ADVANCED MATERIALS IN JAPAN

Source Book 1992

COMLINE INTERNATIONAL CORP

Elsevier Advanced Technology
Mayfield House, 256 Banbury Road, Oxford OX2 7DH, UK

Elsevier Science Publishers Ltd
Mayfield House, 256 Banbury Road, Oxford OX2 7DH, England

British Library Cataloguing in Publication Data can be obtained

ISBN: 1 85617 158 2

Nitrides

Other materials

Oxides

Processing and equipment

CHAPTER ONE:
ADVANCED MATERIALS IN JAPAN

The advanced materials industry is expected to be the one which will carry Japan through to the 21st century as a technological and economic world leader. The Japanese Government and the private sector are currently concentrating on a variety of new materials-related research and development fields.

Private Sector

In the private sector, high growth areas, with regard to the level of research and development activity and market size for new materials, include fine **ceramics**, an area in which Japan can be considered to be particularly strong. Ceramics, such as silicon nitride, are currently being developed for use in bearings, cutting tools, car engines, gas turbines, and other applications — such as reinforcing fibres for use in composites. Research into nitride ceramics was pioneered in Japan by Toyota Motor Corp in 1965.

Research into high-purity aluminium nitride ceramics, which are regarded as excellent materials for the next generation of integrated circuit (IC) substrates, is also proceeding at a fast rate.

Zirconia ceramics are also being increasingly used in industrial applications and engines.

The fine ceramics market is expected to attain a value of ¥6 trillion in the year 2000. Japanese companies are estimated to control 50% of the world market for fine ceramics.

In the **semiconductor** field, demand for gallium arsenide is expected to increase with the spread of broadcast satellite tuners and its use in computers.

Another area currently undergoing rapid development is **engineering plastics**. Domestic demand for such plastics stood at about 520 000 t in 1990. Although Japan has tended to lag behind the rest of the world in basic research on engineering plastics, research into more cost efficient methods of producing super engineering plastics, such as polyphenylene sulphide, is proceeding at a fast pace. Prices for such materials, which also include polyetheretherketone (PEEK), currently stand at ¥10 000/kg. However, some of the more exotic materials, such as some types of liquid crystal polymers, can cost as much as as ¥10 000/g. **Polymer alloys** are another high growth area drawing the attention of materials manufacturers.

The market for **composites**, such as carbon fibre reinforced plastics (CFRP), is growing at an annual rate approaching 20%. Japan is traditionally strong in the field of carbon fibres. Nippon Carbon Co Ltd was the first to commence production of polyacrylonitrile-based (PAN-based) carbon fibres in 1962 (using research and development from the Government Industrial Research Institute, Osaka). High elastic modulus fibres were first produced by Toray Industries Inc in 1971, with the production technology being exported to the USA in 1978. CFRPs are already finding numerous aerospace and military applications, and the trend is likely to spread to the automotive business.

Research and development into **shape memory alloys** and their applications is another area being focused on. Here the Japanese believe they are world leaders in mass production technology and commercial applications. Furukawa Electric Co Ltd pioneered development of shape memory alloys in Japan. The company is said to have developed a nickel–titanium alloy at the same time (in the 1950s) as the US Navy. When the US Navy's patent right expired in 1972, and its application patent right expired in 1982, the company began full scale development of shape memory alloys.

Manufacturers, led by Furukawa Electric, and users of shape memory alloys are currently conducting joint studies into processing technology which will realize practical applications. Currently, shape memory alloys are employed in frames for eye-glasses, brassieres and orthodontic applications. The former two uses account for half of the domestic shape memory alloy market, estimated at ¥1.5–2 billion.

Another area in which the Japanese are highly advanced is **microfibres**. Domestic makers now boast microfibres with thicknesses of less than 0.3 denier, compared to US and European makers who are producing 0.5–1 denier fibres. With research and development progressing at its current pace, the general feeling is that 0.1 denier products will be available by the latter half of 1992. The Japanese industry believes it is two to three years ahead of its foreign competitors.

Research into **hydrogen storage alloys** and their applications in storage batteries and chemical heat pumps is also being conducted. Hydrogen storage alloys are expected also to find applications in residential air conditioning units and industrial cogeneration systems.

Battery manufacturers are developing high-capacity **nickel–hydrogen** storage batteries as replacements for nickel–cadmium batteries which are damaging to the environment.

Superalloys — heat-resistant, corrosion-resistant alloys which use iron or some other low cost metal as a base material along with nickel, cobalt, titanium, etc. — are another active research area.

The Japanese are are also highly advanced in research into **substitutes for natural materials**. Such research is focusing on substitutes for ivory, used in large quantities for official seals, and hawksbill turtle shell, used in traditional crafts.

Associations set up by Private Enterprises and Government

Private sector companies, as well as possessing their own research establishments and conducting their own individual research, also participate in joint research and development projects coordinated by their respective industry associations. Activities of associations are overseen by the Ministry of International Trade and Industry (MITI).

The Association for the Progress of New Chemistry (APNC) is an association formed by private enterprises to promote joint research between its members.

On 1 October 1990, the APNC set up a **Nano-Composite Functional Materials Research Committee** to investigate the possibilities for creating new functional composite materials with nanometre precision. The committee is currently investigating five major areas including solid phase, liquid phase, gas phase and molten-phase techniques.

In May 1991, the APCN formed a group of 38 companies to participate in research and development of new materials using computer simulation systems. The research is initially focusing on the development of **organic magnetic materials** and **functional polymers.** The Association reportedly aims to develop a number of new materials with practical applications by the second half of the 1990s.

The Japan Research and Development Center for Metals (JRCM) is an organization established by metals manufacturers. The Center is involved in conducting working studies and research and development projects into high temperature superconductors, ultra-pure base metals, metal deposition technologies and other metals-related research.

Other private sector associations and organizations involved in new materials-related projects include the **Bioindustry Development Center**, the **Japan Plastics Technology Center**, the **Japan Society of Newer Metals** and the **Cryogenic Association of Japan**.

Public Sector

Research and development in Japan is currently funded 75% by the private sector and 25% by the Government. The major players in research and development into new materials, besides private enterprises, are the Agency of Industrial Science and Technology (a department of the powerful MITI), the Science and Technology Agency (STA) of the Prime Minister's Office, and the Ministry of Education. There is a high level of interaction between the respective groups with numerous "third-sector" companies (quasiprivate operations controlled by government organizations) having been established to promote research and development.

Their activities are guided and coordinated by the Council for Science and Technology (CST), the supreme advisory body to the Prime Minister with regards to establishment of general science and technology policy, establishment of long-term research goals, and formulation of basic guidelines for the promotion of specific research areas to attain these goals.

The direction the country is heading regarding research and development into new materials can be gauged by the contents of the government-sponsored projects currently under way.

Agency of Industrial Science and Technology

The Agency of Industrial Science and Technology's (AIST's) basic policy is: "To provide an environment in which the vitality of the private sector — the foundation of national technological development — can be maximized, and to promote joint development programmes between industry, universities and the public sector, in those areas where a high degree of financial and development risk is involved and where the private sector alone finds it difficult to carry out projects on its own."

In line with its policy, the AIST conducts large-scale research and development projects, energy-related projects, as well as the Research and Development Project of Basic Technologies for Future Industries. All research and development projects are carried out in cooperation with national laboratories, academic institutions and private industry.

Descriptions of the projects currently under way are given below.

(i) **National Research and Development Program** (popularly known as Large-scale Project)

The National Research and Development Program was launched by AIST in 1966, in order to undertake risky large-scale research and development under government coordination by commissioning research and development to third parties. The programme's first three projects were a super-high performance electronic computer, a desulphurization system, and a magnetic hydrodynamic generator.

At this time, a time of high economic growth, most key technologies were still being imported from overseas. The Government saw the need to promote original research and development activities in the industrial sector. The Large-scale Project was inaugurated to meet this need. Its primary goal was to unite national laboratories, academic institutions and private industry, in order to raise the country's standard of technical expertise. Two projects focusing on the development of new forms of energy were started in 1974 (Sunshine Project) and in 1978 (Moonlight Project). These two projects gained independent status during the energy crisis of the 1970s.

There are nine projects currently under way, with two at the planning stage. Of these projects, those related to new materials have only surfaced recently, the two currently in progress being:

a) **Advanced material processing and machining systems** (1986–1993; budget for fiscal year 1991 is ¥3.159 billion).

   AIST has seen the need for highly functional components which do not exhibit a degradation in functionality, even under highly corrosive atmospheres and high pressures and temperatures. Some rare metals have the desired properties but the cost is too prohibitive for them to be applied to mass production.

   The object of this project is to develop techniques for depositing thin films of environmentally friendly rare metals which possess the desired properties onto the surfaces of less expensive base materials. Such techniques are likely to use excimer laser and ion beam techniques, capable of depositing monoatomic films.

b) **Fine materials from marine organisms** (1988–1995; budget for fiscal year 1991 is ¥1.773 billion).

   This project's goal is to develop production methods for new fine chemicals from marine bioresources. Such fine chemicals will find applications in coating materials for underwater structures, pigments, dyestuffs, moisturizing materials, surfactants for cleaning up oil spills, and anti-fouling agents. The following research and development will be conducted:

- basic technologies for using marine organisms;
- production technologies for useful marine organism-based chemicals;
- use of biofunctional marine organisms; and
- development of supporting systems (including research hardware and a biofunctional marine organism database for general use).

A new project in the pipeline, **Advanced Chemical Processing Technology**, will aim to develop technology for fabricating materials such as **functionally gradient materials**, and **highly-aligned polymers**. The project will include development of advanced chemical processing methods including chemical vapour infiltration, purification by laser-induced electric field, and polymer preparation by photo-chemical processing.

The results of all the programmes, whether realized by national laboratories, private industry or academic institutions, become the property of the AIST. It is AIST policy to grant a licence to technologically- and financially-qualified applicants on payment of a fee. To facilitate technological transfer between the AIST and private enterprises, the Japan Industrial Technology Association (JITI), a non-profit organization, was established as an affiliate. JITI possesses the exclusive rights for all the technology possessed by AIST.

Since October 1988, in order to facilitate the full use of the private sector, the Large-scale Project has also involved participation of the AIST-affiliated **New Energy and Industrial Technology Development Organization** (NEDO, see below) which assigns various types of research and development under the programme to private corporations. The results are then shared between NEDO and the companies which develop them.

(ii) **Research and Development Project of Basic Technologies for Future Industries**

This project deals with revolutionary basic technologies. The fields covered by the project are new materials, biotechnology, and next-generation electronic devices. A parallel development system in which several companies carry out the same research and development using different methods is employed. The research programmes are of ten years duration, and are subdivided into a number of periods of a few years, with interim objectives set for each period. The most suitable development methods are chosen during the course of the programme and emphasis is shifted to them.

The project is divided into 14 categories. Research and development programmes currently under way in the new materials area include:

a) **Photoactive Materials** (1985–1992; budget for fiscal year 1991 is ¥432 million).

This programme involves the development of photoactive materials and their applications in high density optical memory devices and micro-disks for home and office computers.

b) **Non-Linear Photonics Materials** (1989–1998; budget for fiscal year 1991 is ¥523 million).

In this programme, research and development of optically non-linear materials for use in optoelectronic components and other applications is being conducted.

c) **Organosilicon Materials** (1991–2000; budget for fiscal year 1991 is ¥44 million).

Organosilicon materials which are heat-resistant, have high strengths and display excellent electrical properties will be developed in this programme. The research and development will be commissioned to the private sector through NEDO. Ten companies will participate in the project, including Toray Industries Inc, Shin-Etsu Chemical Co Ltd and the US company Dow Corning. All Japanese companies involved in the project are members of the **Organosilicon Chemistry Association** (an association established by industrial and academic concerns). The project will entail a total cost of ¥16 billion

d) **High Performance Ceramics** (1981–1992; budget for fiscal year 1991 is ¥1.222 billion).

Heat-resistant, corrosion-resistant, high hardness, high heat efficiency, dimensionally stable fine ceramics for use in nuclear power generation, new energy, and aerospace applications are being developed in this programme. The results are expected to have a large impact on energy conservation.

e) **High Performance Materials for Severe Environments** (1989–1996, budget for fiscal year 1991 is ¥1.699 billion).

This programme's object is to develop ultra-high heat-resistant metal compounds and composites, and carbon fibre- and carbon-based composites which retain their strength at high temperatures. The materials will be used in aerospace, nuclear power generation and new energy applications.

f) **Superconductivity** (1988–1997; budget for fiscal year 1991 is ¥2.179 billion).

Superconductive materials and devices for use in power generation, electronic devices, high performance computers, space systems, transportation and medicine will be developed in this project. Development of technology for practical cooling systems which will enable application of the materials at liquid nitrogen temperature will also be conducted.

Other projects currently under way include development of **synthetic metals** (polymers which are lightweight, as conductive, and as corrosion-resistant as metals) as substitutes for copper and aluminium electric wires and cables. The results of the research are likely to be extended to superconductors.

Previous projects have included **synthetic membranes for new separation technology, high performance plastics**, and **advanced alloys** with controlled crystalline structures.

(iii) **Moonlight project**

This project has the goal of advancing general energy conservation technologies. It was established in 1978 as a breakaway from the Large-scale Project. The project subsidizes industry, etc., for research and development projects in energy conservation as well as promoting international research cooperation and standardization. Current new materials-related projects include:

a) **Development of Ceramic Turbine** (1988–1996; budget for fiscal year 1991 is ¥1.773 billion).

The object is to develop a ceramic turbine with a thermal efficiency of 40–45% when running on natural gas or methanol.

b) **Superconducting Technology for Electric Power Apparatus** (1988–1995; budget for fiscal year 1991 is ¥3.14 billion).

This project aims to develop a superconductive generator with a power output in the 70 MW range. A third sector company, the Engineering Research Association for Superconductive Generation Equipment and Materials (Super-GM) was set up as part of the project. It is made up of 16 corporate members including electronics-related companies (Toshiba, Furukawa Wire) and electric power companies.

In addition, there are other new materials-related AIST projects which exist as separate entities. These include:

a) **Development of components for use in superconductive energy storage systems (1991–1996; budget for fiscal year 1991 is ¥160 million).**

In order to facilitate the earliest possible realization of such systems, research and development into components of superconductive energy storage systems (superconducting coil, AC/DC converter, permanent current switches and DC interrupters) will be conducted.

b) **Development of environmentally friendly materials** (1991–1996; budget for fiscal year 1991 is ¥160 million).

This involves the development of micro-organism-based, protein-based and other natural polymer-based plastics and biodegradable plastics. Basic research for this project began in 1989 as a three-year study at the Government Industrial Research Institute, Osaka.

c) **Development of unmanned space experiment system** (1986–1993; budget for fiscal year 1991 is ¥5.637 billion).

This is the development of a system for low gravity experiments in new materials, electronics and biotechnology. Once developed, the system will be capable of conducting experiments of several months' duration. The system is scheduled to be launched at the end of the fiscal year 1992 using the H-II rocket. Development is being conducted in cooperation with the STA's **National Aerospace Laboratory** and the Ministry of Education's **Institute of Space and Astronomical Science**.

d) **Monetary assistance for research and development activities of private sector**.

Parallel to the research and development projects it sponsors, the AIST supports important technological development projects in the private sector through conditional loans and investment in research and development companies by way of the **Japan Key Technology Center** (JKTC, an organization affiliated with MITI and the **Ministry of Posts and Telecommunications**) and NEDO. A fiscal 1991 budget of ¥26 billion has been set for such projects.

One of the projects the JKTC is participating in is free electron laser research. A **Free Electron Laser Research Center** is to be constructed by a third sector company called the Free Electron Laser Engineering Co (FELEC) at a cost of ¥8.6 billion. The facility will be completed by the fiscal year 1993. FELEC was set up by the JKTC, the Institute for Laser Technology, and 12 domestic firms, including Mitsubishi Electric Corp, Sumitomo Electric Industries Ltd, Kansai Electric Power Co Inc, Hitachi Ltd and Mitsubishi Heavy Industries Ltd.

A seven-year research project at the Center, aimed at building an industrial-use free electron laser system and developing applications for it, started in 1991. In its first year, the project was financed with ¥150 million from JKTC and ¥64 million from the other participants.

The **New Energy and Development Organization** (NEDO) was established in 1980 by the AIST and the private sector. Its initial role was to take over the running of the Sunshine Project which had reached the stage where construction and operation of large-scale test plants was necessary.

In addition to participating in the Large-scale Project, NEDO also establishes third sector companies in association with private industry in order to provide research facilities financially out of reach of private enterprise. These facilities are then opened to universities, and the public and private sectors. NEDO has invested in the **Ion Engineering Research Center** in Kansai Science City (see later) to the tune of ¥2.4 billion in the fiscal year 1991 (¥2.2 billion in 1990). The Center is run by the **Ion Engineering Center Corp** (IECC), a third sector company. It is equipped with expensive ion implantation and deposition systems and electron microscopes.

The Center is also part of a group which has constructed a zero gravity experiment facility, the Underground Micro-G Experimental Center, at a defunct coal mine in Kamisunagawa-cho, Hokkaido.

Independent Research at Agency of Industrial Science and Technology Institutes

The AIST's Tokyo-based national research institutes moved out to Tsukuba Science City (Ibaragi Prefecture) in the 1970s. Since then, a number of industries have follow suit.

The **National Chemical Laboratory for Industry** (NCLI), the **Research Institute for Polymers and Textiles** (RIPT) and the **Electrotechnical Laboratory**, among others, are located in the city. AIST also operates a number of regional research facilities in Hokkaido, Nagoya, Osaka, and four other prefectures.

Four institutes currently up for reorganization are the NCLI, the RIPT, the **Fermentation Research Institute**, and the **Industrial Products Research Division**. The four will be combined into two, one specializing in materials science and one in life sciences.

Research that takes place at the AIST's research facilities is mainly concerned with basic and exploratory technologies which are expected to form the foundation of technological developments beyond the year 2000.

New projects implemented in the fiscal year 1991 included:

- Basic Studies on Advanced Materials from Polyaromatic Hydrocarbons (at NCLI);
- Synthesis of Biodegradable Polymers using Carbon Monoxide (at NCLI);
- Research on Key Technology for Optical Functions of Organic Thin Films (at GIRI, Osaka);
- Processing of Micro Multi-Layer Composite Materials (GIRI, Nagoya);
- Preparation of Novel Carbon Materials based on Molecular Design (**National Institute for Resources and Environment**, until recently called National Research Institute for Pollution and Resources);
- Fabrication Technology of Functionally Superplastic Material by Powder Forming Process (at GIRI, Kyushu);
- Research on Highly Refractory $MoSi_2$ Composite Ceramics (at GIRI, Kyushu);
- Characteristics of Whiskers for Advanced Composite Materials (at GIRI, Shikoku);
- Studies on Polymer Metal Cluster Composites (at RIPT);
- Studies on Technology for Controlling the Function of Polymeric Materials (at RIPT);
- Study on Reaction Mechanism and Advanced Material Production in Specific Reaction Fields (at NCLI); and
- Manufacturing Technique for High-Performance Metal Material Sheets (at **Mechanical Engineering Laboratory**).

The Science and Technology Agency

The Science and Technology Agency (STA) of the Prime Minister's Office is also actively engaged in new materials and other research. Research and development is following a parallel line to that of the AIST, although the projects it is engaged in do not necessarily have immediate industrial applications. Projects initiated by the STA come about as a result of reports released by the **Council for Aeronautics, Electronics and Other Advanced Technologies** (an advisory body to the Director-General of the STA).

The STA promotes and finances research recommended by the council through the following:

1) **Special coordination funds for the promotion of science and technology**

   These funds are directed to projects involving basic research into the latest technologies, those that require participation from a number of organizations, those promoting links between industry, academia and the public sector, and those that promote international cooperation. The fiscal 1991 budget for the funds is ¥10 billion. Projects funded under the scheme tend to run for five years, with total budgets ranging between one and two billion yen.

New materials-related projects started in 1981 with a project which involved the establishment of basic technology for controlling the formation of surfaces and interfaces for highly functional materials. Projects already completed include:

- basic research into superconductivity and ultra-low temperature; development of a tunable high-output laser;
- laser processing techniques;
- creation of new materials under zero gravity conditions; and
- techniques for analysing and evaluating highly functional materials using new beam sources.

Those currently under way include:

a) basic research into the purification of rare metals;

b) development of functionally gradient materials (with an object of attaining a material with a heat resistance of 1700°C and able to endure a temperature gradient of 1000°C, 1987-);

c) generation, measurement and use of ultra-high vacuums (1988-);

d) research into the generation and use of vacuum ultraviolet radiation;

e) research into the use of host–guest reactions in the synthesis of new functional materials (1990-);

f) basic research into material interconnection (1990-); and

g) an international project involving development of experimental evaluation techniques for new materials.

2) **Intelligent materials**

   As a result of a report released by the Aeronautics, Electronics and Other Advanced Technologies Council in November 1989, the STA has adopted a policy that promotes research into intelligent materials. The **Intelligent Materials Forum** (IMF) was established in May 1990 with representation from the public and private sectors, and academic institutions.

The IMF's role is to promote technological cooperation between the government, academia, the private sector and research laboratories through study groups and symposia. The First Intelligent Materials Symposium was held in March 1991 and an international symposium, the International Conference on Intelligent Materials (ICIM-92), in April 1992.

In October 1991, the STA began research into materials with intelligent structures, materials which are capable of perceiving changes in environmental conditions and responding to them, at two of its research establishments, the **National Research Institute for Metals** (NRIM) and the **National Research Institute for Research in Inorganic Materials** (NIRIM). NRIM will conduct research and development into composites in which foreign molecules are implanted into molecular- and atom-sized sites of bulk materials. NIRIM will investigate the synthesis and sintering of zirconia particles. Research into mesophase materials, in which microclusters of 3–1000 atoms or molecules are immobilized on the surface of a solid, will be conducted at the **Institute for Physical and Chemical Research**.

3) **Exploratory research for advanced technology programme**

The STA also has a policy which supports young innovative researchers working in new fields that may have no immediate applications but are likely to lead on to practical uses in a few years. The policy is carried out through the **Exploratory Research for Advanced Technology** programme (ERATO) by the Research Development Corporation of Japan (JRDC), an organization affiliated to the STA. The JRDC selects suitable research topics and allots a sum of approximately ¥60 million over a three-year period to finance it. The research is usually carried out by a team of about 20 young researchers drawn from industry and academia. The fields chosen give an insight to where Japanese research and development in heading.

An example of this is **nanotechnology**. The JRDC has supported a number of such projects in the last few years, and now the AIST is planning to launch a major project in this area.

Current research includes methods of manipulating micron-size particles using infra-red laser light. The new materials-related research field to be covered in fiscal 1991 was control of molecular and crystal structure. Research into the discovery of new phenomena and functions at the mesoscopic level will especially be promoted, with research into intelligent materials also likely. Two projects just starting are the development of materials with free pi electrons (such materials are likely to possess unusual optical and electronic properties) and development of organometallic catalysts for use in synthesizing optical isomers.

The JRDC first supported research into new materials with a project that commenced in 1977. Participants in the project, including Sony Corp and Hitachi Ltd, carried out research and development into amorphous metals.

4) **Superconductivity research multicore project**

This project was started in 1988 with the object of realizing technology to allow the practical application of superconductive materials. National research institutions, and facilities operated by other organizations, are assigned specific research topics and act as core centres for that topic. National and international research and development cooperation on all levels is promoted via the cores. The core organizations include the three STA research institutions mentioned in the above section on Intelligent Materials, plus the **National Space Development Agency of Japan** (NASDA) and the **Japan Atomic Energy Research Institute**.

The STA is also actively following a policy of promoting the employment of foreign researchers in the research and development organizations of Japanese companies. The STA established an **International Science and Technology Exchange Center** in 1990 to promote the policy.

Ministry of Education

The Ministry of Education, in addition to providing funding for the everyday running of public universities, also offers university researchers **Grants-in-Aid for Scientific Research**. Grants totaled ¥526 billion in the fiscal year 1989. These include "special distinguished" grants to individual researchers which can amount to US$1–2 million over a three- to five-year period.

Twelve such grants were awarded in August 1991, two of which were related to new materials. One was a project to develop **microactuators** using photo-fabrication techniques, the other for **high temperature oxide superconductors**. Many Grants-in-Aid are currently being provided by the Ministry for basic research into next-generation electronic devices, including superconductive devices.

Trends in the Japanese New Materials Industry

The Japanese new materials industry was originally centred on a number of electronics, speciality metal, ceramics and other firms whose core businesses were directly related to the field.

However, with the rise of the yen in the mid-1980s companies which were finding their core businesses floundering in the wake of foreign competition sought to diversify their operations. Steelmakers, heavy engineering companies, mining companies, pharmaceutical manufacturers and glass and cement producers, among others, ventured into the new materials business with the objective of producing high value-added products to make up for their lost revenue.

Diversification into New Materials

Key examples include Nippon Steel, which has been diversifying into biotechnology, electronics, composites for aerospace and ceramics, and Mitsubishi Heavy Industries, which is taking part in a wide range of aerospace and defence-related projects involving the development of advanced composites. Other examples include textile maker Teijin, which has diversified into chemicals and pharmaceuticals and high-performance fibres, and Asahi Glass, which is moving into functional glasses such as electrochromic and photochromic materials.

Economic Changes Force Reviews of Corporate Strategy

The stock market boom of the mid-1980s allowed companies to obtain cheap capital to diversify and fund new ventures. However, some firms lacked the necessary marketing ability to be competitive in such a high-growth field as new materials. Many simply waited for one innovative firm to make a move, then jumped on the bandwagon with a similar product.

This trend resulted in an oversupply in some markets for new materials. For example, the market for monocrystalline silicon, supplied by only a handful of companies in the early 1980s, such as Osaka Titanium, was suddenly being fought over by some ten suppliers. The case is the same for gallium arsenide with ten manufacturers, including Dowa Mining, fighting for a market only 10% the size of the monocrystalline silicon one.

Some companies which diversified in the 1980s are now finding new materials markets over-competitive and are turning back to their original core businesses. A good example is steelmaker Nippon Steel which, in 1990, was forced to sell off Toho Kinzoku, a subsidiary set up by the firm in 1987 to produce sintered carbide tools. The subsidiary over-concentrated on the development of new materials and failed to develop a sufficient line-up of competitive products. The subsidiary's line-up of tools was a mere one-tenth that of its rival Mitsubishi Materials (formerly Mitsubishi Metals). Another steelmaker, NKK, was also one to suffer, withdrawing from a monocrystalline silicon production venture in 1989. Another example is Nippon Steel which decided to re-enter the aluminium business, after an absence of 12 years, partly because automakers are shifting to the use of aluminium alloy parts to reduce the weight of their vehicles.

Japanese companies are also beginning to struggle under the burden of higher interest rates, and are facing difficulties in raising equity finance with the Japanese stock market in its current state. Expansionist moves of the order of the latter half of the 1980s are thus not expected in the short term.

Overseas Expansion

Nonetheless, more and more Japanese new materials-related companies are setting up operations overseas or are acquiring stakes in overseas firms. A number of Japanese new materials suppliers have, or are planning to, set up production operations and joint ventures overseas. Examples include:

- Denki Kagaku Kogyo which has established a joint venture in the USA, together with the Austrian firm Rath, that will produce ceramic fibres; and
- Mitsui Petrochemical Industries which is planning to establish a European production base for thermoplastic elastomers by the end of 1992.
- These moves are partly the result of shifts in market demand and domestic labour shortages and high labour costs in Japan.

Unification of the European Community in 1992 has also seen a flurry of activity with Japanese manufacturers establishing joint ventures and overseas production bases. Most of the bases are for the production of mainline products. However, some companies have already formed specialist new materials ventures, such as Ube Industries which has formed a joint venture in Europe with Japanese trading firm Marubeni for producing and marketing plastics composites.

The current rapid changes occurring in the political system of Eastern Europe are expected to result in opportunities for Japanese manufacturers of such products as steel and textiles to expand their markets and sources of raw materials, especially from Japan's resource-rich neighbour, the Commonwealth of Independent States (formerly the Soviet Union). However, it is still too early to assess the potential for Japan's new materials companies.

Conclusion

The continued economic growth of Japan has brought with it a rapid change in the research and development environment. Expectations of society towards science and technology are on the rise, with people demanding a higher standard of living, parallel to the country's economic power. The government of Japan is already looking ahead to the next century, and studying an integrated science and technology policy that will keep Japan at the forefront of research and development, as well as providing its citizens with "a more fulfilling life". New materials are certain to play a leading role in any future initiatives.

CHAPTER TWO
METALS AND ALLOYS

Aluminium and aluminium alloys

Kobe Steel to accelerate development of automotive aluminium alloys

In response to automakers' attempts to reduce vehicle weight, Kobe Steel Ltd plans to expand the activities of its automotive aluminium alloy development groups.

The company plans to combine its aluminium-panel development group and its vehicle-related-materials development group into a single unit. The integrated group will develop lightweight materials for use in engine and suspension parts. The group will also develop appropriate production line technology.

Other Japanese steel producers, including Nippon Steel Corp, NKK Corp, and Kawasaki Steel Corp, are also making efforts to develop lightweight materials for automobiles.

Aqueous aluminium paste for automobile paint

Aluminium paste for use in aqueous automobile paints has been developed by Toyo Aluminum KK. Aluminium has previously not been used in aqueous paint due to its reactivity with water.

The paste is based on 15 μm diameter aluminium particles which have a water-resistant agent adsorbed onto their surface. The water-resistant agent consists of organic and inorganic trace additives. A surfactant is also added to make the particles compatible with water. The company plans to price the paste 50% higher than aluminium pastes made for organic paints.

Now that aqueous paints are coming into use by car manufacturers, the company plans to produce 20–30 types of aqueous aluminium paste.

Aluminium alloy for cars

A hard type of aluminium alloy for use in the outer panels of cars is available from Sumitomo Light Metal Industries Ltd of Tokyo.

The alloy consists of aluminium, silicon, magnesium, and trace elements. Unlike standard aluminium alloys, when it is heat treated at some 200°C in a baking and painting process, the tensile strength produced is 26–28 $kg.cm^{-2}$, approximately twice that of standard alloys, the company claims.

Use of the alloy enables the thickness of the outer panel to be reduced to 0.7–0.8 mm, while standard alloys require a thickness of some 1 mm. As a result, total aluminium alloy use can be reduced by approximately 30%, according to Sumitomo Light Metal Industries.

The material will be priced 10–29% higher than standard alloys.

Iron and steel

Iron and steel industrial standards to adopt SI units

From 1 January 1991, all Japanese Industrial Standards (JIS) for iron and steel have been in Système International (SI) units.

The JIS Committee decided to adopt SI units in August 1985. The iron and steel industry, in advance of others, has now decided to use SI units in all of its standards.

Durable steel for bearings

Bearings five times more durable than conventional products can be made with a high-performance steel, "MT21", according to the material's developer, Nachi-Fujikoshi Corp of Tokyo, Japan. The company plans to market bearings made of MT21 to car manufacturers.

The increased durability of the material has been achieved by increasing the chromium content and adding manganese, although the company has not disclosed the precise composition. Bearings made of MT21 can also withstand temperatures of up to 150°C, 30–40°C higher than conventional products, the company says.

They are expected to cost about 25% more than existing products. The company has already delivered samples to major car makers, and it expects sales of MT21 bearings to reach ¥1 billion a month. This would account for about 20% of Nachi-Fujikoshi's total bearing sales.

Hard stainless steel

A stainless steel with a hardness of over 700 H_V on the Vickers scale has been developed by Nippon Metal Industry Co Ltd of Tokyo, Japan. This is 100 H_V harder than the martensitic stainless steel, "440C", the company claims.

Nippon Metal's "NTK H-1" is an austenitic stainless steel containing more than 3% silicon which makes it resistant to tempering. A molybdenum content of up to 1.5% makes the steel more corrosion-resistant than conventional "SUS304" stainless steel. Furthermore, low temperature annealing, using a 50% reductant, gives an alloy with high yield strength, according to the company.

Nippon Metal has begun sample shipments of NTK H-1 to Japanese appliance and car makers, for use in springs, at a rate of 50 tonnes a month. The steel costs ¥450 000–500 000 a tonne.

Stainless steel with good welding properties

A stainless steel has been made which can support welds with a hydrostatic fracture strength of about 900 MPa, 1.5 times the strength of welds in conventional stainless steel, according to its Japanese developer, Tokyo-based Nippon Yakin Kogyo Co Ltd.

The material is based on the company's "NAS86L" stainless steel, with the addition of unspecified elements. Nippon Yakin initially aimed for a production rate of 3–4 tonnes a month, which it expected to be doubled by the end of the fiscal year 1991.

Strong steel sheet

A steel sheet for the manufacture of cars is being produced by Nippon Steel Corp of Tokyo. The company claims the sheet has 1.5–2 times the tensile strength and ductility of conventional materials.

"Gamma-High-Ten" is an austenitic steel with a high carbon content which is available in thicknesses of 1.4–6.0 mm. Nippon Steel says it has already established a mass production system at its Oita works that can produce 2500–3000 tonnes a month.

Corrosion-resistant ferrite stainless steel

A highly corrosion-resistant ferrite stainless steel has been developed by Nippon Metal Industry Co Ltd of Tokyo. The steel contains only very small amounts of carbon and nitrogen, which aggravate corrosion, plus additional chromium and molybdenum which act to further enhance corrosion resistance.

The steel, "NTR-U20", contains 0.003% carbon, 0.006% nitrogen, 28% chromium and 3.5% molybdenum. The low contents of nitrogen and carbon are achieved by using an improved smelting process, according to the company.

NTR-U20 undergoes almost no stress cracking due to corrosion, and has more than twice the pitting resistance of conventional stainless steel. Nippon Metal has conducted a number of four year-long seawater immersion tests which produced no observable corrosion.

The steel is 2.5 times the price of ordinary stainless steel. The company recently began production of the alloy at its Sagamihara factory, making 50 tonnes a year.

Stainless foil-based catalyst carrier for cleaning car exhaust gas

A catalyst carrier for cleaning car exhaust gas which uses stainless steel foils has been announced by Nippon Steel Corp, in collaboration with Toyota Motor Corp and Nippon Kinzoku Co Ltd.

The carrier uses an oxidation-resistant ferrite-based stainless foil. Whereas conventional ceramic carriers have thicknesses of approximately 150 μm, the new carrier is 50 μm in thickness, thus reducing exhaust resistance. Furthermore, the stainless carrier heats up faster than ceramic carriers, thereby enhancing the cleaning performance. In addition, although slightly more expensive, the new carrier is reportedly more durable than conventional carriers.

Toyota has started to use the new stainless carrier in its "Camry" car model. To supply Toyota, Nippon Steel will produce the new carrier at a rate of several thousand units per month at its Nagoya works.

Ultrathin silicon–steel films

Researchers in Tohoku claim to have created a process for making ultrathin oriented silicon–steel films.

The scientists say their work overturns conventional theories that such film could not be made thinner than 5 μm. The research could have important implications for manufacturers of the tape heads in audio-visual recording equipment.

Loss of energy from the new material is significantly less than for conventional oriented silicon–steel films, or amorphous materials, they say.

High tensile steels for outer panels on cars

Nippon Steel Corp of Tokyo has begun sample shipments of high tensile steel plate, "WEL-TEN80" and "WEL-TEN100N", to car makers for use in vehicle outer panels.

WEL-TEN80 and WEL-TEN100N have tensile strengths of 80 and 100 MPa, respectively, compared with conventional steel plate which has a tensile strength of less than 500 MPa. The use of the new steels will allow door panel thickness to be reduced to less than 0.7 mm, according to the company.

Car makers are increasing their use of aluminium alloys and plastics in order to reduce vehicle weight. Nippon Steel has developed the high performance steels in order to compete with companies which supply non-steel materials to car makers.

Nickel and nickel alloys

Aircraft engine to use cast monocrystalline nickel–cobalt alloy

Ishikawajima-Harima Heavy Industries (IHI) Co Ltd and Ishikawajima Precision Casting Co Ltd (ICC) of Tokyo, Japan, plan to produce aircraft engine parts using a monocrystalline alloy and a high-precision casting technology developed by IHI. This will be the first time that the technology has been used in Japan, the companies say.

IHI and ICC will produce turbine blades, made from a monocrystalline nickel and cobalt-based alloy, for "F100" engines which IHI will deliver to the Japanese Defense Agency. Producing the blades also requires a special, high-precision cooling technology. Only a few manufacturers use this method — under licence from General Electric, Pratt & Whitney, and Rolls Royce — and in the past Japan has had to import these parts from the USA or the UK.

ICC will ship the blades to IHI, and it also plans to develop prototype parts for engines to be developed in Japan. ICC may also export the blades.

ICC expects sales of the blades to total ¥400 million in 1991, increasing to ¥1 billion in two years. Plans are already underway to increase production capacity to meet these goals.

Heat resistant nickel alloy for fission reactors

The Japan Atomic Energy Research Institute (JAERI) in Tokyo has developed a heat resistant nickel alloy which it reports has three times the corrosion resistance, twice the strength and five times the service life of conventional nickel alloys.

The material (which contains manganese, silicon and boron) has applications for use in high temperature gas reactors, the next generation of fission reactor, JAERI says.

Ductile nickel-based heat resistant alloy

Elemental analysis of a nickel-based heat resistant alloy may result in a more ductile material and could lead to a reduction in processing costs, according to The Agency of Industrial Science and Technology's Mechanical Engineering Laboratory.

An X-ray micro-analyser was used to examine the texture of samples of alloy which had been allowed to solidify under different conditions following melting. The alloy analysed was "TMP-3", a material developed in Japan. One sample of the alloy was allowed to solidify while standing still and another sample was agitated while it solidified.

Distribution of phosphorus within the alloy was then analysed. TMP-3 generally contains 0.003% phosphorus by weight, but during undisturbed solidification phosphorus tends to accumulate at the grain boundary, forming a local concentration of 0.1%. It was found that agitated solidification reduces this tendency of phosphorus to accumulate around the grain boundary and this explains why earlier research found that agitated solidification produced alloy which is more ductile at high temperatures.

Alloy which solidified while undisturbed elongated only 11% when stretched at 1100°C, while material which solidified while being agitated elongated 91%.

Other materials

Misch metal to be imported from China

Leading trading house Chori Co Ltd of Osaka plans to import misch metal from China.

The company has acquired sole rights to distribute misch from the Hunan Corp of the National Nonferrous Metals Import/Export Corp of China and its Japanese subsidiary Kinyu. Chori will import 1500 tonnes a year, almost equal to Japan's total demand.

Misch metal is a semi-processed mixture of rare earths, used in the production of nickel-based alloy power cells, which Japanese battery manufacturers are starting to make on a large scale. These cells are less polluting and have twice the storage capacity of cadmium cells.

Chori will price the misch metal at about ¥1000 a kilogram.

Pure base metals target for research and development

The development of ultra-pure base metals and high temperature semiconductors is the aim of the Tokyo-based Japan Research and Development Center for Metals (JRCM), according to its operation plan for the fiscal year 1992.

JRCM says it will give priority to investigating the development of ultra-pure base metals, such as iron, copper and aluminium. It will survey the current situation regarding purities, the evaluation of physical properties, and current research.

JRCM also plans to establish the New Steel Production Technology Research Promotion Department which will investigate the possibility of developing a third steel production process as an alternative to converter and electric furnace methods. The technical evaluation will be conducted from electrical, chemical, and metallurgical standpoints.

Finally, the JRCM will promote research and development on high temperature semiconductors in order to investigate their quantitative characteristics in detail and to develop applications.

Precious metals

Precision electroforming of platinum

A technology for precision electroforming of platinum has been developed by Tanaka Kikinzoku Kogyo KK. This technology is able to produce small hollow products and has applications in jewellery, ornament formation and glass fibre production, the company says.

Prominent features of the technology include:

- solid ornaments can be manufactured from small amounts of platinum;

- processing of complex designs is possible using 99.99% pure platinum;
- the product is about five times harder than ordinary platinum, with improved abrasion resistance;
- films of over 300 μm in thickness can be deposited; and
- 1 μm thick films can be deposited in two to three minutes, 10–15 times faster than conventional plating techniques.

Development and production of the processing system and related products are being conducted by an affiliated company, Electroplating Engineers of Japan Ltd, and the products are being marketed by Tanaka Kikinzoku.

Processing and equipment

Urethane-based coating agent for metal surfaces

A urethane-based coating agent for the treatment of metal surfaces is being marketed by Dainichiseika Colour and Chemicals Manufacturing Co Ltd of Tokyo.

The company says the agent is coated onto a metal surface which is then heated to 150°C. Two ingredients in the agent react and a film is formed.

The coating agent is available in transparent and coloured forms and a product for treating resin surfaces is also available. The company expects annual combined sales of these products to reach ¥1.2 billion.

Metal coatings for all types of material

A technology which can provide all types of materials with metal coatings has been developed, according to Tokyo-based Monsanto Japan Ltd, which plans to begin marketing products made from the process in Japan.

In the process, the base material is coated with a layer of a special catalyst, and then with copper or nickel using a non-electrolytic liquid phase method. A further coating of another type of metal can then be applied, Monsanto says. An advantage of the method is that it does not require large equipment such as that required for electrolytic plating.

The first product to be marketed will be metal-coated fibres for use in electromagnetic shields.

High performance alloy laser plating technique

A high performance alloy plating technique which uses lasers has been jointly developed by the Government Industrial Research Institute (GIRI), Shikoku, and the Okayama Prefectural Industrial Research Center (OPIRC).

In experiments, several layers of alloy, such as nickel–zinc and cobalt–zinc, were plated onto a steel substrate and irradiated with a 2 kW laser beam for three seconds. The research group claims that this produced a layer of alloy which displayed a peeling resistance of 400 MPa, compared with less than 1 MPa for conventional plating.

The researchers report that their technique is applicable to most alloys under a variety of irradiation conditions.

Production of pig iron by melting reduction

The Japan Iron and Steel Federation of Tokyo plans to invest about ¥10 billion in research into melting reduction methods of pig iron production.

Melting reduction, which involves the direct reaction of coal and iron ore in a furnace, is expected to be widely adopted as an alternative production method. Eight major companies, including Nippon Steel Corp and NKK Corp of Tokyo, are cooperating in the development of this technique.

The federation plans to construct a 500 tonnes a day pilot plant at NKK's Keihin Works in Kanagawa Prefecture. The plant is scheduled for completion in the first half of 1993.

Grinding titanium alloys

An electrical discharge method can grind titanium alloys, aluminium composites and other materials considered difficult to cut, according to its joint developers.

Meiji University's Faculty of Engineering and Honda Engineering Co Ltd say their method involves grinding positively charged workpieces with a negatively charged grinding wheel made of cubic boron nitride. The electric field generated reduces clogging of the tool.

The method is expected to find applications in the processing of car and aircraft parts which are increasingly being made of aluminium and other light composites.

Direct molybdenum plating of aluminium

A method of directly plating aluminium with molybdenum has been devised by researchers at an independent laboratory, Tsuda Kenkyusho.

Aluminium coated with a molybdenum film can withstand temperatures of up to 750–800°C; untreated aluminium melts at 660°C, limiting its industrial applications, the scientists say.

In their method, the researchers modify the surface of the aluminium by immersing it in a solution of a carboxylic acid derivative for five to seven minutes. The aluminium is then immersed in a molybdenum plating bath which has a neutral pH and is at a temperature of 90°C.

Immersion for three minutes and 45 seconds results in the formation of a 200 nm film of molybdenum. This rate of processing is considerably faster than conventional methods and is applicable to continuous plating, they claim.

Furthermore, when the molybdenum film is chemically removed, a pearl-coloured surface is left behind which, depending on the incident angle, will split white light into its component wavelengths.

Friction and abrasion between metals reduced

A technique which reduces the frictional coefficient and abrasion between metals has been developed, according to researchers at the Tokyo Metropolitan Institute of Technology.

The scientists say they apply an external DC current to the metals, which controls the induction voltage (including AC) generated by friction between the surfaces. This has the effect of reducing abrasion by 20–50% and lowering the frictional coefficient by 10–20%.

The group has confirmed that this technique is applicable to 28 types of metals and believes it may also be used with other materials, such as ceramics.

Thick silver plating method

The Kyushu branch of the Government Industrial Research Institute (GIRI) says it has developed an improved method of electrolytically depositing thick silver plate. The method does not require chemical additives which tarnish the silver.

The researchers report that their method involves immersing the article in an aqueous solution of cyanogen, hydrogen carbonate, potassium carbonate and silver cyanate. They report that they have produced untarnished silver plate 2 mm thick.

The researchers aim to apply their technique to carbon–silver and palladium–silver composite plating.

Laser modified surfaces

A metal surface modification technology which involves the application of laser light in a liquid, rather than an argon or nitrogen gas, environment has been developed by a group of researchers (led by Professor Tomio Ariyasu) from Kansai University's Faculty of Engineering.

In experiments, pieces of tungsten, titanium, molybdenum and stainless steel were immersed in silicone oil or boric acid. A pulsed yttrium aluminium garnet (YAG) laser with an average output of 600 W irradiated the metal with a 10 mm diameter beam at a beam angle of 25 milli-radian in pulses of 1 ms duration.

The molten metal was rapidly quenched to form a finely textured surface impregnated with carbon (in the case of silicone oil) or boron (in the case of boric acid). The modified surfaces were four times harder, the researchers claim.

The group also succeeded in measuring the surface temperature of the metal during quenching with a time resolution of 1 ms. These measurements, which are difficult to make during conventional modification processes, allow better control of the laser power and the processing conditions.

Automated metal crystal growth technique

A computer-controlled system uses the floating zone (FZ) method to prepare metal crystals, according to its Niihari-based developer.

The National Institute for Research in Inorganic Materials (NIRIM) says the system uses a computer-controlled high-frequency oscillator to control automatically the heating, melting and subsequent crystallization of a metal sinter.

The FZ method involves the application of heat to one end of an elongated sinter. As the heated portion melts, the heat is gradually moved down the sinter's length. The correct application of heat is reportedly difficult to control because the sinter's melt characteristics change as heat is applied, resulting in a lowered melting temperature.

The newly developed automated system is reportedly applicable to the crystallization of silicon carbide, tantalum, tungsten, molybdenum and lanthanum hexaboride.

Thin stainless slab continuous casting process

A continuous casting process for making thin slabs of stainless steel allows production of several 10 mm thick slabs in a single step, according to its Tokyo-based developers.

Nippon Steel Corp and Mitsubishi Heavy Industries Ltd say the method is an improvement on current techniques which require multiple rolling procedures.

Nippon Steel plans to use the process at its Hikari works where it will establish an integrated production system for making thin steel slabs.

Pieces of cast iron joined without welding

Researchers at the Government Industrial Research Institute (GIRI), Nagoya, claim to have joined pieces of cast iron using frictional pressing, as part of a project to develop metal-joining methods which do not require welding.

GIRI reports that localized heat generated by frictional pressing fused the two pieces of iron together. The fused product had 75% of the strength of the individual components. Microscopic examination of the joint's texture revealed none of the cracking found in a welded joint.

The Institute expects that further research will lead to improvements in bond strength.

Cold forging lubricants

A range of cold forging lubricants contains additives which prevent burning of the lubricants and corrosion of the forging dies, their joint developers claim.

Toyota Central Research and Development Laboratories Inc, Nippondenso Co Ltd, both of Aichi, and Toyota Chemical Engineering, say the "Titan Homer CCD Series" of lubricants is based on mineral oils. The lubricants can eliminate the need for bonding treatments and are half the price of conventional products.

In addition to cold forging and cold processing, the product is expected to be used in processing applications where conventional lubricants cannot be used. Relevant patents have been issued in the USA and are pending in Japan.

The CCD Series lubricants are available in three types: standard, low viscosity, and high concentration. Toyota Trading Co Ltd expects monthly sales of 10 kl. The product costs ¥1000 a litre.

Iron-oxidizing bacteria

A transformed strain of bacteria is capable of oxidizing iron, according to Tokyo-based Dowa Mining Co Ltd. The company now plans to develop a potent ferroxidizing bacterial strain for microbial leaching of metals from low-grade ores.

In collaboration with researchers at the Akita Agricultural College, the company says it isolated the gene encoding the enzyme, oxidoreductase, a species of ferroxidant bacteria.

Ferroxidant bacteria derive their energy by oxidizing iron from the ferrous to the ferric form under the catalysing effect of the enzyme.

Spot welding equipment for galvanized sheet steel

Spot welding equipment for galvanized sheet steel is to be developed by Toyota Automatic Loom Works Ltd of Kariya following a commission from the Research Development Corp of Japan (JRDC).

The equipment will be based on a welding technique developed at Osaka College which involves spot welding two steel sheets separated by a thin film of high density polymer such as polypropylene. The film contains channels of alumina and other

conductive materials through which the current is concentrated.

The result is a strong spot weld with reduced denting or distortion of the sheet steel. The technique is also expected to lead to the development of more compact welding equipment with a lower power consumption than conventional machinery.

Spot welding equipment based on the technique is expected to be used in the home appliance, industrial machinery and automobile industries. The development work is expected to take about three years and cost about ¥360 million.

Welding machine for aluminium alloys

An automatic metal electrode inert gas (MIG) welding machine designed for aluminium alloys is now available from Osaka-based Daihen Corp, a manufacturer of compact transformers and welding machines.

The firm reports that the "Wave Pulse 350" produces cleaner welds at higher speeds than is possible using conventional tungsten inert gas (TIG) welding equipment. The machine uses the wave pulse method which involves varying the current between 100 and 200 times a second. It can achieve welding speeds of up to 80 cm a minute.

The Wave Pulse 350 costs ¥1.1 million. Daihen plans to produce 100 units a month.

Automatic agitator for aluminium smelting

"JET Stirrer" is a fully automatic agitator for use in aluminium smelting, according to its developer, Sumitomo Light Metal Industries Ltd of Tokyo.

The machine is equipped with a cylindrical probe which stirs molten aluminium by repeated intake and discharge. The machine is available in two versions: one model which can be transported using a forklift truck and a fixed version for permanent installation.

A standard version of the movable model costs ¥15 million.

Shape memory alloys

Rock splitting device uses shape memory alloy

A device for splitting concrete blocks and rocks which uses a shape memory alloy has been developed by Tokin Corp of Sendai, Japan.

The "Memoalloy Splitter" consists of three to nine cylindrical sections of a nickel–titanium shape memory alloy (25 mm diameter by 29 mm long) fixed to a heat block which is hinged to two double platens. When the platens are closed, the device forms a wedge which is inserted into a 45 mm diameter hole drilled in the rock to be split.

When heated, the memory alloy cylinders expand 1 mm in one minute, generating a splitting force of 10 tonnes each.

Business and financial information

Mass production of amorphous alloy substitute

Hitachi Metals Ltd of Tokyo plans to begin mass production of a magnetic, crystalline substitute for amorphous alloy.

The company describes "Fine Met" as a soft, magnetic, super-fine crystalline material for use in electronic devices and says that it combines the magnetic characteristics of ferrite-based and cobalt-based amorphous alloys.

Hitachi Metals plans to produce Fine Met at its Yasugi factory in Shimane Prefecture at an initial rate of 10 000 tonnes a month. The material will be priced at ¥2000–3000 a kilogram which is cheaper than cobalt-based amorphous alloy.

Mass production of hydrogen-storage alloy

Japan Metals and Chemicals Co Ltd of Tokyo has begun mass production of metal alloys with hydrogen-storage properties. Such alloys have applications for use in dry-cell batteries.

The company produces hydrogen-storage alloy at an annual rate of 120 tonnes. Japan Metals and Chemicals has been test manufacturing alloy at its Oguni factory in Yamagata Prefecture.

The company claims this move has been prompted by increasing demand for nickel–hydrogen (Ni–H) dry-cell batteries.

Engine-parts production facility

Tokyo-based Mitsubishi Metal Corp has invested about ¥1.5 billion in the construction of a powder metallurgy engine-parts production facility at DIAMET Corp, a US subsidiary.

Mitsubishi has already invested ¥5 billion in the construction of a similar facility at DIAMET. DIAMET uses powder metal technology to produce parts such as piston heads which it supplies to General Motor Corp and Ford Motor Co.

Production was around ¥100 million a month but increasing demand prompted the construction of the new facility which will double the current capacity.

Nippon Steel to supply surface treated steel plate technology to Italy

Nippon Steel Corp will supply technology for the production of surface treated steel plate to a nationally-owned Italian steel maker.

Using the Japanese technology, the Italian company will install a 30 000–40 000 tonnes per month "hot zinc dipping" line at its Genoa plant. Production from the new line will reportedly go to Japanese car makers located in the EC.

Kobe Steel and Alcoa agree to collaborate on joint aluminium project

Kobe Steel Ltd and the US firm Alcoa Co Ltd have agreed to form a joint venture company, KSL Alcoa Aluminum Co, to produce aluminium-can stock in Japan and to collaborate on developing aluminium products for the automotive industry.

To be funded with ¥10 billion, the joint venture's shares will be held equally by the two investors. The new company will be based in Tokyo and is expected to employ about 100 people.

KSL Alcoa will construct a factory on a 10 000 m^2 site adjacent to Kobe Steel's Moka plant in Japan's Tochigi Prefecture. The plant will have a cold rolling line and will be supplied with aluminium hot coils from Kobe Steel's Moka plant which will in turn receive its aluminium ingots from Alcoa.

The joint venture will eventually take over production and marketing of the aluminium-can stock presently being produced by Kobe Steel's overseas smelting projects. With anticipated growth in demand for aluminium, Kobe Steel says it needs additional stable sources of ingot.

Alcoa, which has been increasing its global interests in aluminium-can sheet, sees this venture as a means of penetrating the Japanese market.

Both firms expect to reach a definitive agreement on the establishment of the joint venture company so that production can begin in mid-1993.

The companies may also consider a joint venture to produce aluminium products for the automotive industry in Japan and/or the USA. In addition to supplying a variety of aluminium products for cars, Alcoa has been developing an aluminium 'spaceframe' structure which reduces tooling costs and helps reduce weight.

Coated sheet-steel technology transfer

Kobe Steel Ltd has announced that it will provide Voest-Alpine Stahl Linz GmbH with coated sheet-steel production technology. Representatives from both companies signed a three-year agreement on 19 September 1990 in Linz, Austria.

The agreement calls for Kobe Steel to provide Voest-Alpine with technology for the production of its organic-composite-coated sheet-steel, "Hi-Super DN", which is used in automotive applications. The Japanese company will dispatch technicians from its Kakogawa works to Austria to provide technical and operational guidance.

Voest-Alpine plans to produce high quality sheet-steel for Japanese car makers operating in Europe. Hi-Super DN, which is noted for its anti-corrosion characteristics, is widely used by Japanese car makers, including Nissan Motor Co Ltd which calls it "Durasteel", and Mazda Motor Corp.

Hi-Super DN's outer coating is composed of an electrogalvanized layer of a zinc–nickel alloy, a chromate film, a thin organic-composite coating composed of colloidal silica, organic polymer and organic additives.

Kobe Steel is also involved in a steel-related technology transfer with the UK's largest wire-rod maker, Allied Steel and Wire Ltd. Since 1982, Kobe Steel has been furnishing Allied Steel with wire drawing technology.

Canadian–US joint venture for galvanized steel

NKK Corp, the Canadian company Dofasco Inc and the US company National Steel Corp have signed an agreement to establish a joint venture for the production of galvanized steel using continuous galvanizing line (CGL) technology.

The joint venture, DNN Galvanizing, is to be funded with Can$1.2 million, held 50% by Dofasco, 40% by NKK and 10% by National Steel. The planned joint venture will invest about US$200 million in the construction of a CGL facility in Canada with a production capacity of 400 000 tonnes a year.

The new facility, which is scheduled for completion in January 1993, will produce parts for automobiles which will be supplied to Dofasco and National Steel, who will each receive half of the annual production.

Nippon Mining increases electrolytic copper capacity

Plans to invest ¥1 billion in improvements to copper refining facilities in Hitachi and Sagazeki have been announced by Nippon Mining Co Ltd. The planned improvements will increase the company's total electrolytic copper production capacity from 26 000 to 27 500 tonnes a year.

At its Sagazeki works, the company is improving its converter furnace's gas lines in order to increase the smelting rate and boost capacity to 27 500 tonnes a year. The company is also planning to improve its copper electrolysis facility in order to increase its capacity from 15 000 to 16 500 tonnes a year.

In Hitachi, Nippon Mining has improved its copper electrolysis facility, increasing capacity from 10 000 to 11 000 tonnes a year.

Copper ores imported from the Philippines

Mitsubishi Metal Corp has agreed to supply Atlas Consolidated Mining and Development Corp, the largest mining firm in the Philippines, with a US$30 million long-term loan in exchange for a 15-year guaranteed supply of copper ore.

The contract specifies that Atlas Consolidated Mining will guarantee Mitsubishi the right to purchase 130 000 tonnes a year of ore-concentrate, equivalent to about 38 000 tonnes of ingots worth US$100 million, for 15 years. The copper ore will come from the Toredo mine on Cebu island, which is one of the largest copper mines in the Philippines. Mitsubishi will smelt the ore at its works in Naoshima and Onahama.

The contract will account for 20% of Mitsubishi's total demand for copper, its largest single source. The Toredo mine will supply about 50% of its copper ore production to Mitsubishi.

Aluminium production project in Canada

Kobe Steel Ltd plans to invest US$124 million in an aluminium ingot production project in Canada, the Aluette project. Kobe Steel is participating as a means of ensuring a stable long-term supply of aluminium ingots.

This project, which is being coordinated by Canada's SGF (an investment company run by the Quebec State Government), involves construction of a 215 000 tonnes a year aluminium ingot plant on a 900 000 m^2 site in the Septiles area of Quebec. It will require a total investment of US$900 million, and is expected to begin full-scale production at the end of 1993.

Kobe Steel will receive 29 000 tonnes a year of aluminium ingots from the project. These imports will increase Kobe Steel's reliance on overseas aluminium ingots to 33% of its total requirements.

In addition to SGF, the investors are VAW of Germany, Austria Metal, a Dutch company, Kobe Steel (13.3%) and Osaka-based Marubeni Corp (6.67%). The plant will use technology from the French company Pechiney, and power will be supplied by the Quebec Electric Power Corp.

Sumitomo Metal to increase electrogalvanizing capacity

The production capacity of Osaka-based Sumitomo Metal Industries Ltd's Number 2 electrogalvanization (EG) facility at its Kashima works will increase from 180 000 to 270 000 tonnes a year, the company has announced.

Construction began in 1991, and the ¥5 billion expansion will be completed in 1992.

The Kashima Number 2 EG facility combines an EG line with a painting line. The steel plate is electrogalvanized with a zinc–nickel alloy, "Sumizing Super", and it is then painted with a 1 μm layer of a special rust-inhibiting resin.

Resin-coated steel is in demand for use as car panels.

Sumitomo Metal to streamline steel production

Sumitomo Metal Industries of Osaka plans to invest ¥30 billion in improvements to its steel production facilities which should improve efficiency by about 20%.

The improvements, which will involve the introduction of robots and computer control systems, will help reduce the number of workers employed in the company's steel division to about 22 300 from the present 27 000.

The company also plans to invest ¥8 billion in the introduction of dust filters which should improve working conditions.

Powder production increased

Facilities for the production of metal powders for sintering and cold forging are to receive an investment of about ¥5 billion at Tokyo-based NKK Corp.

By 1993, the company will expand powder production at its Niigata Works from 7000 tonnes a year to 10 000–12 000 tonnes a year. NKK also plans to begin producing finished moulded products by 1993 and, sometime after 1994, the company plans to construct a fully automated factory for the production of metal powder.

NKK is currently marketing ¥500 million a year of metal powder, and expects to achieve sales of ¥10 billion by 1993 and ¥15 billion by 1995.

Titanium ore imports from Malaysia

On 19 November 1990, Sakai Chemical Industry Co Ltd recommenced importing titanium oxide ore from Malaysia.

The company terminated imports of the Malaysian ore in August 1990 because it was unacceptably high in radioactivity due to its content of uranium and thorium.

Since August 1990, the company obtained its titanium ore from Australia and India, but because Malaysian ore is much cheaper, the company has gone to the trouble of developing safety standards which enable it to screen Malaysian ore for its content of radioactive elements.

The company determines the radioactivity of each batch of ore and only imports ore which is acceptably low in radioactivity.

NKK to produce titanium pipe

A subsidiary of Tokyo-based NKK Corp, Kawasaki Kokan Co Ltd, plans to purchase a titanium pipe production facility from Nippon Mining Co Ltd, also of Tokyo.

In 1983, Nippon Mining built a titanium pipe production facility costing ¥1 billion at its Hitachi factory which produces 400 tonnes a year. NKK's subsidiary plans to acquire this facility at a cost of ¥300–400 million, and move it to its new Toyooka factory, which is scheduled to begin operating in 1992.

Kawasaki Kokan plans to initially produce 200 tonnes a year of titanium pipe which will be supplied to NKK and Nippon Mining. In addition, Nippon Mining plans to purchase a 10% stake in Kawasaki Kokan.

NKK currently obtains titanium pipe from Nippon Mining who manufactures it under contract.

Metal parts processing facilities to be modernized

Hitachi Metals Ltd of Tokyo plans to invest about ¥60 billion in the modification of metal parts processing facilities at its Kyushu and Yasugi factories.

At the Kyushu factory, the company plans to modify its forging facilities to enable automatic production of stainless steel exhaust manifolds. Production of exhaust manifolds is expected to increase to 3000 tonnes a year following the modifications.

At the Yasugi factory, the company plans to modify its hot rolling facilities to enable automatic production of special steel products.

The company also plans to rationalize its research and development, and marketing activities.

Electrolytic zinc production capacity doubled

Production capacity of electrolytic zinc will double at Tokyo-based Mitsui Mining & Smelting Co Ltd's Hikozima factory. Currently the factory produces 50 000 tonnes a year.

In 1978, the company had a total electrolytic zinc production capacity of 271 000 tonnes a year, but reduced this when zinc prices were low. Mitsui Mining says there is now a domestic shortage of electrolytic zinc, prompting it to expand again.

The company says it is also involved in another electrolytic zinc production venture in collaboration with other companies.

Joint venture in USA for Nisshin Steel

Wheeling Nisshin Inc, a joint venture of Nisshin Steel Co Ltd of Tokyo and Wheeling Pittsburgh Steel Co Ltd, plans to construct a continuous hot-dip galvanizing line (CGL) in the USA at a total cost of US$120 million.

Construction of the CGL, which will have a processing capacity of 216 000 tonnes a year, was scheduled to be completed by June 1993. The production line will bring the company's galvanizing capacity to 459 000 tonnes a year.

The project is being funded by a US$46.25 million outside investment, a US$40 million investment from Nisshin Steel, and by bank loans and leasing arrangements.

Wire production facility completed

A wire production facility capable of making 15 000 kilometres a month has been built by Nittetsu Micrometal Corp, a subsidiary of Tokyo-based Nippon Steel Corp.

The facility produces 10 μm diameter wire, which was developed in association with Oki Electric Industry Co Ltd of Tokyo, and a new type of insulated wire, which was developed in association with Nippon TI.

Nisshin Steel to construct blast furnace

Nisshin Steel Co Ltd of Tokyo plans to invest ¥40 billion in the replacement of one of the two blast furnaces at its Kure works in Hiroshima Prefecture.

The Number 1 furnace at the Kure works, which has a capacity of 2150 m^3, will be replaced by a furnace with a capacity of 2500 m^3.

The Number 2 furnace at the Kure works has a capacity of 1650 m^3. The two furnaces produced a collective total of 3.4 million tonnes of crude steel in the fiscal year 1989, according to Nisshin Steel. Replacing the Number 1 furnace will raise the plant's crude steel production capacity to 4.5 million tonnes a year.

Construction of the furnace has begun and operations will begin in the fiscal year 1993.

Nippon Steel to re-enter aluminium business

Nippon Steel Corp of Tokyo plans to return to the aluminium industry by purchasing a 30% stake in Sky Aluminum Co from Mitsui & Co Ltd, also from Tokyo, thus becoming the second largest shareholder behind a third Tokyo-based company, Showa Denko KK.

Nippon Steel, which left the aluminium business in 1978, says it has returned to the industry because of an expected increase in demand for aluminium from beverage and car manufacturers. Major car manufacturers are expected to increase the use of aluminium as part of their efforts to reduce weight.

Sky Aluminum, which has substantial debts, plans to obtain capital from Nippon Steel and Showa Denko for the expansion of its Japanese and overseas operations. Sky Aluminum also plans to open refining operations overseas, and to invest ¥100 billion in the construction of hot-rolling mills in Japan.

Joint venture to process special steel

A joint venture for processing special steel products has been set up by Nissho Iwai Corp of Osaka and Iwasaki Machinery Co Ltd.

The venture, provisionally named the Nissho Iwai Special Steel Processing Center, was established in the second half of 1991. The venture will stock, cut, process and deliver special steel products.

The two parent companies say they have already begun trial operations at Iwasaki Machinery's Togane factory. The trial operations have a cutting capacity of 1500 tonnes a year and the full-scale operations will have a capacity of 2500 tonnes a year.

Dowa Mining increases metal powder production

Tokyo-based Dowa Mining Co Ltd has invested about ¥3 billion in improvements to metal powder production facilities at its factory in Okayama Prefecture.

Modifications to the facility have been completed and it is now operating at full capacity. As a result, the company's metal powder production capacity will increase from 840 tonnes a year to 1440 tonnes a year.

The company says its metal powder is used in manufacturing magnetic memory media.

Daido Steel Sheet completes plating line

An investment of about ¥5.8 billion in the construction of a 20 000 tonnes a month plating line for steel sheet has been made by Daido Steel Sheet Corp of Amagasaki.

The new line is able to plate steel sheet 2.3 mm thick and 1.23 m wide. The facility will allow the company to produce 40 000 tonnes a month of plated steel sheet, 4000 tonnes a month more than at present.

Aluminium–lithium alloy production facilities

Aluminium–lithium alloy production facilities at Yokkaichi, Mie Prefecture, have been completed by Tokyo-based Alithium Ltd.

The ¥200 million facilities, which include a high frequency induction furnace with controllable atmosphere and continuous forging equipment, can batch-produce 200 kilograms of alloy.

In addition, the company plans to invest about ¥400 million in the construction of a smelting furnace with a capacity of four tonnes.

Alithium Ltd is a research venture owned by seven aluminium companies and the Japan Key Technology Center.

Companies cease export of steel to Middle East

Nippon Steel Corp and Nisshin Steel Co Ltd of Tokyo, and Sumitomo Metal Industries Ltd of Osaka have stopped exporting steel to the Middle East, because of the danger to shipping in the region.

However, the three companies plan to continue exporting steel to Iran. NKK Corp and Kawasaki Steel Corp of Tokyo, and Kobe Steel Ltd are expected to follow a similar course of action.

Production of cold-rolled stainless steel sheet

Two Tokyo-based companies plan to cooperate in the production of cold-rolled stainless steel sheet.

Nippon Steel Corp plans to increase production capacity of its sheet from 15 000 to 22 000 tonnes a month, but in order to avoid investing ¥36–37 billion in processing equipment for making the finished products, it will cooperate with Nippon Yakin Kogyo Co Ltd.

In April 1993, Nippon Steel will introduce cold-rolling equipment with a capacity of 6000–7000 tonnes a month at its Hikari works in Yamaguchi Prefecture. The semi-finished production from this plant will be supplied to Nippon Yakin Kogyo for final processing.

In order to process the steel sheet from Nippon Steel, Nippon Yakin Kogyo will install annealing, acid peeling and press-hardening equipment at its Kawasaki works in Kanagawa Prefecture.

The production from these combined processes will be shared by the two companies. The cooperation agreement covers a period of two years starting in 1993.

Nickel smelting factory constructed

An electrolytic nickel factory is to be constructed at the Besshi works in Ehime Prefecture of

Tokyo-based Sumitomo Metal Mining Co Ltd. The company plans to invest about ¥7.8 billion in the project.

The planned factory, which is to be completed in 1993, will have a production capacity of 2000 tonnes a month. Once it is operating, two older factories at the Besshi works which currently produce a total of 1850 tonnes a month will be shut down.

The facility will use an updated electrodeposition method to produce pure nickel from impure nickel chloride. The method is twice as fast as conventional techniques, cheaper to operate, and requires less maintenance of equipment, Sumitomo Metal Mining says.

Zinc galvanizing equipment exported to Italy

A continuous hot-dipping line capable of galvanizing 30 000 tonnes a month of zinc was exported by Nippon Steel Corp of Tokyo to the Italian state-owned company Illva.

Illva plans to install the equipment at its Novi Ligure works in a suburb of Genoa. Production from the plant will be supplied to building materials manufacturers in Italy and Japanese car makers operating in Europe.

The Novi Ligure works produces electrogalvanized and hot-dipped galvanized steel plates at a rate of 50 000 tonnes a month. The extra line will increase this to 80 000 tonnes a month.

This will be the second time that Nippon Steel has exported steel treatment facilities to a European manufacturer. In 1989, Nippon Steel exported an electrogalvanizing line to the Spanish company Ensidesa.

Sanyo Special to produce steel in Canada

Sanyo Special Steel Co Ltd of Himeji and the Canadian steel maker Stelco Inc are negotiating the possible joint production of special steels in Canada for use in vehicle production.

In early 1990, Sanyo Special Steel supplied Stelco with steel bar inspection technology which Stelco plans to apply at its production facility, to be constructed in Hamilton, Ontario.

Stelco posted a loss of Can$196.7 million in 1990 and, according to industry experts, has sought assistance from Sanyo Special Steel to restructure its operations. The Canadian firm has reportedly offered to convert its Hamilton steel bar operation into a joint venture.

The move is in accordance with Sanyo Special Steel's strategy to internationalize its operations.

Joint venture for marketing lubricants

Kyodo Yushi Co Ltd of Tokyo has established a joint venture with E.F. Horton, a US enterprise based in Philadelphia, to market special lubricants in Japan.

Nippon Horton has its headquarters in Tokyo and is capitalized at ¥200 million, 60% held by Horton and 40% by Kyodo Yushi. The company will market special lubricants which are produced by Kyodo Yushi at its Kameyama factory in Mie Prefecture. The lubricants are used in processes such as aluminium rolling, and steel plate cutting and polishing.

Until now, Horton has been marketing its products in Japan through a local branch office, but the new venture will take over and will develop and improve the product line in response to demand.

Sales of ¥350 million are anticipated for the first year of marketing, and annual sales are expected to reach ¥1.5 billion within five years.

Steel production system

A project to develop an iron and steel production system has been announced by five major iron and steel manufacturers who will cooperate with the Japan Iron and Steel Federation (JISF) of Tokyo.

The five manufacturers (Nippon Steel Corp , NKK Corp and Kawasaki Steel Corp of Tokyo, Sumitomo Metal Industries Ltd of Osaka, and Kobe Steel Ltd) have each established a systems development subsidiary in order to participate in the project.

Cooperative venture to develop aluminium plate for use in cars

Aluminium plates for use in cars will be developed, produced and marketed as the result of a cooperative agreement between three Tokyo-based companies.

Kawasaki Steel Corp, The Furukawa Electric Co Ltd and Furukawa Aluminum Co Ltd say the agreement will combine Kawasaki Steel's experience as a manufacturer of steel for cars with the aluminium production expertise and facilities of the Furukawa group.

At first, the agreement will only concern joint research and development operations, and marketing research. However, in about two years Furukawa Aluminum's production of aluminium sheets for cars is likely to have increased from 150 to 500–600 tonnes a month.

At that point, Kawasaki Steel and Furukawa Electric plan to set up a joint venture to manufacture and sell aluminium sheets, and Furukawa Aluminium's production will be transferred to the new company. Total sales are expected to be 6000–7200 tonnes (¥3 billion) a year.

The move has been prompted by an expected increase in demand for aluminium sheets for making lighter, more fuel-efficient cars in the future, the participants say.

Pilot plant for high speed film formation method

A pilot plant for testing a thin film deposition method has been built by NKK Corp of Tokyo at a cost of ¥500–600 million. The method rapidly forms thin films on steel plate, the company says.

The method involves arc plasma deposition of ions under a vacuum of 10^{-5} torr. It produces a thin film at a rate of 1 $\mu m.s^{-1}$, over ten times faster than conventional techniques, according to NKK.

The pilot plant, completed in late 1991, is situated at NKK's steel laboratory in Kawasaki and handles 1–2 tonnes a day. The company says it hopes the technique will be commercially viable within five years.

Production facilities for plates of tungsten or molybdenum

Tungsten and molybdenum plate rolling mills are to be built at the Toyama factory of Tokyo Tungsten Co Ltd, the company reports. The construction will cost about ¥2 billion.

Construction of the first of two planned buildings was started in late 1991. This will have a floor area of about 5000 m^2 and a tungsten and molybdenum plate rolling capacity of 120 000 tonnes a month.

The second building will also have a floor area of 5000 m^2. Once the facilities are completed, the company says its tungsten and molybdenum plate production capacity will total 200 000 tonnes a month.

Tokyo Tungsten also plans to demolish its existing facilities.

Factory for production of aluminium alloy moulds

Omi Goukin Co Ltd says it has constructed a production factory for an aluminium alloy moulding material near its main factory in Himi, Toyama Prefecture.

The ¥1 billion factory has an area of 1500 m^2 and began operating at the end of March 1991. The factory gives the company a total production capacity for aluminium alloy moulding material of 200 tonnes a month.

Omi Gouken says it currently produces about 70 tonnes a month of aluminium alloy moulding material at its main factory.

Ferrite production increased

Increased production of ferrites is planned by Kawasaki Magnecs Co Ltd, the Chiba-based subsidiary of Kawasaki Steel Corp of Tokyo.

By the end of 1992, the company plans to raise the soft ferrite production capacity of its Mizushima Factory in Okayama Prefecture from 150 to 300 tonnes a month. The company also plans to raise the hard ferrite production capacity of its Honsha Factory in Chiba Prefecture from 200 to 400 tonnes a month.

Kawasaki Magnecs manufactures sintered ferrite alloy from ferrous oxide, a by-product of Kawasaki Steel's operations. The alloy contains strontium and manganese, plus either barium carbonate or barium oxide.

Reduced production of sponge titanium

Toho Titanium Co Ltd of Tokyo plans to cut its production of sponge titanium by 10%.

The company says it currently produces 880 tonnes a month of sponge titanium at its Chigasaki factory in Kanagawa Prefecture for export to customers in the USA. However, two US titanium producers have recently increased their production capacities, resulting in an excess of titanium sponge on the US market.

Nippon Light Metal expands aluminium production

Nippon Light Metal Co Ltd of Tokyo, in cooperation with Alcan Aluminum Ltd of Canada, is expanding its production of aluminium parts for use in cars for Japan and Thailand. The move is expected to encourage car makers to use more aluminium to reduce vehicle weight.

Nippon Light Metal has begun sample shipments of aluminium–magnesium alloy and aluminium–magnesium–silicon alloy sheets in Japan, and will construct a facility for mass production at its Nagoya plant. The company expects the sheets to be used in cars by 1995.

The company foresees total demand for aluminium alloy from Japanese car makers to increase to 190 000 tonnes a year by 1995, from 65 000 tonnes in 1989.

The company's Thai subsidiary, Nikkei Thai Aluminum Co Ltd, will increase its output of aluminium alloy ingots from 6000 to 12 000 tonnes a year by September 1991, and to 36 000 tonnes a year in the future. Nikkei Thai is held 75% bv Nippon Light Metal, with the remainder belonging to Alcan. The company supplies its output to the Thai plants of Japanese car makers.

Sumitomo merges aluminium alloy subsidiaries

Sumitomo Corp of Osaka merged its two aluminium alloy subsidiaries, Kansai Light Metal Co Ltd and Sumi-Al Alloy Co Ltd, on 1 October 1991.

The new company, Summit Aluminum Ltd, is capitalized at ¥100 million and will be the second largest operation in the aluminium alloy industry, Sumitomo claims. Summit Aluminum will have a production capacity of 120 000 tonnes a year (9% of the market).

Kansai Light Metal is a manufacturing and sales company dealing in aluminium alloys. Sumi-Al Alloy is an aluminium alloy manufacturer. The combined yearly sales of the two companies in the fiscal year 1990 was ¥31 billion.

Sumitomo says that the merger will improve efficiency in research and development, production and sales.

Kobe Steel establishes joint venture in Venezuela

A joint venture for refining iron and steel has been formed by Kobe Steel Ltd, in association with the Venezuelan national steel company, Corporation Venezolana de Guayana (CVG).

The venture has been capitalized at ¥4 million, held 51% by Kobe Steel and 49% by CVG. A feasibility study is to be completed on a proposed two million tonnes a year crude iron refining facility. The planned facility will use Kobe Steel's production technology and cost US$1.5 billion. A business plan will be completed in early 1992, the Japanese company says.

Production from the planned factory will be supplied to customers in the USA and Venezuela, giving Kobe Steel the chance to enter the US crude iron market.

Yokohama Rubber increases sales of aircraft parts

Sales of aircraft parts totalling ¥21.992 billion in the fiscal year 1990, 1.7% higher than the figure for 1989, have been reported by Yokohama Rubber Co Ltd of Tokyo. Aircraft parts accounted for 7.7% of the firm's total sales.

Sales of defence-related equipment reflected budget cuts and increased only marginally, the company says. However, demand from the commercial sector was strong.

Higher production rates for commercial transports at Boeing meant that Yokohama sold more honeycomb panels, lavatory modules, drinking water tanks, and other equipment. Further contributing to the increase were sales of metal parts for the "H-2" rocket's "LE7" engine, and honeycomb panels for cars and trucks, and for industrial applications.

Yokohama Rubber expects sales in the fiscal year 1991 to increase by 0.5% to ¥22.1 billion. Plans for major new products include:

- parts for the "SH-60J" and "UH-60J" helicopters;
- parts for the "H-2" rocket;
- parts for the "T700" engine;

- a flexible tank for commercial transports; and
- electromagnetic radiation shielding to be used in "Patriot" missiles and the "EMIS-B".

Yokohama Rubber has also developed a lavatory module for the "B737" aircraft. Modules for 20 aircraft have been built, and about 30 aircraft modules are planned, the company reports.

Japan's major steel makers report drops in profits

In their financial reports for the fiscal year 1990, four out of five of Japan's leading steel makers reported a decline in profits (see Table 2.1).

The five firms are: Nippon Steel Corp, NKK Corp and Kawasaki Steel Corp of Tokyo, Kobe Steel Ltd, and Sumitomo Metal Industries Ltd of Osaka. Only Kobe Steel reported an increase in profits.

The decline in profits was blamed on the increased costs of raw materials, labour and transportation, which more than offset higher sales. These problems are also expected to result in lower profits for the current fiscal year (see Table 2.2).

Kobe Steel invests in joint venture with USX Corp

The Japanese company Kobe Steel Ltd plans to invest US$400 million in plant and equipment in USX/Kobe Steel Co Ltd, a joint venture which was established in equal partnership with USX Corp in 1989.

Kobe Steel plans to invest the money over the next five years. US$200 million will be invested before 1992, and a further US$200 million will be invested in early 1992.

The money will finance:

- rebuilding of USX/Kobe's No.3 blast furnace;
- modernizing the coil handling facilities at the bar mill;
- constructing coke screening facilities;
- modernizing the mandrel mill at the No.4 seamless pipe facility; and
- constructing a metallurgical testing facility.

Table 2.1: Five leading Japanese steel maker's results for the fiscal year 1990.

Company name	Sales	Pretax profits	After-tax profits
Nippon Steel	2 608 283 (1.4)	160 920 (-20.5)	67 061 (-30.9)
NKK	1 326 832 (3.9)	50 371 (-36.6)	23 315 (-24.4)
Kobe Steel	1 321 409 (8.2)	56 530 (0.9)	23 380 (1.5)
Kawasaki Steel	1 185 435 (6.6)	80 529 (-22.2)	38 107 (-24.0)
Sumitomo Metal	1 156 852 (3.3)	73 596 (-23.1)	47 631 (56.3)
Total	7 598 811 (4.0)	421 946 (-21.8)	199 494 (-13.9)

Table 2.2: Five leading Japanese steel maker's forecasts for fiscal year 1991.

Company name	Sales	Pretax profits	After-tax profits
Nippon Steel	2 600 000 (-0.3)	120 000 (-25.4)	n.a (n.a)
NKK	1 340 000 (1.0)	32 000 (-46.5)	n.a (n.a)
Kobe Steel	1 340 000 (1.4)	58 000 (2.6)	30 000 (28.3)
Kawasaki Steel	1 180 000 (-0.4)	55 000 (-31.7)	35 000 (-8.2)
Sumitomo Metal	1 160 000 (0.2)	46 000 (-37.5)	26 000 (-45.4)
Total	7 620 000 (0.3)	311 000 (-26.3)	-- --

USX/Kobe Steel, whose main products are steel bars and seamless pipes, also produces 2.7 million tonnes a year of semi-refined black copper. In 1990, the company had total sales of US$786 million.

Sumitomo Metal to commence production of rare earths

Sumitomo Metal Industries Ltd of Osaka intends to enter into production of rare earth elements.

The company has begun construction of a special rare earth elements plant at its steel mill in Wakayama Prefecture. Construction costs are expected to total ¥900 million.

Sumitomo Metal plans to import ore containing rare earth elements from the US company Molycorp Inc and produce a total of 17 rare earth metals. The company also plans to produce rare earth alloys.

Products will be marketed by Sumikin Molycorp Inc, a Sumitomo Metal and Molycorp joint venture set up in January 1990 to market Molycorp products in Japan. Within three or four years, Sumitomo Metal aims to sell ¥5 billion worth of rare earth elements a year.

Sumitomo Metal steps up gold mining

Sumitomo Metal Mining Co Ltd of Tokyo plans to step up gold mining in Hishikari, Kagoshima Prefecture.

Sumitomo Metal currently mines 100 000 tonnes of ore a year, producing six tonnes a year of gold. With the expansion of the facilities, the company plans to raise its mining capacity to 150 000 tonnes a year.

The company will invest about ¥1 billion to establish additional facilities at the mine. Construction was scheduled to begin in June 1991, with completion by March 1992.

Sumitomo Metal wins order for continuous casting facility from China

Sumitomo Metal Industries Ltd of Osaka has won an order for a continuous billet casting facility from China.

The order, from the Yunming Casting Co in the city of Yunming in Yunnan Province, is for a facility with a capacity of 100 000 tonnes a year and it will include equipment for casting both ordinary and special-grade steels.

The value of the order has not been disclosed, but it is estimated to be in the region of ¥500–600 million.

Alcoa Aluminum constructs rolled aluminium plant

KSL Alcoa Aluminum Co Ltd (KAAL), a joint venture of Kobe Steel Ltd and the US-based Aluminum Co of America (Alcoa), has begun constructing its own plant for the production of rolled aluminium, in Moka City, Tochigi Prefecture, Japan.

The venture, which was established to produce and market aluminium-can stock, is constructing a 9500 m^2 plant on a 39 000 m^2 site adjacent to Kobe Steel's Moka plant. The new plant will be equipped with a German-made tandem cold rolling mill and associated finishing equipment.

When completed in early 1993, the plant will have a production capacity of 180 000 tonnes a year. It will use ingots supplied by Alcoa, and will supply the rolled product to Kobe Steel which will use it to manufacture drink cans.

CHAPTER THREE
CERAMICS

Carbides

Silicon carbide fibre withstands 1800°C

A silicon carbide fibre that displays no loss in strength up to 1800°C has been developed, according to the Japan Atomic Energy Research Institute (JAERI) of Tokyo.

The heat resistance was realized by reducing the oxygen content of the silicon carbide fibre from 10% to 0.5%. JAERI says it reduced the oxygen content by using electron beam irradiation rather than thermal oxidation during the infusion treatment.

The fibre is produced by melt-spinning polycarbosilane (PCS) at 320–350°C. JAERI then irradiates the fibre under vacuum. A fibre of 15–20 μm thickness requires 7–15 million grays to achieve infusion, the Institute says. Finally, the fibre is heated in helium at 1500°C, to convert it into a ceramic.

Self-lubricating cemented carbide

A self-lubricating cemented carbide has been developed and will be mass produced by Nippon Tungsten Co Ltd of Fukuoka.

The company claims the material is the first of its kind. The carbide has a hardness (HrA) of 80.5 and a bending strength of 0.81 GPa.

Nippon Tungsten makes the material by dispersing 15% boron nitride, a solid self-lubricating material, into a powder mixture containing tungsten carbide (1.5 μm in diameter) and 6.5% cobalt. The mixed powders are then sintered.

Refractory silicon carbide fibres

A research group headed by Chief Engineer Seguchi of the Takasaki Radiation Chemistry Research Establishment of the Japanese Atomic Energy Research Institute is claiming to have developed a silicon carbide fibre which does not lose strength on heating to temperatures as high as 1500°C.

The fibre is produced by firstly baking the polymeric precursor fibre under vacuum using an electron beam, followed by firing in an inert gas.

Carbon, graphite and diamond

Epitaxial diamond growth on nickel substrate

The successful epitaxial growth of diamond on a nickel substrate establishes firm prospects for the production of large-area diamond films, according to the Science and Technology Agency (STA) of Tokyo.

Workers at the STA's National Institute for Research in Inorganic Materials (NIRIM) and their colleagues from Onoda Cement Co Ltd used a single crystal nickel substrate, over which they circulated a 0.5% methane gas mixture at about 880°C. This produced epitaxial growth of diamond crystal on the nickel, the first time this has been achieved on a substrate other than cubic boron nitride, the researchers claim.

The group has yet to produce a single crystal film of diamond, but NIRIM has confirmed that epitaxial growth is possible on a single crystal cobalt substrate. In addition, NIRIM reports that a single crystal copper substrate may also be used in view of its similar crystalline structure.

Graphite for fission reactors

A graphite material has been developed by the Tokyo-based Japan Atomic Energy Research Institute (JAERI), in association with Toyo Tanso Co Ltd. The material has applications for use in high temperature gas reactors, the next generation of fission reactor, the partners say.

The material is made from pure graphite crystals, less than 0.1 mm in size, isostatically pressed with a rubber press. This process forces the crystals into an isotropic configuration, reducing shrinkage caused by neutron bombardment during service by one-half to one-third, the developers claim.

Further, because of its fine particle size, the material has a tensile strength of 20 MPa, twice that of conventional graphite, and its high purity doubles the corrosion resistance.

Heat resistant graphite is used in reactor-core structures such as fuel-rod sleeves, fuel blocks, and support posts. However, conventional graphite shrinks and cracks under neutron bombardment, and corrodes at high temperatures.

Diamond coating material

A diamond coating material with improved abrasion resistance and bonding properties has been developed by Osaka Diamond Industrial Co Ltd of Sakai.

The material is polished, using specially-developed equipment, to a mirror finish with a surface roughness of 5 nm and it can withstand a pressure of 6 MPa without peeling, the company claims.

Osaka Diamond expects the material to have applications in the production of semiconductors and optoelectronic components, and is considering collaborating with other companies in order to commercialize the product in the near future.

Superior graphite

The ability to produce sheets and blocks of polycrystalline graphite which has physical properties nearly identical to those of single-crystal graphite, but at a much lower cost, has resulted from research at the Matsushita Research Institute, Kawasaki.

The key to the success of the work was the use of poly(p-phenylene-1,3,4- oxa-diazole), "POD", as the raw material. While normal polymers yield poor quality amorphous carbon products on carbonization (heating to high temperature), POD remains essentially unchanged until about 1000°C. Then it turns into a black solid with a crystal structure resembling that of graphite, but containing nitrogen atoms. This has been called heterographite. Further heating to about 3000°C, combined with pressure, results in the formation of supergraphite.

In addition to the ability to form polycrystalline graphite films, by stacking sheets of the initial POD film, it is also possible to produce blocks of supergraphite since the layers bond together.

Applications for the material are thought to be very extensive. They range from the aerospace and nuclear power industries to automobile engines and gaskets for chemical plants. However, the first application has proved to be as the diaphragm in tweeters (high frequency loudspeakers). The material apparently eliminates some of the distortion found when other materials are used.

Electronic ceramics

Lead zirconate titanate withstands high pressure

A novel powder processing technique produces a lead zirconate titanate (PZT) piezoelectric ceramic which can withstand a pressure of 50 MPa, 2.5 times greater than conventional materials, according to its Kadoma-based developer.

Matsushita Electric Industrial Co Ltd says it used its own powder processing methods to form a mixed fine powder 0.2 μm in diameter. The powder, consisting of lead oxide, zirconia and titania, is then sintered at 1140°C.

The sintering temperature is 150°C lower than that used with conventional ceramics, where the starting powder is usually of 1 μm in diameter, the company says.

Thin films of lead zirconate titanate

A group of researchers from the Tokyo Institute of Technology's Faculty of Engineering claims to have used metal organic chemical vapour deposition (MOCVD) to form a lead zirconate titanate (PZT) thin film.

The research group has developed a novel organometallic compound which is used as the raw material in the MOCVD process. MOCVD has not previously been applied to the production of PZT because the lead has been difficult to vaporize, but the development of the organometallic compound has solved this problem, the scientists claim.

MOCVD produces a smooth film at a deposition rate of 30 nm a minute, compared with the sputtering method which can cause damage to the substrate and has a deposition rate of no higher than 2 nm a minute.

The group claims that the MOCVD method can also be used to form a PLZT (lead lanthanum zirconate titanate) film.

Electrical conductivity of zirconia-based solid raised

Adding lead zirconate titanate (PZT) raises the electroconductivity of a zirconia-based solid electrolyte, according to the Osaka branch of the Government Industrial Research Institute (GIRI).

Compared with liquid electrolytes, solid electrolytes are more stable and easier to handle. Solid electrolytes have potential for use in power cells and oxygen sensors, but have a low conductivity which currently makes their use impractical.

The Institute improved an electrolyte's conductivity by adding 60–70% PZT to an electrolyte powder which was then formed into a solid by sol-gel techniques. Compared to the unmodified solid electrolyte, the new material revealed a 100-fold increase in conductivity at 100°C. However, as the temperature increased the improvement over the conventional material was reduced and at 400°C there was no difference.

Method for making large pieces of lead lanthanum zirconate titanate

The National Institute for Research in Inorganic Materials (NIRIM), based in Ibaraki, claims to have developed a method of fabricating large pieces of lead lanthanum zirconate titanate (PLZT).

NIRIM researchers report that they have formed PLZT ceramic discs measuring 31 x 2.5 x 10 cm by mixing the powdered components in nitric acid followed by alkali precipitation. The hydroxide powder produced is washed and fired under pressure to yield a ceramic.

The method can produce a 5 cm diameter ceramic disc at a cost of about ¥8000, compared with ¥10 000 for conventional methods, according to NIRIM.

Nitrides

High performance silicon nitride turbine rotor

A high temperature ceramic turbine rotor for use in gas turbine engines is the result of collaborative research.

Mitsubishi Motors Corp of Tokyo and NGK Insulators Ltd from Nagoya say the turbine is made of silicon nitride and can operate for up to one hour at 1410°C, a significant improvement over their earlier design which worked for about six minutes at 1300°C. The top speed of the turbine is 700 $m.s^{-1}$ and fuel consumption is comparable to that of diesel engines.

The two firms plan to develop a fully operational ceramic gas turbine in about ten years.

Hybrid ceramic sustains 1000°C temperature gradient

Japan Metals and Chemicals Co Ltd of Tokyo claims to have developed a hybrid ceramic which can withstand a temperature gradient across it of about 1000°C. The material is an intimate mixture of boron and silicon nitrides and is produced by pressure sintering.

Micro-structure of silicon nitride ceramic observed

A means of observing the micro-structure of a silicon nitride ceramic has been developed at its National Institute for Research in Inorganic Materials (NIRIM), says The Science and Technology Agency. A material's strength as well as its resistance to heat and thermal shock are closely related to its micro-structure.

The Institute's technique involves treating a sample of ceramic with hydrogen plasma and then examining the surface with a scanning electron microscope.

Silicon nitride ceramic consists of fine particles of silicon nitride mixed with an amorphous glass-phase. When the ceramic is etched with hydrogen plasma for 3–10 hours, the silicon nitride is dissolved away, leaving the glass phase behind as an irregular surface. This surface was observed with a scanning electron microscope in order to reveal the glass phase's particle size, shape, composition and distribution.

Use of a transmission electron microscope could reveal finer structure within the ceramic, but would be unable to show the larger structures revealed by this technique.

Heat resistant amorphous silicon nitride

Tonen Corp Ltd has developed an amorphous silicon nitride which it says exhibits heat resistance up to 1700°C, a tensile strength of 2500 MPa, and a modulus of elasticity of 200 GPa.

The company makes the material by mixing trimethyl boride with perhydropolysilazane to make polyborosilazane which is then sintered at 1700°C to finally produce silicon nitride with 93% conversion efficiency.

Improved boron nitride–silicon nitride composite

A boron nitride–silicon nitride composite combines high elastic strength with a high resistance to heat shock, according to its Tokyo-based developer.

Japan Metals and Chemicals Co Ltd reports that its material has an elastic strength of 400–600 MPa, compared with 300 MPa displayed by conventional boron nitride–silicon nitride composites. Further, the material can withstand heat shock to 1250°C.

The company produced the composite using novel technology based on normal pressure sintering techniques. Heat-shock resistant composite materials are conventionally produced using hot-press methods which result in a material with a low elastic strength which is hard to mould, Japan Metals and Chemicals says.

Other materials

High strength mullite

A high strength porous ceramic made from needle crystals of mullite, an aluminium silicate compound, has been developed by researchers at Saga Prefectural Institute.

The ceramic has heat resistance to 1700°C in air. According to the scientists, this is partly because the mullite crystals do not contain alkaline impurities which easily undergo glass transition.

The mullite ceramic has a small thermal expansion coefficient, a stable porosity ratio and pore size, and displays resistance to impact, heat, acid and alkali.

The material was made from Kaolin clay which was fired at 1550–1650°C and crushed into particles measuring several millimetres across. These particles were treated with acid or alkali followed by washing with water to obtain needles of mullite 3–5 μm in length and 0.5–1 μm in thickness.

The mullite was press moulded under a pressure of 1770 MPa and fired in air at 1550–1700°C for two hours to yield the porous ceramic. The material does not require a binder because the silica, aluminium, and oxygen constituents chemically recombine at high temperatures and consequently bond (sinter) the mullite crystals together.

When fired at 1650°C, the ceramic was created with a pore size of 0.54 μm, a porosity ratio of 48.1%, and an average flexural strength of 52 MPa. Strength and porosity were reduced when fired at lower temperatures.

High strength cermet

A ceramic–metal (cermet), "CX-90", which has the same hardness and 1.5 times the strength of conventional materials of this type has been developed, according to claims by Dijet Industrial Co Ltd.

To produce the cermet, the company uniformly mixed ceramic and metal powders under a nitrogen atmosphere and then sintered the mixture. The company plans to produce 100 000 CX-90 drill bits a month which will cost ¥1270 each.

Synthetic kaolinite

Synthetic kaolinite has been developed by the Government Industrial Research Institute (GIRI) in Nagoya, in collaboration with the Artificial Clay Research and Development Association.

The material is a thin hexagonal crystal less than 1 μm in diameter and made from silica and alumina. The synthetic kaolinite has a similar structure to the natural material and exhibits high dispersion and plastic properties, the researchers claim. In addition, because the contents of iron and titanium are extremely low, the synthetic kaolinite has a beautiful white colour after sintering.

The researchers expect that the material will prove to be useful for moulding. The association will commission other companies to produce the material, but it will retain responsibility for marketing the synthetic kaolinite which costs ¥30 000 per kilogram.

Machinable ceramic with high flexural strength

A machinable ceramic which exhibits a flexural strength nearly twice that of conventional products has been developed by Mitsubishi Electric Corp of Tokyo.

The company says the material has a flexural strength of 180 MPa, 1.8 times that of standard ceramics.

The material is made by press moulding a mixture of mica and diopside crystals at a temperature in excess of 1000°C. This produces a ceramic composed of fibrous and plate-type crystals.

Because the fibrous crystals, which have a uniform particle size of about 10 μm, are formed during the material's fabrication, they are integrally bound with the plate-type crystals. The small particle size of the crystals and the uniform, integrated nature of the material's structure results in the increased flexural strength.

In tests, the surface of the machined ceramic had a roughness of 0.4 μm, between 1.5 and five times smoother than conventional machined ceramics.

Multi-pore ceramic

"Suncera" is a porous ceramic which has a larger surface area than conventional ceramics plate, and is lightweight and gas permeable, reports Asahi Chemical Industry Co Ltd of Tokyo.

The company has begun test marketing the product which it makes by fusing nylon or acrylic fibre with alumina, zirconia or mullite. Firing of the fused composite then produces a ceramic with pores of diameter 1–2 mm. The process reportedly allows control of the number of pores and their shape.

Asahi Chemical has priced the material at ¥1500 for 15 cm^2 and is expecting to sell it to electronics manufacturers who will use it for component baking shelves.

Support for plant tissue culture

A ceramic support for use as a plant tissue culture medium in the seedling industry is part of Tokyo-based Nippon Steel Chemical Corp's plan for business diversification, the company says.

Most seedlings which are mass produced by tissue culture develop from calluses into plantlets on a solid agar medium. Unfortunately, the procedure is labour intensive as each developing plantlet must be transferred individually from one agar medium to another containing different nutritional and hormone components to promote growth and differentiation. Moreover, transfer leads to root damage which restricts the yield to only 50% of the developing calluses.

To overcome these problems, Nippon Steel Chemical developed the ceramic medium which supports the developing plantlet in a culture solution. This allows the solution to be changed without removing the plant.

In experiments, the company used the ceramic support to produce carnation, chrysanthemum and other plants by tissue culture and achieved a yield equivalent to that obtained by conventional agar culture. Moreover, production costs were markedly reduced.

The company says the ceramic support may allow tissue culture of ginseng plants which have not been readily cultured on agar media to date.

Oxides

Catalyst for hydrogen production

A catalyst for the production of hydrogen from naphtha and water vapour has been developed by Idemitsu Kosan Co Ltd of Tokyo.

Unlike conventional processes, which are catalysed by nickel-based materials, the new catalyst can be used at normal pressure. Furthermore, it uses only one-third of the water vapour consumed by conventional processes, the company claims. The reaction temperature for the new process is 750°C.

The catalyst is based on zirconium oxide powder. Idemitsu Kosan impregnates the uniformly-sized pores of the oxide with 0.5% ruthenium plus several other inorganic catalysts.

Because the catalyst can also be used with propane, kerosene and methanol, in place of naphtha, the company says it has potential applications for phosphoric acid fuel cells.

Rapid production of fine powders of metal oxides

Rapid production of fine powders of metal oxides is possible using a method developed at Saitama University, researchers there claim.

Using the method, a fine powder of a metal oxide is produced in 1–5 minutes. The particle size of the powder can be controlled to between 5 nm and several microns, according to the scientists.

The researchers use a plasma heating system that consists of a microwave generator and an evacuation unit. A metal hydroxide or carbonate is placed in a glass tube, which is evacuated. The tube is then irradiated with microwaves, generating a plasma that produces the fine powder.

In experiments, when 10 grams of magnesium hydroxide was placed in the tube and irradiated with

100 W, 2.45 GHz microwaves for five minutes, a magnesia fine powder with a specific surface area of 372 $m^2.g^{-1}$ was produced.

Magnesia powder is used as a catalyst in benzaldehyde condensation reactions. Reportedly, the powder produced by the new technique had six times the catalytic activity of conventional materials produced by heating in an electric furnace.

Processing and equipment

Uniform lead–magnesium–niobium oxide particles

Uniform particles of lead–magnesium–niobium (Pb-Mg-Nb) oxide with an average diameter of 0.32 μm have been synthesized by researchers at the Tokyo Institute of Technology's Faculty of Engineering.

The composition contains lead, magnesium, niobium and oxygen in the ratio 3:1:2:9. The material has the same structure as barium titanate ($BaTiO_3$) but has superior dielectric properties, according to the scientists.

The group, led by Professor Mizutani, uses an ultrasonic humidifier to generate a mist of 2 μm particles from a solution containing niobium oxalate, and lead and magnesium nitrides. The mist is heated to 700°C, causing components of it to undergo reactions and form 0.32 μm Pb-Mg-Nb oxide particles.

A ceramic is formed by sintering the particles at 750°C. In this way, the group aims to form a high performance dielectric material.

Sol-gel technique produces porous glass

A sol-gel technique has been used to produce a porous glass with 2 nm pores by a research group at the Government Industrial Research Institute (GIRI), Osaka.

The researchers say they used a tetraethoxy-silane (TEOS) and zirconia sol to which they added polyethyleneglycol (PEG). They hydrolysed the sol in an aqueous solution containing ethyl alcohol and hydrochloric acid, and then polymerized, dried, and heat treated it to obtain the silica–zirconia-based porous glass.

The material contains about 30% zirconia in the glass skeleton, and has a constant pore size of 2 nm. According to the researchers, the volume and surface area of the pores can be controlled by adjusting the amount of PEG without changing the pore size. Furthermore, when porous glass with a pore size of 5 nm is produced, there is no carbonization of the residual organic substances that normally occurs during firing.

The group has produced a double-layered film consisting of a porous ceramic with a pore size of 0.2 μm that is coated with the newly-developed material. The film is expected to be used as a separation membrane.

Mass produced diamond-coated cutting tools

A method of mass producing diamond-coated cutting tools has been devised by Nachi-Fujikoshi Corp, the Tokyo-based company claims.

Nachi-Fujikoshi says it has been able to produce the tools on a large scale as a result of developing a substrate suitable for coating with diamond and finding the ideal gas mixture for the deposition.

Using a diamond-coated insert produced using the method, it was possible to cut high-silicon aluminium alloy continuously for over three hours at a speed of 300 metres a minute, with a depth of cut of 0.2 mm and a feed rate of 0.15 mm per revolution. In experiments, a drill and an end-mill were able to cut the alloy 10–30 times more efficiently than was possible using an ordinary cemented carbide tool, the company reports.

Nachi-Fujikoshi expects to be able to produce drills and end-mills cheaper with its new method. Throw-away tips are expected to be about the same price.

Nachi-Fujikoshi anticipates growing demand for diamond-coated cutting tools as car makers are increasingly using high-silicon aluminium alloy, carbon fibre reinforced plastics and other difficult-to-cut materials in order to reduce the weight of their vehicles. The company has begun large-scale production of these tools, and marketing will be targeted at car makers, home electric appliance manufacturers and firms in the aircraft industry.

Silicon nitride thin films to protect liquid crystals

A low temperature thermal chemical vapour deposition (CVD) technique has been developed by researchers in Tokyo. They say it has applications for the formation of thin protective films of silicon nitride over liquid crystal displays.

The scientists, from the Tokyo Institute of Technology's Faculty of Engineering, heat a glass substrate to 350°C and then expose it to trisilane and hydrogen azide gases; 1.8 cm^3 a minute of trisilane and 2.0 cm^3 a minute of hydrogen azide, under a pressure of 1 torr. This produces a silicon nitride film on the substrate at a rate of 0.3 nm a minute.

The film has a dielectric resistance of 80 V, about twice that of films formed by plasma CVD, they claim. Furthermore, the films have no pinhole defects, which are characteristic of films formed by the latter technique.

Silicon carbide produced from rice husks

Biotechnological methods can extract silicon carbide from rice husks, according to a research group at the Industrial Research Center of Fukuoka Prefecture (IRCFP).

The group's method entails mixing the pulverized husks with an alkaline solution, then subjecting the mixture to pressure cooking at 30 atmospheres. Next, enzymes and micro-organisms digest cellulose and other organic compounds in the treated mass, and the residue is condensed and sintered to yield silicon carbide.

According to a spokesperson for the group, scanning electron microscopy shows that the product consists of needle-like crystals of silicon carbide.

Ceramic mould for die-cast aircraft parts

Ceramic moulds could soon have applications in the production of die-cast components for aircraft, according to the Mechanical Engineering Laboratory in Ibaraki, part of the Tokyo-based Agency of Industrial Science and Technology (AIST).

The Laboratory says it has completed tests using a newly developed technique, the "Constant

Temperature Die-Cast Method", to produce ceramic moulds for such uses. AIST expects the method to find applications in the production of parts such as die-cast nickel alloy seats.

The researchers say they used an electrical discharge machine to produce a mould with a diameter of about 25 cm made from sialon. The sialon can withstand casting up to 1000°C and is eventually expected to replace conventional molybdenum alloys for this application. Another advantage of the sialon mould is that casting does not have to take place in a vacuum, as with molybdenum alloy versions, the scientists say.

The casting method was developed using a supercomputer to analyse the behaviour of sialon during processing of the mould and during casting.

Etching silicon carbide substrates

A method for etching silicon carbide substrates might ultimately produce gratings which can efficiently diffract X-rays from a synchrotron orbital radiation source, according to the researchers at Tokai University who developed the technique.

The method uses a 249 nm excimer laser to etch a substrate which is placed in an atmosphere of chlorine trifluoride gas. In an experiment, pulsed laser beams hit a silicon carbide substrate 18 000 times, creating a circuit pattern 4 μm wide and 0.1 μm deep.

The scientists say that the boundary lines of the grooves must be made sharper before the substrate can be used as a diffraction grating. They plan to adjust the laser output and make other improvements to make this possible.

High speed cutting

A method of cutting hard and brittle materials at high speeds has been developed by the Government Industrial Research Institute (GIRI) in Kyushu.

The Institute says it uses a rotating ceramic tool which vibrates at 40 kHz. The method produces a smoother cut with increased precision when compared with conventional techniques and is 5–10 times faster than using a diamond-coated cutting tool.

GIRI plans to use the same principle to develop tools for boring, shearing and grinding operations.

Brazing technology produces diamond—tipped cutting tools

A brazing technology for producing diamond-tipped cutting tools has been developed by workers at Nisshin Diamond Manufacturing Co Ltd, in collaboration with Ryukoku University and the Industrial Research Institute of Shiga Prefecture.

The complete process takes 20 minutes and uses facilities costing less than ¥1 million. This is quicker and cheaper than conventional methods which take about three hours and require an electric vacuum furnace costing several tens of millions of yen, the researchers claim.

The process involves sandwiching a pellet, made from 50% titanium and 50% carbon powder, along with brazing materials, such as silver, copper and titanium, between a diamond cutting tip and the metal tool. Heat is then applied to the pellet, causing the titanium and carbon to react with each other and generating temperatures of 3000°C in the process. This heat melts the brazing materials, creating a bond between the diamond and the metal with a strength of 200 MPa.

Reportedly, the technology can also be used to bond ceramics to metals and is expected to have applications in the development of new engines.

Particle size controlled by jet spray method

Jet spraying of a supercritical solution produces synthetic fine particles of a controlled size, according to a research group, led by Professor Kunio Yoshida, from the University of Tokyo.

Jet spraying involves blowing a supercritical solution through a nozzle under pressure. This pushes the fluid past the supercritical point and solid fine particles are produced. The research group has discovered that the particle size becomes smaller as the concentration of the solution decreases.

In experiments, the scientists prepared several aqueous solutions of silica in different concentrations and blew these supercritical solutions out from a 0.1 mm diameter nozzle. When the silica solution had a concentration of 194 parts per million (ppm), particles with an average size of 0.39 μm were obtained. A 109 ppm solution produced particles with a size of 0.31 μm.

Spherical, mono-dispersed, hydrated alumina particles

A uniform precipitation method produces spherical, mono-dispersed, hydrated alumina particles which are suitable for sintering, according to researchers at Saga University.

The scientists mix 0.001 moles of aluminium sulphate with 0.03 moles of urea, dissolve the mixture in water and heat it to 98°C to form hydrated alumina. They retrieve the alumina by sedimentation or centrifugal separation, and then dry it at 105°C for one day.

The drying process shrinks the particles by 20% without disrupting their shape. Alumina particles of a uniform shape and 0.5 μm size are obtained, they say. The shorter the growth time and the lower the aluminium sulphate concentration, the smaller the size of the particles formed.

High strength ceramic coating technology

Mitsubishi Electric Corp of Tokyo says it has developed a technology which uses a carbon dioxide (CO_2) laser to form a peeling-resistant ceramic thin film at high speed.

In the process, a piece of ceramic is placed in a vacuum vessel along with the substrate. The ceramic is irradiated with the laser as oxygen is slowly introduced to the chamber, at a rate of 0.16 $\mu m^3.s^{-1}$, resulting in the formation of a ceramic vapour which deposits itself upon the substrate as a thin film.

The process does not require a high substrate temperature and is reported to form a film which is ten times stronger than conventional methods.

In experiments, a silicon oxide thin film was formed on stainless steel and silicon substrates which were held at about 200°C. The thin film was formed at a rate of 1 μm a minute.

The technology could be used to coat the surface of mechanical parts with a high quality ceramic thin film which improves their resistance to heat and abrasion.

Aluminium nitride synthesis

A vapour phase deposition technique which forms a fine aluminium nitride powder is the result of research at Saga University.

The aluminium nitride powder is formed by heating an aluminium adduct of aluminium chloride and ammonia ($AlCl_3$–NH_3) to 300–400°C in a nitrogen atmosphere. When the vapour is mixed with ammonia, the crystals are deposited.

Provided the ammonia concentration is four times that of the adduct, the crystals grow radially in the form of tetrahedra. If the ammonia concentration is 10 times higher, spherical particles are formed.

An aluminium nitride film is formed by mixing the adduct vapour with hydrogen, causing the adduct to deposit itself onto an alumina substrate. The alumina substrate is heated from beneath, resulting in decomposition of the adduct and growth of the aluminium nitride film.

So far, five types of film, each comprised of a different particle type, have been formed depending on the position of the substrate.

Aluminium nitride has high electrical insulation, high heat and corrosion resistance, and high thermal conductivity. It has potential applications for producing aluminium alloys, cermet and integrated circuit substrates.

Layered titanium dioxide/tin oxide thin films

Taiyo Yuden Co Ltd, based in Tokyo, has announced that it has developed a method for producing alternate layers of titanium oxide and tin oxide on glass substrates.

No details have been released other than the layers are typically 1–2 μm thick, they are formed at a processing temperature of 500°C, and the titanium dioxide crystal is rutile.

High speed growth of tantalum oxide thin films

The successful development of a high speed process for producing tantalum oxide thin films has been announced by the National Chemical Laboratory of Industry, based in Ibaraki Prefecture.

The process is based on the use of excimer lasers combined with chemical vapour deposition and operates at a temperature of just 130°C. The speed of film growth is stated to be twice that for conventional production of tantalum oxide thin films.

The National Chemical Laboratory reports that if a high energy excimer laser is used the thin film produced will be crystalline, while if a low energy laser is used the film is produced in an amorphous state.

Aluminium nitride whisker processing

The Faculty of Engineering at Niigata University claims to have developed an efficient processing route for aluminium nitride whiskers.

The process involves mixing 5–10% of ammonia with nitrogen gas and aluminium vapour at 1500°C. Under suitable conditions (not specified), aluminium nitride whiskers up to 1 mm long can be grown in about 10 minutes, while whiskers 15 mm long are formed after an hour. The efficiency of the process has been put at 40%.

Titanium dioxide crystals mass produced using floating zone method

Nittetsu Mining Co Ltd has established a facility for mass producing large titanium dioxide (TiO_2) crystals. The new facility employs a floating zone (FZ) technique developed with the assistance of the Science and Technology Agency's National Institute for Research in Inorganic Materials (NIRIM).

The facility produces TiO_2 crystals using the FZ process, which uses focused light from a halogen lamp to melt a metal sinter. The molten sinter then cools to form a single crystal with a diameter of 25 mm. The company reports that its technique produces crystals which have superior qualities compared to crystals produced by the Berouli method.

Initially, the company plans to sell the crystals for ¥450 000 each.

Technology developed for production of highly stable aluminium nitride whiskers

Researchers from Niigata University's Faculty of Engineering have developed technology for producing highly stable aluminium nitride whiskers.

The process involves mixing 5–10% ammonia with nitrogen gas which is then reacted with vaporized aluminium at about 1500°C. The reaction produces a powdered product and the desired whisker product. In experiments, 1 mm long whiskers were produced by the reactor in 10 minutes and these grew to 15 mm in one hour. A reported maximum of 40% aluminium nitride can be converted into whiskers.

Whiskers soaked in water for five days displayed a decomposition ratio of less than 4%, 1/20 of that displayed by the powder. When exposed to a temperature of 1000°C, oxidation of the whiskers was less than 3% compared with 40% for fibrous aluminium nitride.

Seikei University group develops synthesis method

A new method for synthesizing ceramics, the "localized reaction method", has been announced by researchers at Seikei University. Using this method, the group has synthesized a ceramic wire 10–100 μm in diameter and several centimetres in length.

The method uses a hollow thread, made from polymer, metal, or ceramic, as a micro-reactor. Alkoxides are introduced into the tube and then hydrolyzed by water or atmospheric moisture. The hydrolyzed product is removed from the tube, then dried and fired at between 300°C and 1200°C. The group reports that it has used the method to produce ceramics such as lead titanate zirconate, barium titanate, strontium titanate, manganese ferrite, barium titanate and manganese ferrite.

In addition to wires, the method can be adapted to form spherical materials and regularly-structured integrated ceramics.

High-strength bonding of boride–nickel cermet with steel

High-strength bonds between boride-nickel cermet and steel have been made by Asahi Glass Co Ltd.

In experiments, cermet with a hardness of 1000 $kg.mm^{-2}$ was bonded with a segment of SUS 420J2 steel measuring 30 x 2 x 15 μm. The bond produced by the process had a strength of 1100 MPa which is twice as strong as bonds produced by silver soldering. A 50 μm reaction layer was found to exist between both materials but no substantial changes in texture were noted around it.

The process involves polishing and degreasing the surfaces of both materials, placing them together and heating them in an atmosphere of 0.2 torr or less.

Bonding method for ceramics

A method for rapidly bonding ceramic surfaces has been developed by Daihen Corp of Osaka.

The method involves the application of a fluoride-based paste to the ceramic surfaces which are then stuck together and heated. The bond is cemented by passing a 0.6 A current through it for a period of 6–8 minutes.

In experiments, two alumina ceramic specimens measuring 15 x 15 x 20 mm were joined together at their 15 x 15 mm surfaces. The bonded ceramic exhibited a flexural strength of 220 MPa at room temperature, 70% of the ceramic specimens' individual strengths, according to Daihen.

Adhesion technique

A technique for bonding ceramics has been devised by Daihen Corp of Osaka, in collaboration with the Government Industrial Research Institute, Osaka.

The researchers coated potassium fluoride onto two pieces of silicon nitride ceramic which were then placed together and pre-heated to 800°C with a gas flame. They then applied 1000–5000 V across the two pieces for 10–20 minutes, resulting in complete adhesion. The bond has a high heat resistance and reportedly exhibits a bending strength of up to 425 MPa.

Daihen has begun sample marketing products fabricated with the method and expects to begin full-scale marketing in the near future.

Ultrasonic fusing of aluminium and ceramics

A line of computer-controlled ultrasonic machines designed to fuse metal and ceramic parts will shortly be marketed by Ultex Corp.

Known as the "Computer Assisted Welding (CAW) Series", the machines use 20 kHz and 40 kHz ultrasonic vibrations and apply pressure to fuse metal and ceramic parts in only a few seconds, according to the company.

Ultex plans to target marketing of the machine to the research and development laboratories of electrical equipment, car and semiconductor manufacturers. The 20 kHz version costs ¥2.5 million, and the 40 kHz version costs ¥6 million.

A Japanese semiconductor manufacturer has already placed an order for the machine.

Conventionally, a metal with strong affinity to both aluminium and the ceramic is used to form a joint between the materials, but such a process is very time-consuming.

Laser cuts alumina ceramics

A processor equipped with an yttrium aluminium garnet (YAG) laser for cutting alumina ceramics will shortly be marketed by Hoya Corp of Tokyo.

The "Laser Scriber 250" is intended for cutting alumina ceramics used to make substrates for hybrid integrated circuits (ICs). The system is designed to minimize the occurrence of micro-cracks on the cut surface, a problem common when using conventional carbon dioxide lasers, Hoya says.

The YAG laser has an output of 500 W and a wavelength of 1.064 μm. It can bore holes in substrates 0.635 mm thick at a speed of 10 $mm.s^{-1}$. Holes as small as 60 μm in diameter can be made.

In addition to alumina, the system can also cut aluminium nitride and silicon nitride which are beginning to be used to make IC substrates. The Laser Scriber 250 costs ¥40 million.

Electric furnace can operate up to 2600°C

An electric furnace which can generate stable temperatures up to 2600°C has been developed by the National Research Laboratory of Metrology in Ibaraki, according to the Tokyo-based Agency of Industrial Science and Technology (AIST).

The furnace's heat source consists of a specially developed tungsten mesh through which an electric current is passed to generate heat. This is surrounded by an array of reflectors made of molybdenum and tungsten to prevent heat loss by thermal radiation.

The furnace was developed to help measure the thermal resistance of materials with potential uses in high temperature environments — such as space planes, nuclear reactors and gas turbines. Samples are maintained at an extremely low pressure, and the furnace is equipped with a water cooling system to prevent overheating and improve temperature stability, AIST says.

Testing and standards

Ultrasonic apparatus tests fatigue

An apparatus which uses ultrasonic waves to conduct fatigue experiments on ceramics has been developed at the Science and Technology Agency's National Research Institute for Metals (NRIM).

A rod-shaped sample is fixed to a duralumin mount using an epoxy resin. A crack is inserted in the sample which is then irradiated with ultrasonic waves at a frequency of 15–25 kHz.

Changes in the crack are recorded by a displacement sensor. More precise observations can be made with a microscope.

NRIM says conventional hydraulic servo methods can only generate about 20 vibrations a second, and involve complex sample-mounting procedures.

Standard sample of sintered alumina marketed

Standard samples of sintered alumina are for sale from the Fine Ceramics Center of Nagoya.

"Refaceram-AL1" is 10 cm wide, 10 cm long and 1 cm thick. It has an alumina purity of 99.7% or more, a density of 3.92 $g.cm^{-3}$, and a bending strength of 350 MPa, the Center claims.

The material costs ¥24 000 for the general-use standard and ¥2000 for the laboratory-use standard. The Center predicts sales of 1000 sheets in the first year of marketing.

Raw material powder for the ceramic is produced by Sumitomo Chemical Co Ltd of Osaka, and the final product is mass-produced by Asahi Glass Co Ltd in Tokyo.

Research into the mechanical processing of fine ceramics is increasing, but the results from different laboratories cannot be compared as there is no standard, the Center says.

Corrosion rate of glass measured

Natural glass protected from weathering corrodes at a rate of about 1 μm in 2000 years, according to the Power Reactor and Nuclear Fuel Development Corp (PNC) of Tokyo.

Conducted in relation to studies on the disposal of high-level radioactive waste, the research considered a wide range of natural glasses in order to determine how stable these materials are over a long period of time.

Fine particulate basalt glass collected from a 9000 year-old formation in Oshima was observed to be corroded to a depth of about 5 μm from its surface.

Electron microscopic observation revealed three layers of material within the corroded portion, consisting of smectite, iron hydroxide, and amorphous hydrate. Below this depth the glass was unchanged. The sample was found at a depth of 2–3 metres below the ground's surface.

Samples collected from other formations and from around Mount Fuji displayed the same rate of corrosion.

Estimating life of ceramics at high temperatures

A technique for estimating the life expectancy of ceramics at high temperatures is the result of work by a research group from Tohoku University (led by Assistant Professor Toshimitsu Yokobori).

The group has devised relatively simple numerical expressions to calculate a ceramic's estimated life. Data, including the basic characteristics of the ceramic plus the temperature and mechanical stress which it will undergo, are variables in an expression which calculates the rate of crack formation. Integration of the expression leads to an estimation of the ceramic's life.

The technique can also be applied to metals which exhibit a high ductility.

Business and financial information

Joint venture for fibre production in USA

Denki Kagaku Kogyo KK of Tokyo and the Austrian firm Rath Co Ltd have established a joint venture in the USA, Rath Performance Fiber Co.

The new company will produce "Alusen", a ceramic fibre used to reinforce metals. Denki Kagaku is also cooperating with Rath in the construction of an Alusen production factory in Europe.

Alusen is a short fibre consisting of more than 80% alumina and less than 20% silica. Denki Kagaku reports that Alusen has a heat resistance of 1200–1700°C, which is greater than that of conventional ceramic fibre. The Japanese company currently produces 200 tonnes a year of Alusen at its Oumi factory in Niigata Prefecture.

Mass-produced ceramic turbocharger rotors

Kyocera Corp of Kyoto has begun mass-producing ceramic turbocharger rotors at its Kokubu plant (Kagoshima Prefecture) at a rate of 10 000 units a year for supply to Toyota Motor Corp.

Toyota is already using the rotors in 2.5 litre engines used in the "Supra" sports car. Kyocera increased production to 30 000 units a month in mid-1991.

The rotor is made of silicon nitride and has a diameter of 5.2 cm. At a later date, the company plans to begin producing rotors with diameters ranging from 3.5 to 13 cm.

NGK Insulators Ltd and NGK Spark Plug Co Ltd are also mass-producing ceramic turbocharger rotors for supply to Nissan Motor Co Ltd.

Ceramic die production to start

The Japanese company Hara Precise Dies Corp, based in Saitama Prefecture, has now completed construction of a manufacturing plant for making ceramic dies. The plant is located in Aomori Prefecture and was built with a budget of ¥400 million. The company is predicting sales amounting to no less than ¥1000 million in two to three years time.

Engine valve from Kyocera

A ceramic engine valve has been developed by Kyoto-based Kyocera Corp, a major ceramic integrated circuit package manufacturer.

Kyocera is currently conducting durability tests and plans to begin mass production of the valve in 1993. Although the price is presently three times that of steel valves, Kyocera expects to reduce the cost before mass production begins.

Kyocera expects the valve business to increase its ceramic-related sales to ¥1 billion a year by 1995, approximately 2.5 times the current level.

Production factory to be constructed

NGK Spark Plug Co Ltd plans to construct a ceramics production factory in Ise City.

The company will invest a total of about ¥2.5 billion in the project, including the purchase of the site. The factory will consist of a one-storey building with a floor area of about 7500 m^2.

The final details of the construction project have not been decided yet, but construction is scheduled to begin in April 1992 and operations in April 1993. The factory will be staffed by 100 employees, and will produce ¥3.5 billion worth of ceramics a year.

Kobe Steel to develop thin diamond films in UK

Plans to start developing thin diamond films, at a research laboratory which it is renting from Surrey University in the UK, have been announced by Kobe Steel Ltd.

The University has recently constructed a laboratory, with a floor area of 1300 m^2, on the site of its research park. The company plans to rent this in order to house its researchers, whose number are being increased from 15 to 30; Kobe Steel already rents one laboratory where it has been developing polymer composites.

The company expects to produce diamond film for sample shipment in the near future.

Kobe Steel wins major order for hot isostatic presses

A ¥400 million order for two of its hot isostatic presses has been received by Kobe Steel Ltd of Tokyo, the company reports. The order is from the Japan Ultra High Temperature Materials Research Center Corp.

One of the presses has a maximum processing temperature of 3000°C, and the other a maximum pressure of 1000 MPa. The two machines were scheduled to become operational in January 1992.

Kobe Steel has received a combined total of 200 orders for its hot isostatic presses, which the firm claims accounts for nearly 90% of the 230 such pieces of equipment now operating in Japan, including both those for research and industrial production.

Increased production of ceramic parts at Kyocera

Kyocera Corp of Kyoto increased its production of silicon nitride turbo rotors to 30 000 units a month by October 1991.

Kyocera began mass production of the rotors in August 1990 when Toyota Motor Corp began using them. Kyocera raised production in anticipation of orders from other car makers.

In the near future, Kyocera will also begin shipping samples of silicon nitride intake valves which enable improved engine response, it claims. This will be followed by mass production. Previously, the company had only been supplying the valves for use in racing cars.

Kyocera hopes to increase sales of ceramic car parts to ¥10 billion a year by 1995.

Government support for ceramics research in 1991

Ceramics featured prominently in the materials-related research and development themes which the Tokyo-based Ministry of International Trade and Industry (MITI) supported in the fiscal year 1991.

Areas for support included:

- superconductive materials;
- ceramic gas turbine technology;
- superconductive power storage systems; and
- silicon-based polymeric materials.

The superconductive power storage project will concentrate on pilot plant experiments and the development of practical applications. The gas turbine project, part of the Moonlight Project, will concentrate on the development of ceramic blades and pipes for use in high-efficiency (over 42%) co-generation systems which burn natural gas and methanol.

MITI has also announced plans to construct four research centres which will each concentrate on a single topic, such as ultra-high temperature materials.

Glass substrate factory constructed

Nippon Electric Glass Co Ltd of Ohtsu has constructed a non-alkali glass substrate production factory at Wakasa Techno Valley in Fukui Prefecture.

The new factory required a total investment of about ¥3.5 billion and has a floor area of 10 800 m^2.

CHAPTER FOUR
COMPOSITES

Carbon–carbon composites

Carbon–carbon composite for Japan's space shuttle

Fuji Heavy Industries Ltd of Tokyo has asked the National Space Development Agency of Japan (NASDA) to adopt its carbon–carbon composite in the production of an unmanned space shuttle currently under development and called "HOPE".

On commission from NASDA, Mitsubishi Heavy Industries Ltd, Hitachi Ltd and Nissan Motor Co Ltd of Tokyo, and Kawasaki Heavy Industries Ltd of Kobe are involved in developing carbon–carbon composites.

Fuji has been independently conducting development of carbon–carbon composites and ceramics for use in HOPE and other aerospace equipment, in anticipation of joining the HOPE project. The company says it has succeeded in forming a composite which is durable up to 1800°C.

If allowed to enter the HOPE project, Fuji expects to work with Nissan, one of its major shareholders. Fuji and Nissan are already cooperating on development of the Japanese Experimental Module (JEM), a manned space station, scheduled for launch in 1998. Scheduled launch for HOPE is 1999–2000.

Fuji using carbon–carbon technology from Aerospatiale

Carbon–carbon composite technology developed by Aerospatiale of France is being used at Tokyo-based Fuji Heavy Industries Ltd's material production facility in Utsunomiya.

The technology is being used to produce carbon fibre fabric which has obliquely interwoven warps. This has a higher peeling strength than two-dimensional fabric and is easier to form into desired shapes than three-dimensional forms, Fuji claims.

The company also plans to use two other carbon–carbon composite technologies supplied by Aerospatiale: one for increasing the density of carbon by repeated carbonization and another for improving carbon's resistance to oxidation.

Applications for carbon–carbon composites include nozzles for use in solid fuel rockets. Fuji also expects the materials to be used in the Japanese space shuttle, "HOPE", scheduled for commercial use in 1999–2000.

Integrally moulded three–dimensional carbon–carbon composite

A carbon–carbon (C/C) composite made from carbon fibres which are three-dimensionally entangled and moulded into a dense texture has been announced by Nippon Steel Chemical Co Ltd and Nippon Steel Corp.

The C/C composite reportedly features a shear strength of 1.7 kg.mm^{-2}, which is more than 1.5

times greater than conventional products, a tensile strength of 5.8 $kg.mm^{-2}$, over three times greater than conventional products, and has twice the oxidation resistance of conventional products.

Conventional C/C composites, produced by laminating sheet-formed prepregs, tend to delaminate under stress.

The companies plan to market the composite for use in high- value-added products. As a first step, they are marketing nuts and bolts for high-temperature furnaces, as well as the unformed C/C base material.

Oxidation-resistant carbon–carbon composite

Applying molten silicon carbide to the fine pores on the surface of a carbon–carbon composite vastly improves the material's oxidation resistance, the joint developers claim.

Kobe-based Kawasaki Heavy Industries Ltd and the National Aerospace Laboratory of Chufu say the matrix-modified composite survives temperatures of 1700°C without suffering peeling, oxidation or corrosion. Conventional materials display oxidation at temperatures above 1300°C.

Carbon–carbon composite

A carbon fibre reinforced carbon composite which is suitable for use in space has been developed at Tokyo-based Nippon Oil Co Ltd, the company claims.

So far, Nippon Oil says it has produced a gas jet nozzle of complex shape and a 30 mm diameter cylinder which can withstand a temperature gradient of 1000°C across it.

High strength carbon–carbon composite

Joint research between Nippon Steel Chemical Co Ltd and the Nippon Steel Corp, both based in Tokyo has resulted in a carbon–carbon fibre composite which is claimed to have 1.5 times the bending strength and three times the tensile strength of existing similar materials. The composite gets its great strength from the homogeneous, three-dimensional arrangement of the carbon fibres in the carbon matrix, the companies say.

Nippon Steel Chemical, who are to market the composite, adds that probable applications will include electronic devices and nuts and bolts for high temperature furnaces and spacecraft.

Oxidation resistant carbon–carbon composite

Joint research between Kawasaki Heavy Industries Ltd, based in Kobe City, and the National Air Space Laboratory, part of the Science and Technology Agency, Tokyo, has resulted in a carbon–carbon composite which is claimed to have oxidation resistance at temperatures up to 1700°C.

The secret of the success is the introduction of a highly oxidation resistant form of silicon carbide into the carbon matrix. A fairly porous carbon/carbon composite is initially produced and silicon carbide is subsequently impregnated into the pores.

Ceramic matrix composites

Silicon nitride–silicon carbide with plasticity

A composite of silicon nitride and silicon carbide which is said to have plastic properties that greatly increase the ease with which the material can be processed has been developed in Nagoya.

The Government Industrial Research Institute (GIRI), Nagoya, produces the material by hot-press sintering a powder mixture of 80% silicon nitride (0.5 μm in diameter) and 20% silicon carbide (0.2 μm in diameter). The resulting composite has a high strength and toughness in addition to a high plasticity which allows it to expand by 50% at 1600°C, GIRI says.

Conventional silicon nitride–silicon carbide composites have high strength and heat resistance but are hard to process.

Ceramic–polymer coating

A ceramic–polymer coating, "Ceranode", which is resistant to water, chemical corrosion, heat and thermal shock has been developed by Toa Planning Co Ltd in Tokyo.

Ceranode is composed of ceramic-based particles, made from zirconia, silica, zinc and graphite, dispersed in a polymer matrix. It can withstand temperatures of up to 1800°C, whereas conventional coatings can function up to 600°C, the company reports.

In April 1991, Toa Planning began by producing 10 tonnes a month of Ceranode, planning to increase this to 40–50 tonnes a month within three years.

Silicon nitride composite displays strength at high temperature

The Technical Laboratory of Nippon Steel Corp, based in Tokyo, claims to have developed a silicon nitride-based composite which retains a strength of about 600 MPa at 1400°C and to possess a toughness of 9.1 $MPa.m^{3/2}$.

The material is a composite of silicon nitride containing an unspecified amount of silicon carbide whiskers and titanium nitride particulates and has been produced using slurry-based processing. The key to the success in developing this composite has been attributed to the degree of homogeneity attained in the slurry via appropriate control of pH during processing.

High strength ceramic composites

Noritake Co Ltd, based in Aichi Prefecture, claims to have developed high strength, fibre-reinforced ceramics.

The company has reinforced ceramic matrices with a three-dimensional woven carbon fibre cloth to produce materials which, it says, maintain their high strengths up to 1400°C.

Fibres

Elastic carbon fibre

A coil-formed, pitch-based carbon fibre which can expand and contract like a rubber band has been made, according to its joint developers.

The Government Industrial Research Institute (GIRI), Kyushu, Koa Oil Co Ltd of Tokyo and Tanaka Kikinzoku Kogyo KK say expansion and contraction of the fibre can be repeated several tens of thousands of times with no change to its structure. The fibre has a tensile strength of 700–1200 MPa, comparable to that of general purpose, linear carbon fibres.

The fibre is a composite of isotropic and anisotropic pitch. These are raw materials used for making general purpose and high strength, high elasticity carbon fibres, respectively.

The two types of pitch are spun into a composite fibre. 1000 fibres are then bunched together before being fired at 300°C in an oxygen atmosphere for 10–20 minutes, and then in a nitrogen atmosphere at 1000°C for a short time.

Because the isotropic and anisotropic constituents shrink at different rates, the fired fibres become coil-formed, with a small pitch of 1–2 mm.

Fibrous graphite with novel crystalline structure

A fibrous graphite with a novel crystalline structure results from the use of a pentacarbonyl-iron catalyst during synthesis, says Mitsubishi Kasei Corp of Tokyo.

The compound is composed of stacked carbon crystals arranged in a ribbon-like formation. This contrasts with conventional graphite whose component crystals are arranged in a planar structure. The material was produced by a method which involves heating carbon monoxide and hydrogen to 700°C in the presence of a pentacarbonyl-iron catalyst.

Fibrous crystals several microns long, 0.1–0.7 μm wide, and 0.01–0.2 μm thick are produced. The company reports that the iron compound catalyst is important in the formation of the structure.

Functionally gradient materials

Large, stable tungsten–copper composite

A prototype of a large electron beam target has been fabricated from a tungsten–copper composite by Toshiba Corp of Tokyo.

The target measures 80 x 80 x 25 mm, and is a seven-layer structure in which the transition from one material to the other is gradual with no discontinuities at the interfaces, a so-called functionally gradient material (FGM).

Large FGMs tend to suffer from warping, cracking, or peeling due to imperfect sintering, according to Toshiba. However, the company has established technology to control shrinkage of the individual layers during sintering, thereby achieving fabrication of a large, stable FGM.

Compared with two-layer tungsten–copper materials, the target has improved heat transfer properties which result in a surface temperature which is 80 K cooler during operation. The new material is also 50% stronger.

Fabrication of the FGM uses open hot isostatic pressing (HIP) treatment of the tungsten sinter to remove closed pores which tend to disturb heat transfer.

Toshiba has also formed a small tungsten–copper FGM using a sintering impregnation technique which results in a material whose texture becomes denser towards the middle. This technique was found to be effective for fabricating high capacity heat sinks.

Zirconia–stainless steel

A powder spray stacking method has been adapted to fabricate a zirconia–stainless steel whose composition gradually varies across its interface, a so-called functionally gradient material.

According to the researchers from Tohoku University, because there is no discontinuity at the interface, the resulting material exhibits three times greater resistance to thermal shock than zirconia alone.

In the powder spray stacking method, a mixed slurry of metallic and ceramic powders is sprayed through a nozzle onto a hot substrate where it drys into a film. The ratio of powder components in the mixture is continuously varied by computer-controlled feeding of the ceramic component.

The thickness of the functionally gradient layer can be varied from 150 μm up to several millimetres.

Actuator uses functionally gradient material

Tokyo-based Yokogawa Electric Corp says it has fabricated a prototype actuator from a functionally gradient material (FGM).

FGMs are composites in which the transition between the different materials is gradual. The actuator device, measuring about 50 mm long, 14 mm wide, and 0.8 mm thick, consists of alternating layers of piezoelectric lead titanate, which expands and contracts according to the voltage applied, and dielectric nickel lead niobate, which is not sensitive to voltage. The two materials are stacked in vertical and horizontal layers and are intermixed at the interfaces, thus avoiding delamination.

In tests on the actuator, AC voltages of different waveforms were applied. Due to the properties of the FGM, the surface deformed and warped to generate a progressive wave which generated motion. The direction of movement was reportedly changed by varying the voltage and waveform.

Metal matrix composites

Lightweight fuel tank

A lightweight pressure vessel for use as a fuel tank in cars operating on natural gas has been developed as the result of collaboration between three Tokyo-based companies.

NKK Corp, Tokyo Gas Co Ltd and Tokyo Gas Engineering Co Ltd say the vessel has a weight of only 22 kg, half that of conventional ferrite-based vessels. It has a capacity of 31 litres, a charging pressure of 20 MPa and can withstand pressures of up to 76 MPa.

The vessel is made from aluminium alloy lined with epoxy resin-impregnated glass fibre which has been thermoset.

Because of the weight reduction, NKK claims that use of the vessel can increase fuel efficiency by 5%.

High flexural strength nickel trialuminium

A nickel trialuminium (Ni_3Al) composite with a high flexural strength has been developed, according to claims by the National Defense Academy and Shikoku Chemicals Corp.

The joint developers say that the Ni_3Al compound displays a ductility of 0.6% at room temperature and a heat resistance in excess of 1000°C. However, the researchers aim to increase the heat resistance to 1200°C, allowing the material to be used in turbo-chargers and other applications.

Nickel trialuminium is an intermetallic compound, similar to titanium aluminium and niobium trialuminium, which is expected to be used in aerospace applications. Conventionally, intermetallic compounds have suffered from brittleness.

Automotive aluminium alloy composite

A special aluminium alloy composite for automobile engine parts has been developed at Kobe Steel Ltd.

The composite is made from the company's proprietary aluminium alloy, reinforced with silicon carbide particles several microns in diameter. Compared with conventional aluminium engine parts, components moulded from the material display several times greater resistance to corrosion, greater strength at high temperatures, and have a 30% lower coefficient of thermal expansion, Kobe claims. In addition, the composite has a modulus of elasticity equal to that of stainless steel.

Lightweight, high strength composites

The Hyogo Prefecture Industrial Experimental Station is to launch a joint three-year research project with five steel makers to develop lightweight, high strength composites.

The participants and their projects are:

Kobe Steel Ltd of Kobe — development and evaluation of composites containing aluminium;

Kawasaki Steel Corp of Tokyo — melting and precision casting technology for titanium–aluminium alloy;

Kawasaki Heavy Industries Ltd — processing technology for titanium–aluminium intermetallics;

Sumitomo Metal Industries Ltd of Osaka, Mitsubishi Heavy Industries Ltd of Tokyo, and the Hyogo Prefecture Industrial Experimental Station — development of high strength dispersed aluminium ceramics.

Several universities, including Osaka University and Kyoto University, will also be involved. The first year of the project will be funded by an initial investment of ¥30 million from Hyogo Prefecture and ¥10 million from each of the steel companies.

High strength titanium–aluminium

A high strength titanium–aluminium composite containing metal nitride exhibits an elastic limit of 1000 MPa, 2.5 times higher than that of titanium–aluminium alone, according to researchers from the Osaka Prefecture University.

The group reports that at present the ductility of the material is about half that of titanium–aluminium, but the researchers expect to improve this.

The sintered product is a composite structure consisting of a titanium–aluminium matrix, interspersed with approximately 30% Ti_2AlN and Ti_3Al. The interspersed nitride confers the greater resilience on the material, the scientists say.

To make the composite, the researchers blended a mixture of titanium and aluminium powders in stoichiometric ratio. They then moulded the mixture under a pressure of about 140 MPa, de-aerated it under vacuum in a quartz tube, and baked it in a nitrogen atmosphere at 1273 K for about five minutes. The resulting sinter was arc melted in an argon atmosphere to obtain a 40 gram ingot.

Other materials

Tough cermet alloy

A mechanical alloying technique has been used to synthesize a tough cermet alloy, according to the Government Industrial Research Institute (GIRI) of Nagoya.

GIRI says the titanium carbide–nickel alloy it produced exhibited a Vickers hardness of over 1070, with no cracks around the indentation.

The mechanical alloying system used by the Institute consists of a vibration ball mill, with stainless steel balls about 10 mm in diameter. GIRI researchers charged titanium, carbon, nickel and molybdenum powders in the mill, and titanium carbide was synthesized and alloyed.

The scientists say an amorphous phase formed in the powder after it had been mechanically alloyed for 200 hours. The powder was then sintered at low temperature to yield an alloy containing fine particles of titanium carbide.

The alloy is expected to be used in miniature drills for printed circuit board fabrication and in dot matrix printer heads.

Die to extrude engineering plastic sheets and pipes

A die made of composites for extruding engineering plastic sheets and pipes has been developed by Kobe Steel Ltd. The company claims that the die is 10–20 times more resistant to corrosion and wear than its competitors.

The inner surface of the die is coated with copper nitride and a layer of powder consisting of a ceramic and a nickel-based alloy, using a hot isostatic pressing (HIP) method.

The company expects to begin sample shipments imminently.

Strong gypsum composite

A composite based on gypsum, crushed fibre-reinforced plastic and an aqueous solution of sodium alginate has been developed at the Government Industrial Research Institute (GIRI) in Shikoku. The Institute claims that the composite has a compressive strength of up to 5 kPa and a bending strength of up to 2.3 kPa.

The composite is fabricated by kneading the raw materials into a slurry and pressure-moulding. After removal from the mould, the material is left for 12 hours, and then dried for two days at 45°C.

When the slurry is moulded under a pressure of 1 kPa, the compressive strength is 3.2 kPa and the bending strength 2.3 kPa. However, when the moulding pressure is raised to 6 kPa, the compressive strength increases to 5 kPa, but the bending strength drops to 0.8 kPa.

GIRI is now examining reductions in composite strength caused by increases in the fibre-reinforced plastic content or by water absorption.

Fabrication of ceramic–metal pipe

The Nagoya branch of the Government Industrial Research Institute (GIRI) says it has succeeded in using a "planetary ball die" of its own development to fabricate a ceramic–metal composite pipe. Ceramic-lined pipes are valued by the chemical industry for their durability and resistance to corrosive liquids and gases.

The fabrication method involves forcing a metal pipe containing a ceramic pipe into a planetary die. The ball bearings contained in the die do not exert pressure on the entire circumference of the pipe at the same time, thereby preventing breakage.

The outer metal and inner ceramic pipes are bonded together as the metal shrinks on cooling. The process reportedly avoids breakages due to thermal stress, and the processing force is half to one-third that for a rigid body die so there is no danger of fracturing the pipe.

When this type of composite pipe is fabricated using conventional techniques, the inner ceramic pipe tends to break or is insufficiently bonded.

Polymer matrix composites

Composites developed for fighter planes

The Japanese Defense Agency has reportedly decided to commission Mitsubishi Rayon Co Ltd and Toray Industries Inc of Tokyo to develop composites for use in the main wing and tails of "FSX" fighter planes.

The FSX will have large integrally moulded composite panels made from carbon fibre and polymer resin. The lower exterior panel, which measures 1.5 × 6 m and incorporates the wing and the spar in a single unit, will be moulded by Toray from prepreg supplied by Mitsubishi.

The upper exterior panel of the wing, as well as the tail, will be moulded by Fuji Heavy Industries Ltd, also of Tokyo, from prepreg developed by Toray. The whole plane will be assembled by Mitsubishi. The General Electric Co will also be fabricating wings for the FSX using materials supplied by Mitsubishi and Toray.

Prepreg for manufacture of aircraft primary structures

A prepreg for use in the manufacture of primary structural materials for aircraft is available from Yokohama Rubber Co Ltd of Tokyo.

The company says its prepreg is based on carbon fibre and an epoxy resin. In experiments, the prepreg was used to make a laminated carbon fibre-reinforced plastic (CFRP). Reportedly, at 121°C, the CFRP had a tensile strength of 540 MPa, a compressive strength of 1.69 GPa, and an interlayer shear strength of 830 MPa. Furthermore, the problem of plasticization during extended use was eliminated.

Composite aircraft bolts

Carbon-fibre reinforced plastic (CFRP) bolts for use in civil and military aircraft are to be developed by Kawasaki Heavy Industries and Fukuvi Chemical.

The use of CFRP has been increasing in aircraft production. However, since CFRP parts are currently joined with metal fasteners they are susceptible to galvanic corrosion. The new bolts are designed to overcome this problem.

Phenol resin for prepregs

Plans to develop phenol resins for use in prepreg production have been announced by Sumitomo Durez Co Ltd, a joint venture of Tokyo-based Sumitomo Bakelite Co Ltd and the US company Occidental Chemicals Corp.

Sumitomo Durez, which currently produces 4500 tonnes a year of phenol resin at its Shizuoka factory, plans to improve the adhesive and processing properties of the material so that it can be used to make prepreg. Usually prepreg is produced from epoxy resin because it is more adhesive and easier to process than conventional phenol.

The prepreg is expected to be used in the production of structural composites for aircraft.

Processing and equipment

Carbon fibre floc coatings

A method of electrostatically floc coating activated carbon fibre (ACF) has been devised by Wako Engineering Co Ltd of Tokyo. The company says it is usually difficult to coat ACF uniformly in this manner because of its electrical conductivity.

ACF fibre 0.8–1 mm in length and 10 μm in thickness can be coated to a density of 20 000–50 000 pieces of floc/cm^2, according to Wako Engineering.

The method involves preparing the coating tank in order to avoid the build-up of static charges. The metallic electrode is coated with an insulator, residual static charges in the tank are dispersed, and the inner surface of the tank is coated with an antistatic agent.

Because ACF has about 20 billion fine pores/cm^2, it can adsorb odours and solvents faster than activated charcoal. Further, the fine pores can be filled with insecticides or anti-mildew agents. ACF may be applied to building materials and clothing because of its fire-retardant and deodorization properties, the company says.

Sol-gel preparation of polyimide/silica

A sol-gel preparation has been used to make a composite composed of silica particles and heat-resistant polyimide, researchers in Tokyo report.

The group, led by Professor Toshio Imai and Associate Professor Masaaki Kakimoto from the Tokyo Institute of Technology's Faculty of Engineering, says the type of composite produced can be influenced by varying the relative silica content. It is possible to form either polyimide containing glass components or glass containing polyimide components. A greater proportion of silica gives rise to larger silica particles; 70% silica results in a particle size of 5 μm.

The production process involves mixing polyamide acid, a precursor of polyimide, with tetraethoxysilane, a precursor of silica.

Reinforcements

Lightweight reinforcement

A lightweight reinforcement which is cheaper than conventional butyl rubber-based materials has been developed for use in car panels, according to Nippon Zeon Co Ltd and Nissan Motor Co Ltd, both of Tokyo.

"Zeotac" is a mixture of thermoplastic and thermosetting resins, and a forming agent. The mixture is coated on glass cloth laminated with aluminium foil. Zeotac is 1 mm wide and weighs only 1 $kg.m^{-2}$, the companies report.

When the coated panel is heated on a baking paint line to 140°C, the Zeotac increases 7–8 times in size and hardens, resulting in good reinforcement, anti-vibration, and heat insulation properties.

Nissan is already using Zeotac in some of its cars. The material costs ¥2000 a square metre and Nippon Zeon is aiming at annual sales of ¥2 billion in five years.

Testing and standards

Structural test facility

A structural testing facility for composites has been completed by Mitsubishi Heavy Industries at its Nagoya Aerospace Systems Works.

The facility has a large environmental control chamber that can provide temperatures from –75°C to +150°C and relative humidities from 20% to 95%. There are also eight, 10 tonne actuators which each have a stroke length of 30 cm, and a 100 tonne fatigue test frame with a 15 cm stroke length.

Composite structures will be tested under a variety of loads, temperatures and levels of humidity at the facility, Mitsubishi Heavy Industries says.

Ultrasonic system for inspecting fibre reinforced plastics

The Research Center for Advanced Science and Technology of the University of Tokyo and Hitachi Construction Machinery Co Ltd have jointly developed an ultrasonic system for inspecting fibre-reinforced plastics (FRP) and other materials.

The system is expected to find applications in aerospace research to measure the effects of shock tests on FRP and multi-layer laminated structures. Defects can be determined with a high rate of accuracy by measuring changes in the speed and rate of attenuation of ultrasonic waves travelling through materials.

Hitachi Construction Machinery plans to begin marketing a commercial version of the system in the near future.

Business and financial information

Toray to boost carbon fibre production in France

By June 1992, Toray Industries Inc of Tokyo plans to double carbon fibre production capacity at its French subsidiary Société de Fibre de Carbone SA.

The company says it will invest FF270 million in expanding the subsidiary's annual capacity from the current 340 tonnes to 700 tonnes.

Toray is currently constructing a new carbon fibre production facility at its Ehime plant in Ehime Prefecture, Japan. When both facilities are completed the Toray group's total carbon fibre production capacity will be 2950 tonnes a year.

The company claims that the expansion is largely in anticipation of increased demand for carbon fibre by aircraft manufacturers. A portion of the increased production will be exported to the USA.

Dainippon and Hexcel establish composite company

Dainippon Ink & Chemicals Inc of Tokyo and US company Hexcel Corp have established a joint venture for producing composites used in aerospace applications.

The joint venture, called DIC Hexcel, is located in Ishikawa Prefecture and funded with ¥600 million. In December 1991, DIC Hexcel began producing honeycomb structured materials and prepreg sheets.

Carbon fibre production factory to be built

Mitsubishi Rayon Co Ltd plans to invest ¥5 billion in the construction of a 500 tonne a year carbon fibre production factory at its Toyohashi works.

The company currently produces carbon fibre at its Ohtake works in Hiroshima (150 tonnes a year) and at its Toyohashi works (350 tonnes a year). Construction of the new factory began in late 1991, completion is scheduled for late 1992, and operations will begin in spring 1993.

Following completion of the factory, the company plans to increase its exports of carbon fibre from the current 10–15% to 30% of production. Carbon fibre-based material is in demand in Europe and the USA for airplane construction.

Nylon-based composite produced

Full-scale production of a composite based on Nylon 6 has begun at Ube Industries Ltd.

The company will produce 1200 tonnes a year of "Ube Nylon 1015C" at its Ube Chemical plant in Yamaguchi Prefecture. The material is composed of Nylon 6 and silicate and was developed in association with Toyota Motor Corp and Toyota Central Research and Development Laboratories Inc.

The composite has higher heat resistance and about 70% greater elastic modulus than conventional nylon materials, but its specific gravity, 1.15, is only marginally larger than that of conventional Nylon 6, 1.14.

Ube is supplying Ube Nylon 1015C to Toyota for use in engine-related parts. The company also plans to supply it to resin film makers and is aiming at annual sales of ¥350 million in 1993. The material costs ¥700 per kilogram, some 15% more expensive than conventional Nylon 6.

UK joint venture for production and marketing of plastic composites

A joint venture for the production and marketing of plastic composites based in London, UK, has been established by Ube Industries Ltd in collaboration with Marubeni Corp of Osaka.

European Technological Composite (ETC), is capitalized at about ¥500 million, held 70% by Ube Industries and 30% by Marubeni. The joint venture plans to construct a 6000 tonnes a year plastic composites production facility which is scheduled to begin operating in April 1992.

Nikkiso to build aircraft parts plant

Tokyo-based precision products maker Nikkiso Co Ltd plans to construct a plant in Ishikawa Prefecture for aircraft parts made of carbon fibre reinforced plastic (CFRP).

Scheduled to begin operating by the end of 1993, the ¥5 billion plant will produce "Cascade", a CFRP composite used in jet engines and in advanced medical devices, such as artificial kidneys. Nikkiso's Cascade is reported to be 15% lighter than that of its rivals.

Once the plant is completed, Nikkiso will be able to produce Cascade worth ¥6 billion yen annually, approximately three times its current production level.

¥2 billion glass fibre fabric factory

Chukoh Chemical Industries Ltd of Tokyo has invested about ¥2 billion in constructing a factory in Matsura City which produces the company's glass fibre fabric.

The three-storey factory has a floor area of 3790 m^2 and began operating in 1991. The company's fabric is a glass fibre cloth coated and baked with a fluoro-resin.

Carbon fibre-reinforced plastic process factory

Techmate Co Ltd of Osaka has invested about ¥340 million in constructing a processing factory for carbon fibre-reinforced plastic in Hakodate, Hokkaido Prefecture.

Mass produced preforms

Plans to mass produce preform mats, which provide the reinforcement for plastic moulded products, have been announced by Chugokukogyo Co Ltd of Hiroshima.

From January 1992, the company plans to produce 50 000 tonnes a month of preform mats at its Kure No.2 factory. Half of this production will be supplied to other companies.

Chugokukogyo aims to sell mats worth ¥100 million in the first year of marketing.

Teijin and Araya Industrial to produce composites

Teijin Ltd and Araya Industrial Co Ltd, both of Osaka, say they plan to cooperatively produce moulded composites on a commission basis.

Teijin is investing some ¥500 million in the construction of a mass-production plant at its Iwakuni factory in Yamaguchi Prefecture. Teijin is currently producing composite bicycle rims, developed jointly with Araya, using a test facility at the Iwakuni laboratory.

Teijin will initially supply 400 racing bicycle rims a year to Araya. The rims, which consist of a urethane core reinforced with carbon fibres, weigh only 240 grams, which is at least 14% lighter than those made from aluminium alloy, the companies claim. The planned factory will produce 10 000 rims a year.

Araya will carry out post-processing of the rims, such as the drilling of spoke holes, and it will use them in bicycles which it began marketing in April 1991.

CHAPTER FIVE
ELECTRONIC MATERIALS

Batteries, capacitors and fuel cells

Solid electrolyte for secondary cell electrodes

A crystallized glass which is suitable for use as an electrode in lithium secondary cells has been developed by a research group, led by Professor Hosono, from the Nagoya Institute of Technology.

To produce the electrode, a titanium–phosphoric acid complex salt is mixed with other materials, melted in a crucible, and then rapidly quenched to produce a crystallized glass. The glass is made porous by treatment with acid to dissolve its calcium phase; pore sizes are 0.2 μm.

The porous glass has a specific surface area of 40–70 $cm^2.g^{-1}$ and the optimum composition has an electrical conductivity of 5×10^{-5} $S.cm^{-1}$ at room temperature. Because the glass has a high conductivity, the addition of liquid lithium electrolyte to the pores produces an electrochemical reaction which facilitates charge and discharge of the cell.

Porous electrode has higher capacitance

A porous electrode material has been developed by NEC Corp of Tokyo which the company claims it has used to fabricate an electrolytic capacitor with a rating of 5.5 V and 1000 F.

The material, which can be formed into any desired shape, is made from two types of phenolic resin which are mixed, press-formed, and then carbonized by heating to 500–800°C in a nitrogen gas atmosphere. The resulting material has numerous fine pores measuring from between several angstroms to one micron in diameter, giving it a specific surface area of 1500 $m^2.g^{-1}$.

An electrolytic capacitor can be formed from this material by immersing it in an electrolytic fluid. A conventional 1 F capacitor measures 3 × 3 cm, but the new material has been used to fabricate a 5 × 5 cm, 8 mm thick device which can store up to 100 F. Another version, measuring 7 × 7 cm and 10 cm in height, can hold 1000 F.

Low temperature, solid electrolyte fuel cell

A prototype solid electrolyte fuel cell has been fabricated by researchers who claim it can operate at a lower temperature than conventional devices of this type.

The researchers, led by Associate Professor Yotaro Yamazaki from the Tokyo Institute of Technology, say the cell is comprised of a stabilized zirconia electrolyte film which is 17 μm thick and is inserted between the anode and cathode, all of which rests on a nickel mesh.

The nickel-based cathode is directly formed on the electrolyte film by chemical vapour deposition. Once the film is deposited on the cathode it is oxidized and reduced at high temperatures. This produces a uniform film with numerous fine pores in the cathode to facilitate the passage of hydrogen ions.

In experiments, the cell was operated at about 700°C to generate as much as 17 $mW.cm^{-2}$. Conventional solid electrolyte fuel cells operate at around 1000°C, the researchers say.

High capacity polymer battery

A polymer battery which has 10 000 times the capacity of existing types has been developed, according to a joint research group from Osaka.

The Kansai Electric Power Co Inc and Sumitomo Electric Industries Ltd say their polymer battery is compact, has an output of 106.9 W and a capacity of 855 W.hours.

The polymer battery has a polyaniline anode and a lithium–aluminium alloy cathode in a propylene carbonate electrolyte. The anode is fabricated from a stainless steel mesh onto which polyaniline is deposited by electrodeposition. The

research group produced the high capacity battery by improving control of the current to the stainless mesh in the anode.

The battery is expected to be used for power storage and in electric cars.

Solid electrolyte fuel cells

The development of solid electrolyte fuel cells is the aim of a cooperative research project.

Tokyo Electric Power Co Inc, Kansai Electric Power Co Inc of Osaka, Chubu Electric Power Co Inc of Nagoya, and the Fine Ceramics Center say they will concentrate on fuel cells which use a nickel oxide–magnesium oxide (NiO–MgO) fuel electrode. The electrode, which is manufactured using a novel baking process, has better polarization characteristics than its conventional rivals.

The researchers report that their work may result in the development of methods for the efficient generation of electricity from methane gas.

Conductive polymers

Electroconductive polymer films synthesized

A chemical vapour deposition (CVD) technique has been used to synthesize a thin electroconductive film of polythiophene at the University of Tokyo.

A research group, led by Assistant Professor Naoki Tojima of the Faculty of Engineering, says it used bithiophene, in place of the conventional thiophene, to prepare the polythiophene which was then deposited on a substrate as a thin film with a uniform thickness. The research group is now trying to raise the electroconductivity of the deposited films and to develop applications.

Conventionally, polythiophene thin films have been produced using electrolytic polymerization techniques, the scientists say.

Integrated circuits and printed circuit boards

Conductive paste

A copper-based conductive paste has been developed by Nippon Oil & Fats Co Ltd of Tokyo. The paste is intended for use in the fabrication of circuits on printed circuit boards (PCBs), and of electromagnetic shielding.

The paste consists of copper particles uniformly distributed throughout a polymer. It contains special additives to increase its adhesiveness to PCBs, and to prevent oxidation of the copper. The copper paste has superior soldering properties to silver-based alternatives, Nippon Oil and Fat says. It can form tracks as narrow as 0.2 mm.

Paste for shielding applications costs ¥25 000 per kilogram, and paste for soldering applications costs ¥35 000 per kilogram. Nippon Oil & Fat has already begun to ship samples of the paste and expects to attain sales of ¥12 million in the first year of production. The company hopes to increase annual sales to ¥167 million within five years.

Currently, the Japanese market for conductive pastes has total sales of ¥680 million. In five years, this is expected to grow to ¥1.6 billion.

Integrated circuit is resistant to heat

A silicon integrated circuit (IC) which functions at 370°C has been fabricated by a research group at the Toyota Technological Institute. Furthermore, the group estimates that the IC can function normally at up to 450°C.

The researchers say they doped the silicon with high levels of arsenic and boron. The higher impurity concentrations stopped current leakage, a cause of malfunctions in ICs at high temperatures.

Heat-resistant ICs have potential applications in cars and aircraft. They can also be used in control systems and sensors in rockets, nuclear reactors and geothermal power generators.

Field effect transistor uses conductive organic material

A field effect transistor (FET) made from an electroconductive organic material has been fabricated by the Research Development Corp of Japan's Sakaki Quantumwave Project.

The electroconductive material is an alkyl-substituted oligothiophene that consists of four thiophene molecules with methyl or ethyl groups attached to both ends.

Research Development researchers fabricated the FET by forming two metal electrodes (for the source and drain) on an n-type silicon substrate coated with insulating film. A solution of the alkyl-substituted oligothiophene was then coated onto the electrodes to form the channel of the FET.

In experiments using a gate voltage of 10 V and a drain voltage of 30 V, a current of 70 nA was generated. Furthermore, the FET had an electron mobility of 100 $\mu m.s^{-1}$ at a source–drain voltage of 1 V. This value is over 1000 times higher than that obtained using a compound consisting of three thiophene molecules, and is comparable to the value obtained with polyacetylene, which is the most electroconductive polymer developed to date, the Research Development Corp says.

In further experiments using a compound consisting of five thiophene molecules, an electron mobility of 1 $mm.s^{-1}$ was recorded.

Formation technology for capacitors in memory devices

A formation technology that can be used to simplify the structure of the stacked-cell polysilicon capacitors used in 64 Mbit dynamic random access memories (DRAMs) has been devised, claims Tokyo-based Oki Electric Industry Co Ltd.

The technology increases the surface area of a capacitor cell by producing it with a rough surface, through the use of film forming techniques which do not involve lithography. The company reports that using a liquid phase chemical vapour deposition (LPCVD) method, at a temperature of 570°C and in an atmosphere of less than 0.2 torr, yielded the maximum surface area for a capacitor cell grown to a thickness of 0.1 μm.

Oki Electric optimized the three basic parameters of cell growth — film thickness, growth temperature and gas pressure — in order to develop a cell with a roughly cylindrical surface and a capacitance 2.5 times that of a flat film capacitor.

No-clean soldering method

A printed circuit board (PCB) soldering technology, which does not need to be followed by a washing step has been developed by Tokyo-based Fujitsu Ltd. The company says that "Clean Soldering" does not need solvents such as chlorofluorocarbons (CFCs) and trichloroethane.

The technology uses a specially developed flux in a low-oxygen atmosphere, such as nitrogen gas. The flux evaporates after the soldering step and contains few residual elements. Because of this, the flux does not produce residues and no washing is required.

In an experiment, the ion residue of this technology was found to be no more than two-thirds the value specified in the US military standards. This result indicates that the technology can provide insulation resistance and soldering characteristics equal to, or better than, conventional PCB soldering techniques.

Preventing degradation of silicon oxide films

An ashing technology (to remove photoresist) which can reduce deterioration of large-scale integrated circuits (LSIs) caused by the diffusion of sodium from the resists into the silicon oxide insulator films has been developed, reports Fujitsu Ltd of Tokyo.

The method uses an oxygen–water plasma to remove photoresist and avoids direct contact between charged particles and the substrate.

In an experiment, a resist containing a large amount of sodium was used. Reportedly, it was confirmed that there was no diffusion of sodium particles into a silicon oxide film even when the temperature of the substrate was 200°C. Sodium molecules from the resist were found clustered on the surface of the film.

Fujitsu also claims its technology offers other benefits:

- no diffusion of sodium occurs even when the resist removal process takes four times longer than usual;
- the resist can be removed at a rate of 1 μm or more a minute, about twice as fast as in the oxygen ashing method; and

- no charged particles come in contact with the substrate, thus eliminating many of the problems associated with conventional ashing methods.

The company plans to apply its ashing method to the production of advanced LSIs.

Chlorofluorocarbon-free flux application system for printed circuit boards

A flux application system, for use in soldering printed circuit boards (PCBs), which does not use chlorofluorocarbons (CFCs) is being marketed by NEC Shizuoka Ltd.

Reportedly, "Micro-coat HY-10" reduces the amount of flux required by a factor of five compared to conventional systems. The HY-10 contains a flux applicator, a flux tank and spray-nozzle control units.

The Micro-coat HY-10 costs ¥4.5 million. The firm expects to sell ten units in the first year of marketing and 50 units in the second year.

Method for depositing copper foil in multi-layer printed circuit boards

A process for depositing copper foil in high density, multi-layered printed circuit boards (PCBs) has been devised at Kanto Gakuin University of Kanagawa Prefecture.

The process is not electrolytic and gives a higher bond strength than is available with conventional methods, the University says.

A film of 96% copper, 3% nickel and 1% phosphorus is deposited on the copper surface which is then treated with what the University describes as the "black oxidation" process. This improves the bond to the copper and allows holes to be more accurately drilled in the final PCB.

Electrodeposited photoresist

A negative-type, electrodeposited photoresist with a resolution of 30 μm is now available from Toagosei Chemical Industry Co Ltd of Tokyo.

The resist, "Photo Emerge ED", is electrodeposited as a film, which may be as thin as 10 μm. The company describes the adhesion properties as outstanding and expects it will be widely adopted in the production of printed circuit boards.

Nippon Kodoshi develops heat-resistant flexible printed circuit board

Nippon Kodoshi Corp, a manufacturer of insulating paper, has developed "flexible flat connector" (FFC), a three-layer flexible printed circuit board (PCB) that can withstand temperatures of up to 350°C. The company is using the material in a position detector for automatic blinds.

The new PCB does not use adhesive layers between the copper foil and the covering polyimide resin films. This gives the PCB improved heat resistance since it is the adhesives which tend to be sensitive to heat.

Photoresist resins

Mitsubishi Electric Corp has developed a series of photoresist resins. The "COP series" is reportedly four times as sensitive as conventional photoresist resins.

To produce COP, TBTMA, a tin compound, is mixed with methylmethacrylate and benzoyl peroxide and reacted for about eight hours. The final tin concentration in the COP ranges from 5% to 31%.

COP with a tin content of 5-10% has what the firm calls "posi type" characteristics, meaning that a positive image of the deposited COP pattern is produced by curing. COP with a tin content of 15% or more has "nega type" characteristics, meaning that a negative image of the COP pattern is etched in the substrate during curing.

Hitachi develops highly sensitive photoresist

A photoresist for use in the production of black-matrix colour-CRT screens has been developed by Hitachi Ltd of Tokyo.

The resist has 2.5 times the photographic sensitivity of conventional photoresists. The company claims that, because curing the new photoresist requires less photo-energy, a larger width of screen can be cured in a shorter time.

The photoresist is made of acrylamide-diacetone copolymer and an azide-based sensitive material.

Strong lead-frame material

Hitachi Metals Ltd has developed "Tough-Frame", an integrated circuit lead-frame material with higher strength than existing rivals.

Conventionally, beryllium and other elements are added to a nickel alloy to produce a strong lead-frame material. However, this type of material has soldering, etching, and rusting problems. Hitachi Metals has developed a method which uses heat to strengthen the material and does not need additional elements.

The material has a tensile strength of 110 $kg.mm^{-2}$, 80% higher than that of conventional lead-frame materials. The company says that 1 Mbit and 4 Mbit large-scale integration packages can be made much thinner with the new material.

The material is two to four times more expensive than the conventional material that usually sells for ¥1500–1600 a kilogram.

Laser-curing photosensitive material for printed circuit boards

A photosensitive material which is cured by exposure to an argon laser has been announced by Mitsubishi Chemical Industries. The material is useful for forming dry films such as printed circuit boards.

The material is 60–70% binder resin (a copolymer of methyl methacrylate and methacrylate), 30–40% monomer (pentaerythritol triacrylate, PETA), 1–2% of a polymerization initiator, and 1–2% of a sensitizer. Two combinations of initiator/sensitizer have been developed: (1) trichloromethyl triazine/cyanine dyestuff, and (2) bisimidazole compound/ P- dimethyl aminostyryl naphthothiazole.

Because a circuit pattern can be formed directly onto circuit boards made with the new material, the result is claimed to lower production costs and increase productivity.

Ultraviolet sensitive resist

A silicon-containing photoresist which is sensitive to ultraviolet (UV) light from an ordinary mercury lamp and is intended for use in double-layer resist photolithography has been developed at the Tokyo Metropolitan Industrial Technology Center.

The resist is a mixture of a silicon-containing photo cross-linking agent (a material which hardens when exposed to light) and a polymer. It has a silicon content of about 6%.

The resist's UV-light sensitivity and resolution are comparable to those of conventional photoresists which do not contain silicon. In addition, the photoresist reportedly has high resistance to oxygen plasma etching.

Conventional photoresists are sensitive to 300 nm UV light or electron beams. However, the new resist can be exposed with an ordinary mercury lamp which emits 365 nm (i-line) UV light.

The photoresist is still in the experimental stage, but according to the Center, it has potential for use in the fabrication of extremely fine printed circuit board wiring patterns or other high resolution fabrication processes.

Liquid crystals

Ferroelectric liquid crystal with fast response

A ferroelectric liquid crystal (FLC) has been developed by Sumitomo Chemical Co Ltd of Osaka. The company reports that the material functions from –20°C to about +50°C and that it has a response time of 100 ms, or less, at 0°C.

The rotational viscosity is as low as 0.2 to 0.3 Pa.s (at 20°C), one-third that of the viscosity of conventional FLCs. Adding an alignment reagent improves the memory function, Sumitomo claims.

FLCs are being developed by a number of other firms, including Mitsui Toatsu Chemicals Inc of Tokyo and Takeda Chemical Industries Ltd of Osaka, in cooperation with electronic equipment manufacturers.

Ferroelectric liquid crystal

A high performance ferroelectric liquid crystal, which is expected to lead to a 100-fold improvement in the performance of liquid crystal televisions and other devices, is the result of its research, according to Nippon Telegraph and Telephone Corp.

The liquid crystal is a ketone-type organic compound with a molecular weight of about 600. The rate of spontaneous polarity, a measure of the crystal's usefulness in visual display applications, has been enhanced through the addition of molecular groups such as hydroxyls.

As a result, the charge per square centimetre, which is a measure of spontaneous polarization, has been increased to a record 1.1 μC.

Semiconductor processing and growth

Growing silicon films at low temperature

A method of growing single crystal silicon films at a substrate temperature of only 400°C has been developed, according to researchers at Tokyo-based Nippon Telegraph and Telephone Corp's LSI Laboratories.

The process, which uses photo-assisted chemical vapour deposition (CVD) and synchrotron orbital radiation (SOR), takes place under an ultra-high vacuum. SOR is used to irradiate the substrate as the deposition process takes place.

The researchers say single crystals of excellent quality are obtained at a temperature of 400°C. This compares with a temperature of 700°C necessary for the growth of single crystal films when SOR is not used. Such low temperature techniques are necessary for the fabrication of large-scale integrated circuits (LSIs).

Mixed crystal of gallium arsenide and silicon grown

Migration-enhanced epitaxy (MEE) has been used to grow a mixed crystal of gallium arsenide (GaAs) and silicon (Si) at Nippon Telegraph and Telephone (NTT) Corp in Tokyo, the company claims.

In experiments, NTT says it used MEE to grow mixed crystals at relatively low temperatures, resulting in an even distribution of the Si atoms. A high quality mixed crystal was obtained when the growth temperature was 450–500°C and the Si content was 25%.

The researchers formed a superlattice of ten alternating layers of the mixed crystal and GaAs, each 20 nm thick, on an Si substrate. MEE was then used to form a GaAs layer on top of this structure.

Photoluminescence from the top GaAs layer was as strong as that from a layer on a GaAs substrate, strongly suggesting that the superlattice acts as a buffer between the GaAs and Si, the scientists say.

Mass production technology for gallium arsenide wafers

A technology for the mass production of four inch gallium arsenide (GaAs) wafers is under development by Dowa Mining Co Ltd of Tokyo.

The company has been test producing four inch wafers at its factory in Akita Prefecture and has introduced evaluation equipment to check the characteristics of the wafers including composition, heat resistance, and electronic properties. The evaluation system will help improve the quality of the wafers produced, Dowa Mining says.

GaAs wafers are hard to mass produce because of their brittleness.

Mass production of gallium arsenide substrates

A mass production technology for gallium arsenide (GaAs) substrates has been developed by Nippon Mining Co Ltd of Tokyo.

Arsenic deposits tend to cause malfunctions in integrated circuits, so only the portions of the wafer with no arsenic deposits can be used. Nippon Mining says its method leaves only 7000 arsenic deposits/cm^2. Conventional techniques produce wafers with about 100 000 arsenic deposits/cm^2, the deposits being 50–200 nm in size.

In addition, the new method enables production of wafers with improved local flatness and processability because the number of 200 nm pits is only one quarter that seen with conventional processes.

The technology consists of a two-step wafer annealing method. A GaAs ingot is heated to 1100°C under a gaseous arsenic atmosphere at constant pressure and temperature. The ingot is then heat treated at 950°C to create a uniform crystal structure.

Nippon Mining has already begun sample shipments to semiconductor manufacturers, and plans to use the method to mass produce four inch wafers at its Toda laboratory in Saitama Prefecture.

Gallium arsenide production technologies

Several non-ferrous metal companies in Tokyo have released details of techniques they have developed for producing gallium arsenide (GaAs):

- Nippon Mining Co Ltd says it has developed a thermal treatment for GaAs wafers which overcomes some of the material's deficiencies;
- Furukawa Co Ltd claims to have a method for the safe mass production of a liquid form of GaAs; and
- Dowa Mining Co Ltd is trying to develop a mass production technique which will enable pre-determination of the product's electrical properties.

Synthesis of large crystals of cadmium telluride

A technique for synthesizing large crystals of cadmium telluride (CdTe) and associated compounds has been developed at Sumitomo Electric Industries Ltd in Osaka, the company says.

Uniform CdTe crystals are difficult to synthesize because cadmium is too volatile. Sumitomo Electric's technique has been used to grow cadmium zinc telluride and cadmium manganese telluride crystals measuring 50 × 60 mm. The technique reportedly produces crystals of a higher quality than can be produced using conventional methods.

Sumitomo Electric plans to develop mass production techniques and expects to have begun sample marketing the crystal to semiconductor device manufacturers during 1992.

Liquid raw material for gallium arsenide synthesis

High purity tertiary-butylarsine is a safer alternative to arsine gas used in the production of gallium arsenide (GaAs), according to its producer, Furukawa Co Ltd of Tokyo.

Tertiary-butylarsine is a liquid at room temperature (boiling point 50°C) making it easier to handle than gaseous arsine. Furthermore, it is ten times less toxic than arsine, with an air concentration of 0.5 parts per million being tolerable.

The company produces tertiary-butylarsine by chlorinating arsenic trioxide which it then reduces with a tertiary-butyl compound. By improving the production process, it has succeeded in purifying the compound to 99.99999% (seven-nines), Furukawa claims.

Semiconductor manufacturers have already evaluated the material and Furukawa is investing several hundred million yen in setting up a mass production facility.

Molecular beam epitaxy monitoring technology

A technology for monitoring the atomic surface of a semiconductor as it is grown by molecular beam epitaxy (MBE) has been developed by Matsushita Electric Industrial Co Ltd in Tokyo.

Helium ions are accelerated at 1 keV to irradiate the surface of a semiconductor substrate being processed by MBE. The energy of the scattered ions is measured, giving information on the atomic structure of the surface of the semiconductor, as well as surface and point defects. The measurement device is 7 cm in diameter and 150 cm in length.

In a related development, Matsushita Electric says it has discovered that the surface of a sample being processed by MBE can be modified when irradiated by neon and other heavier ions, opening the way for the development of new ion beam assisted epitaxy technologies.

Electroconductive flooring material for clean rooms

An electroconductive, antistatic flooring material based on epoxy resin has been developed by the Yokohama Rubber Co Ltd.

The antistatic flooring material is made for use in clean rooms at factories which produce semiconductors, communication equipment, precision machinery and instrumentation, and biotechnological facilities.

The material, "Smoony Floor", is applied in three layers:

- a primer, which has good adherence to concrete;
- an undercoat; and
- a top-coat.

The individual layers differ in their properties but are all based on an electroconductive epoxy resin developed by the company. The floor material conducts charges away and removes them through ground plates.

Fujitsu Ltd has already purchased the material for use in its new personal computer factory in Shimane Prefecture.

Gallium arsenide on silicon

A technique for growing high quality gallium arsenide (GaAs) crystals on a silicon substrate is the result of their work, claims a group of researchers from the Tokyo Institute of Technology.

The method has reportedly produced GaAs crystals with good surface flatness and an electron mobility of 2240 $cm^2.V^{-1}.s^{-1}$, a figure comparable with bulk GaAs.

The group has developed a method of coating the silicon substrate with a film of dielectric calcium fluoride prior to growth of the GaAs crystal. The calcium fluoride acts as a buffer between the GaAs and the silicon which, because they have different lattice constants and thermal expansion coefficients, are incompatible at their surface interface.

In the method, formation of the thin calcium fluoride film is followed by low temperature (500°C) growth of a thin film (30–50 nm) of GaAs and subsequently high temperature growth of a much thicker GaAs crystal. The initial layer of GaAs is grown as a buffer against migration of the calcium fluoride into the second layer of crystal which is of superior quality because high temperature growth reduces the incidence of crystal defects.

Observing crystal growth processes at atomic level

A system which allows crystal growth of compounds such as gallium arsenide (GaAs) to be observed at the atomic level has been developed by the Optoelectronics Technology Research Corp in Tokyo.

The system grows the crystal and generates an ion beam to observe the growth. The time taken by ions to leave the ion gun, strike the crystal surface and return to the detector is monitored. This provides constant measurements of the crystal's surface topography as it forms.

Furthermore, the types of atom which the crystal is made of can be determined because the reflected ions vary in speed depending on the mass of the atoms upon which they impinge. The company plans to further improve the system for observing atom's three-dimensional positions by applying ions in multiple directions.

In experiments, GaAs thin films were observed as they were grown on an aluminium arsenide substrate using neon ions accelerated through a potential of 3000 V.

Optical chemical vapour deposition reaction mechanisms elucidated

Fujitsu Laboratories Ltd of Kawasaki claims to have elucidated the reaction mechanism of optically catalysed chemical vapour deposition (CVD) in its application to the formation of amorphous silicon thin films.

A group at the company used infra-red (IR) light to monitor the formation of products during the irradiation of disilane gas with 0.16 μm ultraviolet (UV) light. The group claims to have observed that, upon irradiation, disilane is first decomposed into hydrogen and silicon–hydrogen radicals.

These radicals engage in degradative chain reactions and the disilane is finally reduced to hydrogen and monosilane. The chain, up to the point where the thin film is reacted with diborane, involves a total of seven reactions.

This is reportedly the first explanation to date of the gas reactions involved in the optical CVD process.

Selective growth of gallium arsenide

The Electronics Laboratory of NTT Corp has used an ECR-MBE method — a combination of electron cyclotron resonance (ECR) and molecular beam epitaxy (MBE) — to selectively grow gallium arsenide (GaAs) crystals.

The use of ECR produces high density arsine plasma and this, in turn, results in improved crystal growth characteristics. The first step in the process involves depositing a silicon nitride pattern onto a gallium or silicon substrate. Vaporized gallium and arsine-gas plasma are then introduced and single GaAs crystals are deposited on the unmasked portions.

Conventional MBE techniques have tended to deposit polycrystalline and amorphous GaAs on the masked surfaces but this does not occur with the new technique. The technique can also be carried out at low temperatures, thus avoiding thermal stress dispersion. Because the technique provides selective growth and results in flat growth surfaces, regardless of the pattern width and shape, it is expected to be applied in the production of future ultra-fine semiconductor devices.

Semiconductor thin film production techniques

The US-based semiconductor production equipment manufacturer Applied Materials Inc has started a joint research and development project with a group (led by Professor Yasuhiko Ohmi) from Tohoku University. The project will develop thin film production techniques.

The research group is developing new techniques for forming thin films of aluminium or copper on silicon wafers. In particular, the group will develop an interface technique for smoothly connecting crystals of silicon with crystals of metal. Within three years, the US firm aims to apply the technique to the manufacture of high-density intergrated circuit (IC) production machines, such as those which produce 16 Mbit dynamic random access memories (DRAMs).

Synthetic ruby bonding capillary for device assembly

A polycrystalline synthetic ruby which has a hardness and durability comparable with single-crystal ruby provides the basis of an improved wire-bonding capillary for use in semiconductor device assembly, according to its Kita-Kyushu-based developer.

Toto Ltd says the capillary combines the high durability of natural ruby with the cost savings of polycrystalline ruby. Toto claims its capillary is 20–30% cheaper than single-crystal ruby versions and has two to three times the working life of conventional polycrystalline ruby rivals.

The company's polycrystalline ruby is made from aluminium oxide, chromium oxide (0.07–0.3%) and magnesium oxide (0–0.05%). These powders are mixed with an organic binder, injection moulded, sintered at 1080–1350°C, and finally treated by hot isostatic pressing (HIP).

Substrate cleaning and crystal growth methods for semiconductors

Cleaning semiconductor substrates with a hydrogen plasma involves a lower processing temperature than is conventionally required, according to Tokyo-based Nippon Telegraph and Telephone Corp (NTT). NTT also claims to have improved existing methods of growing semiconductor crystals, again at lower than usual temperatures.

The company's researchers clean the semiconductors by heating the substrate with a beam of hydrogen plasma. The technique requires heating gallium arsenide substrates to 300°C and silicon substrates to 400°C. Conventional processing requires baking at temperatures more than twice as high, NTT says.

NTT scientists also say they have used molecular beam epitaxy (MBE) to grow a gallium arsenide crystal at 350°C. Normally, crystal growth by MBE requires processing at 700°C.

Sub-surface analysis of silicon crystals

NEC Corp of Tokyo says it has devised a sub-surface technique which can analyse an encased epitaxial layer of boron within a single crystal of silicon.

NEC says it uses synchrotron radiation to reflect electrons off the underlying boron–silicon layer. Analysis of the reflected electrons reveals that the structure of the boron–silicon layer remains unchanged following deposition of a silicon film.

The company uses molecular beam epitaxy (MBE) to prepare a surface modified silicon crystal in which one in every three silicon atoms on the surface is replaced by a boron atom. A thin film of silicon is deposited on top.

Semiconductors packaged in copper using atomic beams

A method of packaging semiconductors in copper has been developed, according to a research group from the University of Tokyo's Research Center for Advanced Science and Technology.

In experiments, 60 μm square copper films were irradiated with atomic beams of argon. The irradiation took 30 minutes at room temperature under a vacuum of 10^{-7} torr. A 1 mm thick copper or aluminium plate was then pressed against the copper, resulting in a bond with a strength of 100–200 MPa.

Semiconductors packaged in copper have greater resistance to pressure, temperature, humidity and electricity, the scientists say. They are expected to find applications in space and in the atomic power industry.

However, conventional joining techniques require heating and this tends to change the properties of the material at the interface, resulting in an increase in contact resistance.

Conventional superconductors

Superconductive alloy with high critical field

A triniobium–aluminium–germanium alloy superconductor with a critical magnetic field of 40 T has been fabricated, a research group from Tokai University claims.

The material remained superconductive in an external magnetic field of 28 T while carrying a current up to 5000 $A.cm^{-2}$ at liquid helium temperature (4.2 K).

To make the superconductor the group crushed a compound of diniobium–aluminium–germanium in a mortar, mixed this with niobium powder, and heated the mixture to 1000°C, using either an electric furnace or laser irradiation.

The scientists say they have fabricated a tape 5 mm wide, 0.4 mm thick and 2 m long. They have also made a multi-strand wire consisting of seven such tapes.

The wire is expected to have applications in the fabrication of superconductive magnets. The triniobium–aluminium–germanium alloy is expected to be used in superconductive magnets in nuclear fusion reactors and nuclear magnetic resonance (NMR) imaging systems.

Removing residual magnetic fields from superconductors

A method for removing residual magnetic fields from devices and magnetic shields made from superconductors has been developed, according to a research organization in Tokyo. Residual magnetic fields are a major cause of device degradation, the Research Development Corp of Japan (JRDC) says.

JRDC's method uses an argon laser which emits at a wavelength of 488 nm. The beam is focused to a 0.1 mm spot which heats a point on the surface of the superconductor. The heated portion becomes normally conductive and attracts the otherwise trapped residual magnetic fields. The focus of the laser can then be gradually moved to the edge of the superconductor, removing the residual field.

In experiments, superconductive niobium at liquid helium temperature (4 K) was treated and residues of magnetism were completely removed, JRDC claims.

Fine superconductive wire

The Electrotechnical Laboratory of the Agency of Industrial Science and Technology (AIST) has developed a superconductive wire measuring 0.4–0.5 μm in diameter.

To make the wire, researchers added 1–3% manganese to a conventional superconductive material, niobium-3-tin. Superconducting materials have an effective diameter which is the minimum thickness which a wire must have if it is to exhibit superconductivity when alternating current is circulated. The effective diameter of conventional niobium-3-tin wire is so thick (some 10 μm) that the superconductivity cannot be maintained due to heat generation. The effective diameter of the new wire material is 2.1 μm which requires bunching several fine strands together.

The Laboratory plans to continue efforts to develop even finer wires.

High temperature superconductors

Superconductive films of different orientation grown on same substrate

Researchers at Sanyo Electric Co Ltd in Osaka report they have fabricated bismuth-based superconductive films of different orientation on the same substrate.

The researchers grew the films using a molecular beam epitaxy (MBE) process with two types of bismuth-based superconductor. One type contained one calcium atom per unit cell, and the other contained no calcium. The researchers produced films whose orientations were rotated by 45°, with respect to each other.

The technique is expected to find applications in the fabrication of Josephson junctions. The company plans to use elements fabricated using the above process in multi-channel superconductive quantum interference devices (SQUIDs).

Superconductive tape production method

A method of producing bismuth-based superconductive tape, which has a high critical current density (J_c) under an intense magnetic field, has been developed, according to the National Research Institute for Metals (NRIM) in Tokyo.

In the production process, powders of bismuth oxide, strontium carbonate, calcium carbonate, and copper oxide are mixed and heated. The powder is then mixed, first with polyvinyl butyral and then with trichloroethylene, and agitated to form a slurry.

Silver tape is immersed in this slurry to coat it with an oxide superconductor film. The coated silver tape is then heated.

The thickness of the oxide film can be reduced by diluting the slurry. Furthermore, the silver can be formed into various shapes prior to immersion in the slurry, NRIM says.

The J_c was 20 000 $A.cm^{-2}$ at 77 K, under a zero magnetic field, and 10 000 $A.cm^{-2}$ under a 10 T field.

High temperature superconductive wires

The National Research Institute for Metals (NRIM) in Tokyo plans to begin joint research with the Indian National Physical Research Institute on the fabrication of high temperature superconductive wires.

The research will concentrate on bismuth-based superconducting wires which are expected to find applications for use in magnets and power cables.

High temperature superconductive fibre

A high temperature superconductive fibre has been developed by Nitivy Co Ltd of Tokyo, in association with the Nagoya Institute of Technology.

The fibre exhibits zero resistance at 90 K and has a critical current density of 17 000 $A.cm^{-2}$. A current density of 1000 $A.cm^{-2}$ was maintained under a magnetic field of 5 T, the developer claims.

The superconductor is based on yttrium, barium and copper oxides. To make the fibre, acetates of these elements are dissolved in a polyvinylalcohol solution containing special additives. The solution is uniformly mixed, concentrated, and then injected through a nozzle to obtain a raw fibre.

The diameter of the raw fibre can be controlled to between several microns and 500 μm. The raw fibre is then fired in air at 500–600°C, the company reports. Nitivy has already spun the superconductive fibre at a speed of 30–100 metres per minute, but adds that the fibre tends to break during firing.

Large crystals of bismuth-based superconductor

A method for growing large crystals of bismuth-based superconductors with transition temperatures (T_cs) of about 110 K has been devised, researchers in Osaka claim.

The Government Industrial Research Institute (GIRI), in collaboration with Associate Professor Tomoji Kawai at the university, says it has grown large crystals of an oxide of bismuth, strontium, calcium and copper, where the latter elements are in a ratio of 2:3:2:3. Until now, researchers have only been able to synthesize large crystals of bismuth-based superconductors where this ratio is 2:2:1:2 and the T_c is about 90 K.

The crystals are about 1 mm square and have a thickness of 10–100 μm. Synthesis involves placing a 2:2:1:2 phase compound in a crucible along with powdered bismuth, strontium and lead oxides, and large amounts of calcium and copper.

The mixture is baked at about 840°C for a period of between five days and one week. The powders liquify and the surface of the crystal partially melts, allowing calcium and copper to penetrate the crystal matrix to form the 2:3:2:3 phase.

Magnetic shield has medical applications

A large magnetic shield is available which can cover the entire body during the measurement of weak fields emitted from the brain and heart, according to its co-developers.

Nagoya-based NGK Insulators Ltd and Furukawa Electric Co Ltd of Tokyo say the shield is a metal cylinder with a surface coating of a sintered bismuth oxide superconductor which has a high critical temperature.

The cylinder, which has a diameter of 650 mm, a length of 2000 mm, and a surface area in excess of 4 m^2, surrounds the entire body to block any extraneous magnetic fields.

Fluorometric analysis of superconductors

Fluorometric analysis can determine the composition of high temperature superconductors using samples as small as 10 mg, according to the National Research Institute for Metals (NRIM) in Tokyo.

A small sample of the superconductor is melted with some flux, and this mixture is allowed to cool into a glass bead. The bead is then irradiated with X-rays, which are absorbed by the material.

Subsequently, the material emits the absorbed energy in a spectrum which is characteristic of its elemental composition. This enables exact elemental determination of the material's ingredients, NRIM claims.

The type of flux used varies with the type of superconductor being analysed:

- bismuth-based superconductors require a lithium hydroxide–sodium borate flux;
- yttrium-based and lanthanum-based superconductors need a lithium carbonate–sodium borate flux; and
- thallium-based superconductors demand a sodium borate–sodium pyrosulphate flux.

The Institute has developed a melting/casting crucible which it says reduces the time required to prepare the glass bead by about one-third.

Flexible high temperature superconductive wire

A high temperature superconductive wire which has improved resistance to bending stress has been developed, according to Sumitomo Electric Industries Ltd from Osaka.

The wire is composed of 1296 strands of a bismuth-based superconductor, encased in a silver sheath. Researchers at the company's Osaka laboratory placed 1296 fine superconductive strands in a rectangular silver sheath, 0.15 mm thick and 3 mm wide. Each strand is about 2 μm thick and 30 μm wide.

Sumitomo Electric reports the wire can be subjected to bending and deformation without suffering damage to its superconductive characteristics.

The company has fabricated a prototype power cable composed of 22 units of the wire, wound ten times around a 75 cm long, 2 cm diameter support. When cooled with liquid nitrogen to 77 K, a current of 80 A could pass through the cable. The current in each wire was about 4 A, a current density of about 1000 $A.cm^{-2}$.

Making high temperature superconducting wires

A method of fabricating wires from high temperature superconductors has been developed by a research group from the University of Tokyo.

Professor Junji Kihara's team place a small billet of superconductor in a silver sheath about 10 mm in diameter, which is then drawn out into a 0.33 mm diameter wire with standard machinery. This treatment reduces the superconductor to a 0.7 μm powder. The researchers then sinter the wire in order to solidify the superconductive material.

The group has so far failed to remove all of the internal bores in the superconductor and so have not fabricated wires with sufficiently high critical current densities for practical purposes. However, the researchers report that complete removal of these bores may be possible by using hot isostatic pressing.

60 metre long, bismuth-based superconducting wire

A 60 metre long, bismuth-based superconducting wire has been made by researchers at the Osaka laboratory of Sumitomo Electric Industries Ltd.

The research group, led by Chief Researcher Kenichi Sato, says the wire is 0.4 mm thick and 4 mm wide. It consists of a silver sheath covering a 0.1 mm thick core of bismuth-based superconductor.

In experiments, the wire carried a current of 10.5 A at liquid nitrogen temperature (77 K), corresponding to a current density of 2450 $A.cm^{-2}$.

According to Sumitomo, superconducting wires with a current carrying capacity of 100 A must be developed before they will be of commercial interest to electric power companies or heavy electrical firms.

Low temperature formation of superconductive film

Nisshin Electric Co Ltd, jointly with the Government Industrial Research Institute (GIRI), Nagoya, has developed an improved method of forming superconductive films of yttrium barium copper oxide (YBCO).

Conventional methods require the film to be heated in oxygen to a temperature of 700–800°C. The research group reduced this temperature to 500°C by alternately irradiating the film with argon ions and oxygen ions. The group claims to have elucidated the relationship between the crystalline structure and the irradiation current applied. They claim that the crystallinity and critical transition temperature (T_C)of the film are improved with irradiation-current values of up to 1 mA and that irradiation can change the film's crystal axis orientation to the a-axis orientation.

In experiments, the group formed a 1 μm YBCO film with a T_C of 62 K.

Superconductivity confirmed in 2 nm thick film

Nippon Telegraph and Telephone Corp (NTT) of Tokyo says it has demonstrated superconductivity in an yttrium oxide superconductive thin film just 2 nm thick.

NTT researchers claim that this is the first time superconductivity has been detected in such a small sample. They report that experiments demonstrated that the critical transition temperature (T_C) dropped with the size of the film. Films over 100 nm thick exhibited a T_C of approximately 90 K, 10 nm films exhibited a T_C of 70 K, 5 nm films exhibited a T_C of 50 K, and 2 nm films exhibited a T_C of 22–23 K.

The company formed the yttrium-based superconductive films on the surface of magnesium substrates by a finely controlled sputtering process.

Superconductive tape suitable for magnets and simple motors

A bismuth-based superconductive tape has a critical current density (J_C) of 11 000 $A.cm^{-2}$ at liquid nitrogen temperature (77 K) under a magnetic field of 1 T, according to the Osaka-based manufacturer.

Sumitomo Electric Industries Ltd says this J_C is high enough to make it useful for applications in magnets and simple motors. The tape is 0.15 mm thick, 4 mm wide and 4 cm in length. The company plans to develop methods for producing the tape in lengths up to hundreds of metres.

The company made the wire by charging the raw material powders in a silver sheath, followed by drawing and firing.

Superconductor with high magneticity

An yttrium-based superconductor has a magneticity of 2500 gauss, the highest so far achieved, claims the manufacturer, Nippon Steel Corp of Tokyo.

The high level of magneticity is attributable to the increased size of the crystal grains contained within the superconductor, according to Nippon Steel. This is because large grains cause a reduction in the surface area to volume ratio.

The superconductor contains 1-2-3 phase yttrium barium copper oxide crystal grains measuring 3 cm square, three times larger than those found in conventional products. The company produced it by melting the ingredient powders at 1400°C. The melt was then quenched and melted again, this time at 1200°C. Subsequently, the material was held at 1100°C for 30 minutes followed by gradual cooling at a rate of 1°C an hour.

Superconductor research association develops high current wires

The Engineering Research Association for Superconductive Generation Equipment and Materials has developed high-critical transition temperature (T_c) superconductive wires with a high current capacity.

The Association claims to have produced wires which can conduct currents of up to 180 A, which is 18 times the previous record. Two different wire-fabrication methods have been developed. Both methods succeed by forming a uniform crystalline arrangement in a material consisting of bismuth, strontium, calcium, and copper oxides. The two methods are: (1) a rolling method which combines segments of superconductive material with segments of silver and produces a wire which can carry 180 A at –169°C, and (2) an extrusion method which uses a laser to melt the superconductive material which is then drawn into a wire which can carry 139 A at –169°C.

The rolling method was developed by Furukawa Electric Co Ltd, and the laser melting method was developed by Sumitomo Electric Industries Ltd.

Superconductive thin film made using new method

A superconductive thin film with a critical transition temperature (T_c) of 95 K has been produced by a new method by NGK Spark Plug Ltd and researchers from Nagoya University.

In the new process, a semiconductive thin film forms onto the substrate prior to the deposition of the superconductive film. The 70-angstrom-thick semiconductive film is made from bismuth-strontium-copper (2:2:1) oxide and the superconductive 700-angstrom-thick film is formed by the sputtering of bismuth, strontium, calcium, and copper oxides in a ratio of 2:2:2:3. The sputtering process involves alternate exposure of the substrate, at 760°C, to two atom-ejecting blocks of material: bismuth-strontium-copper (2:2:1) oxides; and calcium-copper (1:1) oxides.

Computer-controlled shutters, placed between the materials and the substrate, are opened and closed in order to alternately stack bismuth-strontium-copper-oxygen blocks and calcium-copper-oxygen blocks onto the substrate.

When the high-T_c phase superconductive thin film was formed directly onto the substrate, the T_c was only 60 K. When, with a semiconductor buffer layer, the high-T_c phase was formed to a thickness of 70 angstroms (two crystal layers), the T_c was about 50 K.

Fine superconductive fibre

The "laser heated pedestal method" produces polycrystalline fibres of a bismuth-based superconductor, according to its developers, the National Research Institute for Metals (NRIM) of Tokyo and Sanyo Electric Co Ltd from Osaka.

In the laser heat pedestal method, a carbon dioxide (CO_2) laser is used to melt a small area on the top of a pedestal of sintered superconductor several millimetres in diameter. A seed crystal is then inserted into the melted portion and slowly drawn out as an ultra-fine polycrystal fibre. Because the laser can heat a very narrow area of the pedestal, the researchers have been able to produce fibres with a diameter as small as 200 μm.

The group reports that the laser heat pedestal method is not a practical technique for producing superconductive wire directly because only several millimetres of fibre are produced an hour. However, it hopes to develop a practical technique by combining aspects of the laser pedestal method with other wire formation procedures, such as the doctor blade method.

Superconductor displays electro-optic properties

Researchers at the Applied Electronics Laboratories of Nippon Telegraph and Telephone (NTT) Corp, based in Tokyo, have found that thin films of a yttrium-based superconductor will change the voltage across them when exposed to certain wavelengths of electromagnetic radiation.

The workers deposited a 100 nm-thick layer of the superconductor onto a strontium titanate substrate and applied a voltage across the component at liquid helium temperatures. When exposed to infra-red light with a wavelength of 1.3 μm, the voltage changed by 5 V in 5 ns. Applications for the discovery are still being considered.

Superconductive tape in lengths up to 100 metres

Joint research and development between Hitachi Cable Ltd and Hitachi Ltd, both based in Tokyo, has resulted in the ability to produce superconducting tape in lengths of up to 100 metres.

The product consists of the thallium-based superconductor encased in silver. The manufacturing route involves simultaneous rolling and pressing of the constituent materials.

The tape has a critical current density (J_c) of 10 000 $A.cm^{-2}$ at liquid nitrogen temperatures; however, a maximum value of 22 000 $A.cm^{-2}$ can be produced at the expense of other properties.

Sol-gel route to yttrium superconductors

A sol-gel route to the production of films of yttrium-based high critical transition temperature (T_c) superconductors has been developed by Mitsubishi Cable Industries, based in Tokyo.

The process requires the use of the acetic salts of yttrium, barium and copper in the appropriate ratio (1:2:3). These are mixed together and then dissolved in propionic acid solution. The product is coated onto silver tape 0.1 mm thick and fired at 900°C. The properties of the resultant film are quoted as critical transition temperature (T_c) greater than 88 K and critical current density (J_c) of 6700 $A.cm^{-2}$.

The main advantage of the process is claimed to be the cheapness of the raw materials compared to the more conventional alcoxide precursors.

High critical current density superconductor

A gadolinium–barium–copper oxide superconductor which displays a critical current density (J_c) of 40 000 $A.cm^{-2}$ under a magnetic field as high as 4 T, has been developed, according to claims by Nippon Steel Corp, based in Tokyo. The material is made by a melting process, followed by oxygen annealing.

Meanwhile, Sumitomo Electric Industries Ltd has developed superconducting wire based on the bismuth-series materials which has a J_c of 11 000 $A.cm^{-2}$ at liquid nitrogen temperature under a magnetic field of 1 T. The tape measures 0.15 mm thick by 4 mm wide and is made by sheathing the superconductor in a silver pipe before firing.

Method for synthesizing lanthanum-based superconductor

A method of synthesizing lanthanum-based high temperature superconductors has been devised at the Superconductivity Research Laboratory (SRL).

SRL says, compared with the synthesis method developed recently by the US company AT&T, its process is simpler and cheaper.

The annealing process involves hot isostatic pressing (HIP) under an oxygen partial pressure of 10 MPa for ten hours. The process can accommodate a wide range of annealing temperatures from 970–1170°C and is reportedly more predictable and therefore more practical than previous processes.

The superconductor fabricated by this method exhibited a critical transition temperature (T_c) of 63 K, and the Meissner effect was confirmed. The Laboratory believes that its method holds promise for industrial applications if the material's T_c is raised by 15 K, to liquid nitrogen temperature.

Superconductive tape has high critical current at high magnetic field

A bismuth-based superconductive tape which exhibits a critical current density (J_c) of 140 000 $A.cm^{-2}$ at 4.2 K under a magnetic field of 25 T has been developed at the National Research Institute for Metals (NRIM).

NRIM says it has improved the doctor blade method which it uses to fabricate superconductive tapes. In the past, superconducting powder was mixed with a binder, extruded into a tape, placed onto a silver substrate, and heat treated at a constant temperature of 880°C. However, now the temperature is gradually decreased from 890 to 870°C. This results in well oriented superconductive crystal grains which facilitate current flow and increase the J_c value.

Long fibre of high temperature superconductor

A bismuth-based high temperature superconductor has been formed into a 50 cm long fibre by researchers at the Technological University of Nagaoka.

The superconductor contains oxides of bismuth, strontium, calcium and copper in the ratio 2:2:1:2. To make the fibre, a mixture of these powders is melted at 1200°C and then poured onto a steel plate at 200°C to form a 1 mm^2 glass-like superconductor.

The researchers then heat the superconductor to between its glass transition and crystallization temperatures, before drawing it into a fibre about 150 μm wide, 20 μm thick, and 50 cm long. The fibre can be coated with silver paste and heat treated to form a glass ceramic, with properties similar to those used for optical communications.

The fibre has a relatively low critical transition temperature (T_c) of 45 K, but the group claims that the T_c can be further increased by better control of crystal grain formation. The superconductor can be formed in lengths of up to one metre and is expected to find applications in the formation of composites.

Fibrous oxide superconductor

A polycrystalline fibre, 0.5 mm in diameter, of a bismuth-based superconductor has been fabricated using a laser melting technique at the National Research Institute for Metals (NRIM).

The technique involves melting the tip of an upright rod of superconductor with a carbon dioxide (CO_2) laser. A polycrystalline ceramic seed crystal is brought into contact with the molten tip, and then slowly pulled away. As the ceramic seed crystal is drawn up from the molten superconductor, it draws out a superconductive crystal.

If the extraction speed is less than 5 mm an hour, a fibrous superconductive polycrystal is formed with virtually only one phase. The technique is highly valued because it can produce crystals with a predictable grain size. This will aid research into the relationship between a crystal's fine structure and its critical current density, NRIM claims.

Business and financial information

Epoxy resins for printed circuit boards

Mitsubishi Gas Chemical Co Inc is constructing a factory which will produce epoxy resins for use in the manufacture of printed circuit boards (PCBs).

The company will invest about ¥10 billion in the factory which will have a floor area of 200 000 m^2. Construction of the factory, to be located at Nishi-shirakawa, began in late 1991 and operations should start in late 1992. This is the company's tenth factory.

Nippon Sanso to supply superconductor raw materials

Nippon Sanso KK of Tokyo is to start supplying raw materials for making thin films of high critical temperature (T_c) superconductors using the metal organic chemical vapour deposition (MOCVD) method.

Nippon Sanso expects sales amounting to ¥20 million for the first year of business, rising to ¥100 million per year by the third year. Prices are quoted as ¥6000 per gram for barium, strontium, yttrium, copper and bismuth organic compounds, according to the Nikkei Industrial Daily.

Osaka Sanso and BOC Group to develop gases for semiconductor industry

Speciality gases for use by electronics manufacturers will be developed as a result of cooperation between a leading manufacturer of industrial-use gases, Osaka Sanso Kogyo Ltd, and a major UK gas producer, BOC Group plc.

Under the planned programme, "BOC SPECTRA 100", both companies will integrate and reorganize their electronic gas divisions and will jointly develop gases required for manufacturing semiconductors. Technologies to be developed include equipment for producing ultra-pure nitrogen, and equipment for the transportation and storage of ultra-pure nitrogen, oxygen, and argon.

The two companies plan to market their products and services as a package to semiconductor makers.

Eight inch silicon wafer production plant

Plans to invest ¥30 billion in the construction of a facility which will produce eight inch silicon wafers have been announced by Shin-Etsu Handotai Co Ltd of Tokyo.

The facility is to be built at the company's Shirakawa plant in Fukushima Prefecture and is scheduled to be operating by April 1992. The company claims it is the first mass production facility for eight inch wafers in Japan.

Initially, it will produce several tonnes of wafers a year, rising to 30 tonnes a year by 1993. In February 1991, the company also began producing 10 000, eight inch wafers a month in the USA.

Funding for semiconductors and superconductors

Funding for research in the production technology for superconductive wires and magnetic shields, and the plasma processing of semiconductor wafers, has been announced by the Research Development Corp of Japan (JRDC) in Tokyo.

The superconductive materials research is expected to produce commercial applications for technology developed by the National Research Institute for Metals (NRIM), and by Professor Fueki of the Science University of Tokyo and Professor Kitazawa of the University of Tokyo. The technology to be developed involves the formation of superconductive wires and magnetic shields from composite oxides made from mixtures of the oxides of bismuth, strontium, calcium and copper.

The plasma processing research will further develop a technique that was originally conceived by researchers (led by Professor Mori) at Osaka University. The technology involves slicing a semiconductor single crystal into wafers using a halogen plasma. This method produces none of the deterioration caused by conventional mechanical slicing, eliminates the need for surface polishing after slicing, and generates a higher yield of wafers from each crystal, JRDC says.

Eight inch silicon wafer production increased

By the end of 1991, Komatsu Electrical Metal Co Ltd had invested more than ¥10 billion to increase its production capacity for eight inch silicon wafers.

The company was producing 10 000 wafers a month at its factory in Miyazaki Prefecture, but plans to increase the plant's capacity to 20 000 wafers a month. The company also plans to construct a 5000–10 000 wafers a month production line at its factory in Nagasaki Prefecture.

Expanded production for high electron mobility transistors

Production of high electron mobility transistors (HEMTs) rose from three million to four–five million units a month at Fujitsu Ltd of Tokyo, the company has announced. Fujitsu says the move was an attempt to satisfy an increase in demand for the devices which are used in broadcast satellite tuners.

The company has built a ¥30 billion facility with a capacity of up to five million units a month at its subsidiary, Fujitsu Yamanashi Electronics, in Yamanashi Prefecture, which became operational in early 1991.

Initially, the company used three inch gallium arsenide wafers to produce HEMTs, but moved quickly to four inch wafers.

Nippon Sanso acquires US semiconductor company

Tokyo-based Nippon Sanso KK completed the acquisition of Semi Gas Co Ltd, a US supplier of gases for the semiconductor industry.

The acquisition, through Nippon Sanso's subsidiary Matheson Gas Products Inc, was basically agreed upon in May 1990 with Semi Gas's parent firm, US aerospace company Hercules Co Ltd. However, the deal was suspended in December 1990 because the US Department of Justice alleged that the acquisition would violate anti-trust laws. Several parties in the USA felt that Japanese acquisition of the company, which is at the leading edge of semiconductor gas technology, would harm the national interest.

In March 1991, a district court of Pennsylvania dismissed the legal challenge and the acquisition proceeded. Negotiations were concluded on 10 April 1991 in a contract said to be worth US$20–30 million.

CHAPTER SIX
OPTOELECTRONIC MATERIALS

Adhesives

Adhesives for optical glass

Two kinds of adhesives for use with optical glass which cure when exposed to ultraviolet light have been developed by the Japan Institute of Advanced Dentistry in Tokyo.

The two adhesives, "Optokleb HO2" and "Optokleb MO1," harden after exposure to ultraviolet light for one minute at room temperature, the Institute claims.

Optokleb HO2 has a refractive index of 1.63, the same as optical fibres. Consequently, it can bond optical fibres together without distorting or attenuating the signal. Three grams cost ¥10 000. Optokleb MO1 has an adhesive strength of more than 25 MPa, high water resistance and does not peel, even after immersion in water at 37°C for one week. Ten grams cost ¥15 000.

Adhesive for optical equipment

Daikin Industries Ltd of Osaka has begun marketing "Optodyne UV", an adhesive containing fluorine which cures under ultraviolet light and is for use in the fabrication of optical communications equipment.

The refractive index of the adhesive can be adjusted by manipulating its fluorine content during manufacture, the company says. The product comes in four types with refractive indices of 1.45, 1.48, 1.50, and 1.57. The material costs ¥2 million per kilogram.

Optodyne UV has the following properties:

- 91–94% transmission of 1.3 μm light per 0.1 mm of adhesive;
- adhesion within one minute at room temperature;
- low viscosity;
- adhesive strength of 70–120 $kg.cm^{-2}$;
- resistance to 500 temperature cycles of between –10 and +60°C; and
- resistance to peeling, even after immersion in boiling water for five hours.

The adhesive can be used for bonding quartz optical fibres to optical lenses and prisms.

Devices

Device for optoelectronic integrated circuits

A device which consists of a light emitting diode (LED) formed on top of an array of field effect transistors (FETs) could be used in optoelectronic integrated circuits (OEICs), according to its Tokyo-based developer.

Oki Electric Industry Co Ltd makes the device from gallium arsenide (GaAs), the FETs forming a NAND drive circuit which controls the LED. The prototype measures 4.5 × 3.5 mm.

Using a two-stage, metal organic chemical vapour deposition (MOCVD) process, the company deposited GaAs doped with vanadium on a silicon substrate. On top of this film a layer of pure GaAs was deposited. The company then used standard photolithography techniques to form the array of FETs.

The FETs were then covered with insulating silica, on top of which the LED was formed, made of n- and p-type indium gallium phosphide (InGaP). The LED, measuring 80 μm square, has a threshold current of about 10 mA.

Oki Electric is now developing an experimental integrated circuit with the same structure containing 35 of the LEDs.

Optical amplifier

A semiconductor device capable of amplifying 1500 nm optical signals by a factor of more than 100 and which has an output of 30 mW has been developed, claims Nippon Telegraph and Telephone (NTT) Corp of Tokyo.

The amplifier has an active layer with a multiple quantum well (MQW) structure which contains six pairs of alternating films of indium gallium arsenide (InGaAs) and indium gallium arsenide phosphide (InGaAsP). On both sides of the active layer, between it and the cladding layers, are superlattices consisting of 12 alternating 0.2 μm thick films, each with a different light refraction factor.

For light signals in the 1500 nm band, the device has a maximum amplification factor of 24 dB. However, because the actual maximum optical output is about 30 mW, the amplification factor is limited to approximately ten when used with 3 mW input signals. Further, the bandwidth in which signals can be amplified is 100 nm or more, twice as wide as the bandwidth of conventional amplifiers, according to NTT.

The amplifier could be used to simultaneously amplify 15 different light signals of different wavelengths, providing the wavelengths are all at intervals of at least 6 nm from each other.

Photoelectric device could mimic human eye

A photoelectric device has been produced which may eventually be developed to the point where it can mimic human sight, according to Mitsubishi Kasei Corp of Tokyo.

The device is a Langmuir-Blodgett film consisting of an array of photoelectric domains, which range in diameter from several microns to 100 microns; they have an average diameter of about 50 μm. The researchers claim that several Langmuir–Blodgett films could be stacked onto a glass plate, covered by a protective layer and wired with electrodes to yield a photoelectric device.

If the individual domains differed in their reactive wavelengths, and if they were all individually wired to monitor their electrical signals, then a colour-sensitive eye would be obtained. If a learning function were added, then a perfect simulation of human sight would be possible.

To fabricate the device, Mitsubishi researchers combined pyridine-based porphyrin polymer, which generates electricity when irradiated, with arachic acid, which has insulating properties. The two molecules were immersed in pure water at 30°C, stimulating the porphyrin to form a honeycomb structure. When this procedure was repeated three times, the single molecule of porphyrin self-organized into an array of photoelectric domains separated by insulating portions of arachic acid.

Lasers

Optically controlled power output for laser

Researchers in Shizuoka claim to have controlled the power output of a laser by irradiating it with ultraviolet (UV) light.

The scientists, led by Associate Researcher Koji Sakaguchi from the university's Electronics Engineering Laboratory, say that when a Langmuir-Blodgett film of a ruthenium complex is used as a wavelength converter, the power of the output laser light can be controlled by irradiating the film with UV light.

The group formed the film from a ruthenium complex and a dialkyl- ammonium salt, and stacked 60 such layers on a 1 × 3 cm glass.

In experiments, an yttrium aluminium garnet (YAG) laser's 1.06 μm beam was converted to 0.532 μm by the film and the power of the output was controlled with 0.355 μm ultraviolet.

The researchers were able to vary the output of the laser light to between 50 and 100% of the incident light. In theory the output can be reduced to one-tenth, they claim.

Variable wavelength, solid state laser with high output

A high output, variable wavelength, solid state laser which can emit 100 000 times the energy of conventional lasers has been developed, according to the Institute of Physical and Chemical Research (IPCR) and Mitsui Mining & Smelting Co Ltd of Tokyo.

The laser generates 250–500 nm ultraviolet light using an infra-red excitation of 130–700 nm. Output is 50–100 mJ.

The oscillator crystal is either titanium chrysoberyl or chromium forsterite. The researchers say that instead of using a yttrium aluminium garnet (YAG) laser or non-linear crystals to excite the oscillator, the device uses light from a flash lamp. As a result, energy loss is reduced, enhancing the final ultraviolet output.

The laser is expected to have medical applications, such as in the treatment of cardiac thrombosis.

Silica glass laser source

A laser source based on silica glass has been developed by a research group at Shonan University of Engineering. This light emitting property of silica glass has previously been unobserved, the researchers say.

The group irradiated high purity silica glass with a krypton–fluorine excimer laser. This stimulated the glass to emit light with an oscillation energy of 2–4 eV, at a wavelength of 450–650 nm. The light emissions can reportedly be manipulated so that a single laser wavelength can be obtained.

Detailed investigations revealed that electrons on the outermost shell of the oxygen in the silica glass are excited to a higher energy state by the laser. Light is emitted when the electrons drop to a lower energy state.

Shortest wavelength semiconductor laser

A semiconductor laser with the shortest wavelength to date of any continuous-beam laser emitted at room temperature has been announced by NEC Corp.

The laser measures about 9 mm and consists of multiple thin films of aluminium-gallium-indium-phosphorus semiconductor crystal stacked on a GaAs substrate. It is able to emit a wavelength of 632.7 nm at a temperature of 20°C. Compared with other short-wavelength semiconductor lasers of about 670 nm, the device is ten times brighter to the human eye, making it easier to use in positioning control and measurement applications.

NEC reports that the short-wavelength laser was developed without resorting to increases in the crystal's aluminium content. Crystals containing large amounts of aluminium tend to degrade more quickly.

Liquid crystals

Liquid crystal display using azobenzene compound

A liquid crystal (LC) display incorporating a photochromic azobenzene compound has been developed by the Agency of Industrial Science and Technology's Research Institute for Polymers and Textiles in Ibaraki.

The azobenzene compound, which adopts the cis-form under ultraviolet rays and the trans-form under visible light, produces a high contrast image, the Institute says.

In experiments, an 8 μm LC layer was formed on top of an azobenzene layer deposited onto a glass substrate. When the LC layer was irradiated with linear polarized ultraviolet light, the LC molecules oriented themselves in a direction perpendicular to the direction of polarization, producing an image.

The Institute expects the development to expand applications for photochromic materials.

Lightguide brightens liquid crystal displays

A lightguide plate for use in backlit liquid crystal displays (LCDs) produces devices 30% brighter than usual, according to its maker, the Ohno Research and Development Laboratory Co Ltd.

The company's lightguide is made from transparent acrylic resin which is dissolved with solvents and resolidified under controlled conditions.

The texture of the resin is finer near the light source and coarser away from it, providing a uniform surface brightness in the LCD, Ohno Research claims.

The lightguide can also be formed into thinner plates than conventional products because of its improved light scattering properties.

Liquid crystal polymer developed

Osaka-based Polyplastics Co Ltd has developed a liquid crystal polymer (LCP) in collaboration with the US firm Hoechst Celanese Corp.

"Vectra E Series" is heat resistant to 270°C, about 30°C higher than the conventional Vectra polymers, Polyplastics says. In addition, Vectra E Series has high fluidity and is easy to process.

In the near future, Hoechst Celanese will begin producing the polymer and Polyplastics will handle the marketing. The cost of Vectra E Series will be comparable to that of the conventional Vectra.

Fast response ferroelectric liquid crystal

A simple matrix ferroelectric liquid crystal with fast response has been produced at the New Material Research Laboratory of Tokyo-based Nippon Mining Co Ltd, according to the company.

In an experiment, a liquid crystal display (LCD) made using the material had a switching speed of 20 μs, a wide view angle, and a much higher contrast ratio than that of conventional nematic LCDs.

To fabricate the material, the researchers used a fluorine reagent to oxidize chiral epoxide with a methyl group. This material was then added at a concentration of 10% to phenyl pyrimidine resin and similar substances.

In a separate development, the Laboratory says it fabricated an LCD using a liquid crystal with a different structure which could retain images for more than ten hours.

Materials for recording/Optical disks

Pure target materials for erasable optical disks

Two target materials which are 99.999% pure and intended for use in erasable optical disks have reportedly been developed by Dowa Mining Co Ltd of Tokyo.

One of the materials is an alloy of indium, tin and tellurium, and the other is an alloy of germanium, tin, and tellurium. Dowa has begun sample shipments of disks which are 4–6 inches in diameter.

Elemental composition is variable, depending on the needs of the user. Sample prices are ¥390 000–540 000 for indium-type disks and ¥320 000–460 000 for germanium-type disks.

Durable organic photochromic recording media

An organic photochromic media for optical recording which can reportedly be encoded, read, and erased more than one million times has been produced, according to a group of researchers from Sanyo Electric Co Ltd of Osaka, who made the material working in collaboration with scientists from Kyushu University.

To make the media, 1% of a diallylethene derivative dye is mixed with polymethacrylic acid, and coated onto a substrate. The media adopts a red colour in the recorded state and a yellow colour in the erased state.

Recording on the media is achieved with an argon laser (which emits at a wavelength of 0.458 μm) at high temperatures. Erasure is achieved with an helium–neon laser (wavelength: 0.633 μm). A combination of the two lasers together is needed for reading.

Focus detection device for reading optical disks

Matsushita Electric Industrial Co Ltd has announced the development of an integrated prism/lens focus detection device for optical disk readers.

A focus detection unit for optomagnetic disk readers is generally composed of a lens and two prisms. The new device combines all three functions in a single integrally-moulded component. The glass device measures 4 x 4 x 2 mm and weighs 0.12 g. Half of the device's incident side is spherical and the other half is a flat surface inclined by 2.4 degrees. The unit has two focal positions which reportedly enables it to detect focus errors. The unit's output lens is aspheric which reportedly eliminates cross-track noise.

Sample shipments have begun at a unit cost of ¥10 000–30 000. (*C)High speed, high storage capacity optic material

Nippon Telegraph and Telephone (NTT) Corp of Tokyo has developed an optic memory material which it says is capable of high speed information retrieval and greater storage capacity.

The material is called a photon gated electron transfer system or PHB. It is about 100 million times faster than current devices. Although development is at an early stage, an NTT spokesperson says possible uses include floppy disks for computers. The material could significantly increase the speed of data retrieval — the full page of a newspaper could be retrieved in six ten-thousandths of a second. Further, a five inch compact disk of the material could hold 1000 times the information stored on conventional disks.

Resin for magnetic optical disks

Nippon Zeon Co Ltd has begun mass production of "Zeonex", a polyolefin-based engineering plastic used in magnetic optical disks.

The company has invested ¥3 billion in the construction of a 1000 tonnes a year Zeonex production plant at its Mizushima factory in Okayama Prefecture. The company previously produced Zeonex at a pilot plant at its research and development centre in Kanagawa Prefecture.

Zeonex reportedly has the same light permeability of conventional polycarbonate (PC) resin but has only 10% of the water absorption properties.

Photochromic memory material

A photochromic material which can be used to record and retrieve data using existing semiconductor lasers is now available, according to Mitsubishi Electric Corp of Tokyo.

The company says the material can store data for up to ten years, and can be read non-destructively. Conventional photochromic compounds are sensitive to light with wavelengths of 500–600 nm, but the newly developed material is sensitive to 780 nm wavelengths and so can be used with existing semiconducting lasers.

The material is based on a compound containing a benzothiopyrane-type spiropyrane group. In tests, Mitsubishi Electric recorded data on a disk formed of the material using a 780 nm semiconductor laser with an output of 14–20 mW.

Material for high density optomagnetic recording

The Kuroda Solid Surface Project of the Tokyo-based Research Development Corp of Japan (JRDC)

has developed a material which, it claims, can be used in a high density optomagnetic recording system.

The material is a 50-layer superlattice consisting of alternating 2 nm thick films of europium and manganese–ytterbium. The researchers found that the magnetization intensity of the superlattice increases with temperature to a peak value, and thereafter decreases.

The Kuroda Solid Surface Project intends to exploit this principle to develop an optomagnetic recording system which will use the material.

Non-linear materials

Long-life, non-linear optical material

An organic non-linear optical material has been developed by Hoya Corp of Tokyo which the company claims has a service life of over 100 years.

In durability tests, no deterioration in performance was noted, even after 180 days. Theoretically, the material is expected to be able to maintain its functionality for about 130 years. Existing non-linear optical materials, based on polymethyl methacrylate (PMMA), have a service life of about 90 days, according to Hoya.

The material is based on poly-N-methylphenylacrylamide (PMPA) resin interspersed with nitroaniline. When a voltage is applied to a film 1 μm thick, the material can halve the wavelength of an incident laser beam.

Controlling molecular orientation in optically non-linear organic films

A method for controlling the molecular orientation of optically non-linear organic films by the application of an electric field has been developed, according to researchers at Tokyo University's Research Center for Advanced Science and Technology.

Organic materials display over one thousand times the non-linear optical effect of inorganic materials. However, when a pentacosane-based acid is fabricated into a film the molecules adopt random orientations, and the acid loses almost all of its non-linear optical effect, the scientists say.

The researchers used a 4 kV electric field to restore the molecular orientation of a 25 Å pentacosane-based acid film, fabricated using the Langmuir-Blodgett technique. They now hope to improve the technique to attain a highly efficient, optically non-linear organic material.

Non-linear optical materials under development

Tertiary non-linear optical materials for use with semiconductor lasers are under development by Asahi Glass Co Ltd of Tokyo. The company is trying to develop materials which display these effects at wavelengths of 380 nm or more.

Last year, Asahi Glass, in collaboration with a research group from Nagoya University, has developed a material which it says exhibits significantly enhanced tertiary non-linear optical properties at wavelengths below 380 nm. This material consists of copper (I) chloride particles, 5–10 nm in diameter, which are deposited in a glass vessel.

Non-linear optical film

A non-linear optical film made from a film of polymethyl methacrylate (PMMA) dispersed with fine crystals of polydiacetylene has been developed at the Institute for Polymers and Textiles.

The film has non-linear optical characteristics comparable to polydiacetylene single crystals. Polydiacetylene has a switching time of 1 ps and has potential for applications in the development of optical computers. However, polydiacetylene is difficult to form into thin films and this has been a major obstacle to its commercial use, the Institute says.

The film is made by placing a mixture of diacetylene monomer and PMMA (in a ratio of 9:1) on a glass plate. The plate is then heated to 60°C for 48 hours to produce the polymerized film.

Improved optical crystal

A potassium titanate phosphate (KTP) crystal with improved optical characteristics has been developed by Toshiba Corp, the company claims.

KTP crystal is a non-linear optical material which displays optical second harmonic generation (SHG). For instance, when 1.06 μm wavelength light from an yttrium aluminium garnet (YAG) laser is passed through KTP, the wavelength can be halved.

The resulting YAG laser light, with a wavelength of 0.502 μm, does not deeply penetrate the surface of materials and can be used to smooth surfaces and make relief patterns.

However, if the KTP crystal has oxygen defects the laser energy tends to be absorbed and converted into heat, degrading it. To reduce oxygen defects in the KTP crystal, the research group baked the crystal in an oxygen atmosphere at 800°C for 48 hours, thereby uniformly implanting oxygen atoms into the crystal lattice.

As a result, the transmissivity was reportedly enhanced to about 90%, compared with about 70% for unbaked crystals, and 80% for crystals baked in air.

When 100 W YAG laser output is applied to KTP, conventional crystals tend to crack in about 60 hours. On the other hand, the oxygen-annealed KTP crystal is reportedly expected to withstand continuous irradiation of YAG laser light for 600–1000 hours.

Three inch lithium niobate mono-crystal grown

A resistance heating Czochralski (CZ) method for growing a three inch diameter mono-crystal of lithium niobate (LN) is the result of studies at its Device Development Center, according to Kyocera Corp of Kyoto.

The resistance heating method features indirect heating by ceramic coils which are in turn heated by silicon carbide elements. Conventional methods involve direct heating of the crucible by high-frequency radiation.

In repeated experiments, the company reportedly succeeded in pulling up a mono-crystal 85 mm in diameter and 70 mm in length, using a platinum crucible 130 mm diameter, at a pull-up speed of 5 mm an hour and a crystal rotation speed of 5–15 revolutions a minute.

The equipment used costs less than ¥10 million, 40% cheaper than the high-frequency heating CZ method, Kyocera claims.

Other materials

Optical waveguide uses polymer thin film

An optical waveguide which uses a polymer thin film which has a non-linear optical effect has been developed by the Institute of Physical and Chemical Research (IPCR).

This waveguide is based on a thin film of soluble polythiophene. When YAG laser light shines on the film, through a prism placed on the film, the film exhibits the Kerr effect, in which the refractive index of the film varies with the electric field intensity of the laser light. Conventional polymer thin films have no Kerr effect and thus cannot be used in optical waveguides.

The Institute expects that polymer film optical waveguides will lead to the development of optical memory and optical shutters.

Zinc selenide mono-crystal

A method of producing zinc selenide mono-crystals which have applications in producing substrates for blue light emitting diodes (LEDs) has been developed by Toshiba Corp.

The researchers report that when a 10 μm polycrystalline zinc selenide grain is heated in a selenium atmosphere at high temperatures, it grows into a perfect 10 mm square crystal. When a high purity polycrystal of the same size was placed in a quartz tube and heated in a selenium atmosphere at 1000°C for two weeks, the light-yellow polycrystal was converted into a green mono-crystal.

The mono-crystals produced have the perfect composition which is required for the epitaxial growth of thin film substrates suitable for use as LEDs. Therefore, the development of the zinc selenide mono-crystal is expected to lead to the development of blue LEDs.

Artificial quartz produced with fewer defects

A method of producing artificial quartz using high quality seed crystals reduces the number of defects to one-tenth or less of that found with conventional techniques, according to the Tokyo-based developer.

Quartz oscillator manufacturer Tokyo Denpa says it has developed a method of optimizing production conditions to reduce the number of defects in the seed crystals. Artificial quartz grown from its seed crystals has fewer than ten defect lines per centimetre, compared with 100–400 lines per centimetre in quartz grown from conventional seeds.

The company expects this technology will considerably improve the yield and reduce the costs of producing artificial quartz, especially production of 40–50 MHz crystal oscillators.

Production of high molecular weight polysilane

A simple method of producing high molecular weight polysilane from methanol has been developed, according to researchers in Osaka and Kyoto.

Polysilane is drawing attention as a next-generation optical and electronic material with applications in solar cells and other devices, says Osaka Gas Co Ltd, which collaborated with a research group from Kyoto University, led by Professor Shono.

The method involves the reduction of methanol by the application of 50 V through a magnesium electrode. It can be adjusted to produce polysilane with a molecular weight of between 3000 and 30 000, the developers claim.

The reaction is carried out in an argon atmosphere, free of oxygen and water, at room temperature and normal pressure. Adjusting the reaction conditions, such as the concentration of methanol, enables control of the product's molecular weight.

Conventionally, polysilane with a molecular weight of 3000 is produced from methanol. However, polysilane with a molecular weight in excess of 30 000 is produced from methylphenyl-dichloro-silane in a reaction catalysed by metallic sodium. This method is dangerous and the molecular weight of the product tends to fluctuate.

Zinc telluride superlattice

Researchers from the Tokyo Institute of Technology claim to have produced a superlattice film of zinc telluride using atomic layer epitaxy (ALE). The technology could open the way for the development of blue light emitting diodes with high intensities, they report.

The researchers say they alternately fed zinc and tellurium into a reaction vessel containing a substrate at 230–270°C, over 50°C cooler than that used in conventional methods. The feed rate was 2–15 NS for tellurium (1 NS is the minimum quantity required to cover the substrate with a mono-atomic layer) and 6.6–44 NS for zinc.

The superlattice formed by this method consisted of alternating stacks of zinc and tellurium. Each layer was 50 atoms deep and about 6 Å thick; in conventional ALE, each layer must be 20–30 Å thick, the researchers say.

Alkyl halide crystal grown on gallium arsenide

A single crystal film of sodium chloride (NaCl) has been grown epitaxially on a gallium arsenide (GaAs) substrate by a research group from the University of Tokyo's Faculty of Science.

Professor Atsushi Koma's team used molecular beam epitaxy (MBE) to grow a single crystal film of sodium chloride which was 2 nm thick. The scientists believe their research could eventually enable the formation of materials with promise for optical applications.

The group suggest that epitaxial growth was achieved because NaCl and GaAs have similar lattice constants. The scientists also say they grew a perfect single crystal thin film on potassium bromide.

Because it is easy to grow a different alkyl halide crystal on top of any other alkyl halide film, even if it has a different lattice constant, the group claims it can stack films of sodium chloride, potassium bromide and organic phthalocyanine, in that order, onto a GaAs substrate.

Optical fibre with high transmission capacity

A plastic optical fibre has been developed that has 100 times the information transmission capacity of existing plastic optical fibres, according to a research group from Keio University.

The fibre is made from two types of monomer, methyl methacrylate and benzyl methacrylate, which are polymerized in a cylindrical reactor tube by the team, led by Lecturer Yasuhiro Koike of the Faculty of Science and Engineering.

The polymerization forms a transparent rod about 1.5 cm in diameter which can be heated and drawn out to produce an optical fibre. In terms of propagation loss, mechanical strength and weather resistance, the fibre has characteristics comparable to conventional products, the scientists add.

The fibre is suitable for short range information communication, such as in LAN (local area network) systems.

Permeable, heat-resistant polymer for optical filters

A polymer which is almost totally transparent to light of wavelength 0.85–1.6 μm has been developed by Nippon Telegraph and Telephone (NTT) Corp of Tokyo. The company intends to use the material in filters employed in optical communications equipment.

The polymer is made from a polyimide whose hydrogen atoms have been entirely replaced with fluorine. It has high permeability and heat resistance, and absorbs ten times less water than conventional polymers. The polymer can be fabricated into thin films with surface areas larger than those made from glass, NTT says.

Solar cells

High efficiency amorphous solar cells

A novel p-type semiconductor material allows the production of amorphous solar cells with improved photoelectric conversion efficiencies, claims Fuji Electric Co Ltd of Tokyo.

According to the company, a 1 cm square solar cell incorporating the semiconductor exhibited improved absorption of short wavelength light in the visible region. The conversion efficiency was increased by 0.5% to 12.0%, which is the highest efficiency to date for an amorphous solar cell.

The p-type semiconductor is made using boron trifluoride gas in place of the diborane gas used in conventional methods. The resulting semiconductor contains less boron than conventional products and is therefore more transparent.

The company claims that its p-type semiconductor will enable the fabrication of large area solar cells with high efficiencies. It plans to develop 30–40 cm square cells in the near future.

Producing polycrystalline silicon for solar cells

An electron beam melting technique which produces polycrystalline silicon for use in solar cells has been developed by researchers from Tokyo University's Institute of Industrial Science.

The technique uses silicon oxide feed-stock which is converted to silicon through carbon reduction. Purification is achieved by melting the silicon in an electron beam furnace. The molten silicon is then cast in a 15 × 3.5 cm water-cooled copper mould.

According to the scientists, the carbon content of the silicon was reduced from 75 parts per million (ppm) to less than 1 ppm and the phosphorus content was reduced from 30 ppm to less than 3 ppm. Other impurities, such as boron and iron, were not removed but the group is conducting further research in this area.

Polycrystalline silicon is currently produced from semiconductive silicon scraps. However, supplies for this method are at the mercy of fluctuations in the demand for semiconductors.

Large silicon solar cell has 15.6% conversion efficiency

A polycrystalline silicon solar cell which is 15 cm square has been made with a conversion efficiency of 15.6%, the highest ever achieved for a cell of this size, according to its Kyoto-based manufacturer.

Kyocera and Hitachi have previously achieved a conversion efficiency of 15.7% with a 10 cm square cell, but this is the first time that an efficiency of over 15.5% has been achieved with a 15 cm square device.

The high conversion efficiency was reportedly obtained by:

- reducing the film thickness, in order to enhance the voltage and current;
- changing the electrode formation method from screen printing to photography, in order to reduce the area of the electrode; and
- heat treating the electrode at 700°C, in order to stabilize the molecules in their position, thereby improving the electron flow.

Gallium aluminium arsenide solar cell

A solar cell made of gallium aluminium arsenide (GaAlAs) has been fabricated by a research team from Kagawa University.

Reportedly, the researchers made a 60 × 85 cm solar panel made from 1944 GaAlAs cells. They claim that the solar panel has a total power output of 5 W and that, by changing the way in which the cells are connected, it can produce up to 3000 V. A single cell can produce 1.5 V.

Further, the GaAlAs cell has great longevity compared to silicon cells, which are damaged by ultraviolet light and alpha rays and therefore have a limited service life of 10–20 years.

The researchers expect that the solar cell will be used in applications such as powering satellites, or compact electronic devices.

The group, led by Associate Professor Okamoto, reported their findings to the Fifth International Conference on Solar Light Electric Power Generation.

Amorphous solar cell with zinc oxide electrode

Tokyo-based Fuji Electric Co Ltd has fabricated a prototype amorphous solar cell which uses a transparent zinc oxide electrode.

There are no data on the conversion efficiency of the prototype, but the company says the zinc oxide film has been confirmed to be highly transparent and low in electrical resistance.

In experiments, the zinc oxide film transmitted light with an efficiency of approximately 80%, almost the same as for a tin oxide film (0.8 μm thick) formed by thermal chemical vapour deposition (CVD). However, the zinc oxide had less than one-tenth the electrical resistance of tin oxide film.

To make the prototype, Fuji Electric used a cluster ion beam to deposit high quality zinc oxide film without heating the glass substrate.

The highest conversion efficiency to date for an amorphous solar cell is 11.4%, which was achieved by the company. Solar cells using zinc oxide transparent electrodes are still in the test fabrication state, but the company claims higher conversion efficiencies can be achieved by improving the film formation techniques.

Business and financial information

Amorphous silicon solar cell marketed in Australia

A 20 W amorphous silicon solar cell panel with a conversion efficiency of 6–7% is being marketed in Australia by Showa Shell Sekiyu KK of Tokyo.

In 1989, Showa Shell produced solar cells with a total capacity of 1.4 MW, of which only 10% were made from amorphous silicon. Within five years, the company plans to raise its total silicon solar cell production volume to 10 MW and the proportion of amorphous cells, which are less expensive to produce than the mono-crystalline versions, to 30%.

The company plans to begin selling the 20 W solar panels in Japan. The panels cost ¥800–1000 per Watt.

CHAPTER SEVEN
MAGNETIC MATERIALS

Fabrication of thin films

A Langmuir–Blodgett technique for fabricating thin metal films with magnetic properties is under development at Osaka City University's Faculty of Engineering.

A research group led by Professor Tadao Nakaya claims to have synthesized films from five metal oxide powders:

- magnetic ferrite oxide was formed into a 10 nm diameter film;
- titanium oxide was formed into a 7 nm film;
- barium ferrite was formed into a 50 nm film;
- gamma-ferrite oxide was formed into a 100 nm film; and
- superconductive fine powder was formed into a 200 nm film.

To produce the films, metal powder mixed with toluene and a surface agent, such as oleic acid, were reacted to form carboxyl- metal fine powder. This preparation was then treated with benzene, placed in water and coated onto a substrate of either polyethylene terephthalate (PET) or silicone.

Teijin supplying tape to JVC and Hitachi Maxell

Teijin Ltd of Osaka is supplying its polyethylene naphthalate (PEN) film for use in high performance video tapes to the Tokyo-based Victor Co of Japan Ltd (JVC) and Hitachi Maxell Ltd from Ibaraki.

The PEN film improves the rigidity by 30–50% and yields better heat resistance characteristics than conventional tapes which use polyethlylene terephthalate (PET), according to Teijin. Using PEN, the thickness of a tape can be reduced to 0.8 μm, compared to the 1.5 μm of tapes fabricated using PET.

JVC and Hitachi Maxell are marketing S-VHS and VHS tapes using PEN film which are able to record up to three hours of video in normal mode.

Teijin says it is the only domestic firm producing PEN film, and has a current capacity of 80 tonnes a month at its Gifu plant in Gifu Prefecture. Sales of the material for use in video-cassette tapes accounted for over 50 tonnes a month by December 1990.

In 1992, the firm will construct a ¥15 billion PEN film production facility at its Utsunomia plant in Tochigi Prefecture and will expand its total production capacity to 500 tonnes a month.

Matsushita Electric Industrial Co Ltd of Osaka also plans to commercialize a PEN-based VHS-C video tape capable of recording 40 minutes of video in normal mode, while Tokyo-based Sony Corp is developing a PEN-based 8 mm video tape which will have a greater storage capacity than its current products.

Developing applications for high performance fluid

The development of practical applications for the high performance magnetic fluid "2,300-G" is the aim of a co-operation between the Tokyo-based National Research Institute for Metals (NRIM) and nine private firms.

Applications for 2,300-G, which was originally developed by NRIM, are expected to include use in car suspensions and in the fabrication of vibration-absorbing materials for audio speakers.

The participating companies are Nittetsu Mining Co Ltd, Mitsubishi Petrochemical Co Ltd, Tabuchi Electric Co Ltd, Kyoseki Gijutsu Kenkyusho, NOK Corp, Nippon Pillar Packing Co Ltd, Otsuka Kagaku, Nippon Seiko KK, and Cosmo Sogo Kenkyusho. However, the Institute reportedly plans to recruit more participants.

Mass production technique for single crystals of manganese zinc ferrite

A technique for mass producing single crystals of manganese zinc ferrite has been developed by Nippon Kessho Kogaku Co Ltd, a member of the Tokyo-based Mitsui Mining and Smelting Co Ltd group.

The parent company says crystals with a diameter as large as 80 mm can be produced which have a platinum content of almost zero.

Conventional ferrite single crystals have a diameter of 40–50 mm and contain platinum which diminishes their quality in applications such as video recorder heads. Furthermore, the new technique gives a yield of 40% while conventional methods have typical yields of about 25%, Mitsui claims.

The technique uses special ovens for the fine control of the temperature gradient during crystal growth. The company is currently operating three such ovens with a total production capacity of 5 kg a month.

It expects to market 18 kg of the crystals in the first year and predicts demand for ferrite single crystals will grow at an annual rate of 15%.

CHAPTER EIGHT
PLASTICS

Additives

Additive increases toughness and heat resistance of resin

An additive has been developed which increases the toughness and heat resistance of epoxy resins, according to a research group from Yokohama National University.

The fracture toughness of a resin containing the correct amount of additive is doubled, the scientists claim. In addition, the glass transition point, an indication of heat resistance, is raised from 180°C to 188°C.

The additive, N-phenyl-maleimide-styrene copolymer, is produced by copolymerizing phenyl maleimide monomer with styrene monomer. It has a structure similar to that of engineering plastics.

Tough, heat resistant epoxy resins are required for use in printed circuit boards and composites reinforced with carbon fibres.

Cyclohexymaleimide additive improved

A cyclohexymaleimide (CHMI) agent which confers heat resistance but, unlike conventional CHMIs, has no adverse effect on resins if added in large amounts has been developed, according to Nippon Oil & Fats Co Ltd in Tokyo.

CHMI, a derivative of maleic anhydride, is added during polymerization in order to improve a resin's heat resistance. Conventional CHMI products must be added in large amounts but this makes the polymerized resin hard and brittle. The new product needs to be added in small quantities only because of the addition of a special organic peroxide compound, the company says.

Additive for highly heat resistant epoxy resin

An additive which increases the heat resistance of epoxy resins has been jointly developed by the Tokyo Institute of Technology's Research Laboratory of Resource Utilization and Asahi Denka Kogyo KK.

Epoxy resin containing the additive displays heat resistance to temperatures of greater than 200°C, compared to 130–150°C for untreated materials. The additive is made from isocyanate in a reaction which involves a caesium compound catalyst.

Applications

Plastic fuel tanks in new vehicles in 1992

A plastic fuel tank has been developed by Toyota Motor Corp and Horie Metal Co Ltd, which Toyota says it will begin using in new car models during the first half of 1992.

Horie Metal will produce the tank at a ¥1 billion facility at its Hirota plant in Aichi Prefecture. Production will be at a rate of 3000 units a month.

This will be the first time Toyota has used plastic fuel tanks. Tokyo-based Nissan Motor Co Ltd is already using them in its vehicles.

The tank is made of a polyethylene sheet sandwiched between two nylon layers. It is light and easily formed, the developers say.

Resin for car parts

An acrylonitrile butadiene styrene (ABS) resin for use in car parts is available from Asahi Chemical Industry Co Ltd of Tokyo.

The company says that conventional ABS resins are difficult to apply to blow moulding because of their insufficient viscosities. However, the "Stylac ABS BL Series" has twice the molecular weight and can be used in the blow process.

The Stylac ABS BL Series costs ¥500 per kilogram, which is ¥100 per kiligram cheaper than conventional engineering plastics. Asahi Chemical expects to post annual sales of 2000 tonnes in 1993. The company has already begun supplying the resin to Mazda Motor Corp in Hiroshima, which uses it to make air spoilers.

Polymer membrane for separating chlorides

A group from Meiji University (led by Professor Tsutomu Nakagawa) and Lintec Corp has developed a separation membrane which efficiently and selectively separates organo-chloride compounds from aqueous solutions.

The new separation membrane uses a flexible acrylic- ester/acrylic-acid copolymer. The copolymer is mixed with a special cross-linking agent and is sandwiched between porous films. In experiments, the membrane showed an organic- compound/water separation factor of between several hundred and several thousand. Organo-chlorides such as trichloroethylene and tetrachloroethylene are difficult to separate from water because of their high boiling point and low combustibility.

At present, this membrane has a thickness of 60 μm but the researchers are aiming to develop thinner membranes as well as multilayered membranes for enhanced separation efficiency.

Anti-reflective plastic lenses

Fukui Prefecture's Industrial Technology Center has developed a technology which gives plastic lenses anti-reflective properties.

In the process, a plastic lens coated with a silicon/metal compound is subjected to low-temperature plasma treatment. A porous layer is formed in the silicon/metal coat by an etching action, the metal molecules are decomposed by the plasma treatment, and a low-refractive silicon oxide film is formed. The combination of the low-refractive film and the porous layer forms an anti-reflective film. The researchers report that a 92% transmissivity displayed by the coated lens was enhanced to 98% by the plasma treatment. 98% reflectivity is comparable to conventional multi-coating methods which are much more expensive. The multi-coating method requires six to seven hours processing using equipment costing ¥200–300 million. The plasma treatment method involves equipment costs of only ¥50 million for the same capacity, and the processing time is only one to two hours. In addition, the low-temperature plasma process has no problem with lens warping and allows further processing steps such as dyeing.

Photosensitive polymer

A type of photosensitive polymer which can be used as a negative type photoresist as well as immobilizing micro-organisms has been developed at the Research Institute for Polymers and Textiles.

The polymer is based on polyvinylalcohol (PVA), with hydrophilic pyridiniumylide groups attached. The polymer is soluble in water, but becomes insoluble when irradiated with ultraviolet rays.

The pyridiniumylide group is a complex containing a pyridine structure. When about 10 of these functional groups are attached to one PVA molecule, a transparent light-sensitive polymer is obtained which can be formed into a thick film.

Biodegradable polymers

Biodegradable thermoplastic from wheat protein

A biodegradable thermoplastic has been made from wheat protein (gluten) by researchers at the Hyogo Prefectural Industrial Experimental Station, in collaboration with Nagata Sangyo Co Ltd and Kyoto University of Industrial Art and Textile.

The researchers mixed gluten powder, glycerin, glycol, emulsified silicone oil, and urea which they then dried and heat pressed under a pressure of 150 $kg.m^{-2}$ into a translucent sheet 300 μm thick. The sheet was buried in soil and within four weeks it had completely degraded, one to two weeks faster than other biodegradable sheets, they claim. The increased biodegradability of the plastic stems from the fact that its polymeric structure is partially stabilized by sulphur bonds within the protein which are easily degraded.

A sample of the sheet displayed a tensile strength of 25 MPa, comparable to supermarket bags. Another sample containing larger amounts of urea, employed as a plasticizer, had a lower tensile strength.

Other materials

Thermoplastic polyolefin elastomer

A thermoplastic polyolefin elastomer is now being sold by its developer, Showa Denko KK of Tokyo.

"Oleflex" is resistant to heat, has high tensile strength, is light in weight and displays good weather resistance, the company says. The material may find applications as a substitute for polyurethane-based cross-linking agents.

Oleflex costs ¥500 per kilogram and Showa Denko hopes to achieve annual sales of ¥2 billion by 1995.

Silicone dissolving agent

An agent which uniformly dissolves hardened silicone products that have adhered to metals and ceramics has been developed by Mitsu Kagaku Co Ltd.

According to the company, silicone can be dissolved by immersing the product in the agent at room temperature for one to three days, depending on the thickness of the silicone. The agent has no adverse effects on materials such as aluminium, copper, iron, glass, and ceramics, all of which can be reused after cleaning with trichloroethane or xylene.

The agent costs ¥3800 per kilogram and the company aims to achieve sales of ¥50 million in the first year of marketing. Samples are available.

Polyimide hollow membrane fibres to separate solvents

A polyimide hollow membrane fibre can separate vaporous mixtures of water and organic solvents, according to Tokyo-based Ube Industries Ltd.

The membrane fibre is differentially permeable to different types of solvent vapour. Water will penetrate the membrane fibres 250 times faster than ethanol, 14 000 times faster than isopropylalcohol, 6600 times faster than acetone, and 9500 times faster than ethylacetate. In addition, the fibres have good resistance to heat and solvents, Ube says.

Industrial distillation units, which are currently used to separate solvents, generally use two distillation columns, one of which can be replaced with the hollow membrane fibres. This reduces energy costs by between one-half and one-third, and provides savings on space and installation costs.

Highly flexible polymers

A liquid crystal polyarylate which has a flexural modulus of 52 GPa, 2.5 times that of conventional products, and a hybrid ion cross-linked polymer which has a flexural modulus of 64 GPa have been synthesized at the Research Association for Basic Polymer Technology.

The liquid crystal polyarylate has a molecular structure containing 4,4-diphenyl carboxylic acid, consisting of two benzene rings with carboxyl groups at both ends. A 0.8 mm sheet of this polymer, which was formed by injection moulding, displayed a flexural modulus of 44 GPa, which was further increased to 52 GPa by heat treatment.

The other polymer is based on polyacrylic acid, cross-linked with polyvalent metallic ions. This was compression moulded under vacuum and displayed a flexural modulus of 64 GPa, close to that of aluminium. This isotropic polymer also exhibited improved heat resistance and dimensional stability.

High molecular weight copolymer

A highmolecular weight copolymer of acrylonitrile and vinyl acetate, called 'Porosoft', has been developed for filtration and air-diffusion applications. Its main characteristics are high gas permeability, high affinity for water, and virtually instantaneous absorption of aqueous solutions and organic solvents, according to Spacy Chemical Co of Hachioji.

The material is made by first heating and press-forming an acrylonitrile–vinyl acetate powder. The resulting sinter has pores of 3–50 μm in diameter. In addition, when placed in hot water (70–80°C), it can be moulded easily, even by hand.

The company says it could be used in water purification systems as a prefilter to ion-exchange columns to remove dust and small particles, thus lengthening column life. It could also be used for dewatering: if a water-oil mixture is passed through a pipe made of this material, the water will separate out through the pipe walls, leaving pure oil. Furthermore, air forced through a Porosoft pipe submerged in water diffuses through the walls into the water.

Colour-fast resin-coated polyester

'Perlex' is a resin-coated polyester which does not suffer from the pigment migration problems which are characteristic of other such fabrics, according to its Ishikawa-based developer.

Komatsu Seiren Co Ltd says that resin-coated fabrics, such as waterproofs, have not previously been made from polyester because dyestuffs from the polyester tend to move through the coating resin and transfer to an adjacent fabric.

The company claims to have prevented dye transfer by mixing additives with the coating resin. Komatsu Seiren plans to produce ¥400 million worth of Perlex a year.

Polycarbonate

Polycarbonate sheet

A new polycarbonate (PC) sheet has improved resistance to weathering and abrasion, according to the manufacturer, Mitsubishi Rayon Co Ltd of Tokyo.

The PC sheet is coated with a 1 μm thick layer of acrylic which is integrated by treatment with ultraviolet radiation. The company expects to sell 300 tonnes in the first year, rising to an annual total of 600 tonnes within three years.

Mitsubishi Rayon has also tripled its total production capacity of PC sheet to reach a total annual capacity of 5800 tonnes.

Polyvinyl chloride

Series of polyvinyl chloride resins developed

Four series of functional polymer resins based on polyvinyl chloride (PVC) — the "UE series", "SR series", "SE series" and the "JO series" — have been developed by Shin-Etsu Polymer Co Ltd of Tokyo.

The company reports that:

- the UE series contains an ultraviolet stabilizer and exhibits three times greater weather resistance than conventional PVC compounds;
- the SR series contains a thermoplastic rubber and displays high elasticity and good processing properties, such as adhesion;
- the SE series contains a special resin with a low friction resistance and is very smooth; and
- the JO series contains an olefin resin and flows readily so that it can be injection-moulded at low pressures and temperatures.

The company has begun marketing the four resins to car manufacturers, and expects to achieve sales of ¥1 billion in the first year of marketing.

Polyvinyl chloride sheet

A polyvinyl chloride sheet, "SG Sheet", is now available from Toppan Printing Co Ltd and Shin-Etsu Polymer Co Ltd, both of Tokyo.

The joint developers say the sheet, which Toppan is marketing, can be moulded and expect it to be used for making furniture and other products.

The sheet has three layers and is 0.8 mm thick. The material is made by laminating a transparent sheet onto a polyvinyl chloride substrate, which may be printed with a decorative pattern, and this is further coated with an ultraviolet-curable acrylic resin. The sheet has a mirror-like gloss and is easily processed into curved shapes.

A sheet measuring 1.04 × 1.83 m costs ¥7000 and the two companies also plan to market finished products, including doors and furniture. Toppan and Shin-Etsu expect sales of ¥1 billion in the first year of marketing.

Processing and equipment

Thin polyphenylene sulphide film can be injection-moulded

A polyphenylene sulphide (PPS) engineering plastic has been made which its joint developers say can be injection-moulded into films just 0.3 mm thick.

Toray Industries Inc of Tokyo and Matsushita Seiko Co Ltd of Osaka intend the plastic to be used as an insulating material in the condenser motors of home appliances. Use of the plastic will allow the production of smaller motors, they claim.

Toray was producing six to seven tonnes a month of the plastic, but increased this to 20 tonnes a month. The material costs ¥1800–2000 per kilogram.

Water-soluble polymer is easy to process

A thermoplastic has been produced which can be easily cast into films and mouldings and has five times the dissolution rate in water of conventional products, according to its developer, Dai-Ichi Kogyo Seiyaku Co Ltd of Kyoto.

"PAOGEN" is easily processed and (because it consists only of carbon, hydrogen, and oxygen) it does not release noxious gases, such as NO_x, when it thermally decomposes. A film 30 μm thick dissolves in water in five seconds, the company claims.

The product comes in the form of granules 1 mm in diameter and 3 mm long. It costs ¥3000 per kilogram. The company aims to achieve sales of ¥600 million in the first year of marketing.

Polyester can be rapidly injection-moulded

A polyester is available which its developer, Toray Industries Inc from Tokyo, claims can be injection-moulded ten times faster than conventional polyesters.

To synthesize the material, ethylene glycol and terephthalic acid, the two monomers used to produce polyethylene terephthalate (PET), are copolymerized with diphenylcarboxylic acid in a molar ratio of 5:2:3.

The new polyester has a 10–30% higher glass transition temperature and three to four times the elastic modulus of PET, Toray reports.

Fluorine ion implantation increases abrasion resistance of polyimide resin

The abrasion resistance of a polyimide resin is raised by a factor of ten as a result of implanting it with fluorine ions, according to the Agency of Industrial Science and Technology's Mechanical Engineering Laboratory in Ibaraki.

In experiments, a block of polyimide (30 × 30 × 6 mm) was implanted with $10^{15}cm^{-2}$ of fluorine ions. The implantation energy was 1 MeV and the depth of penetration was 3.5 μm. This process carbonizes the top layer of the sample, the Laboratory says.

Abrasion tests revealed that the implanted sample exhibited only half the friction coefficient and suffered only one-tenth the abrasion of an untreated sample. The method is expected to expand applications for polyimide resin.

Conventionally, the abrasion resistance is increased using reinforcing materials.

Polyphenylene sulphide synthesis

A method of synthesizing polyphenylene sulphide (PPS) has been devised by researchers from Waseda University's School of Science and Engineering.

Waseda University's method takes place at normal temperature and pressure, and produces no sodium chloride (NaCl) as a by-product, eliminating the need for a desalination step, the scientists say. Conventional methods of PPS synthesis involve extreme conditions of 200°C and 20 atmospheres, and produce NaCl.

The new synthesis method uses p-xylene and sulphur chloride as raw materials, dichloro-dicyanobenzoquinone (DDQ) as an oxidizing agent, and a vanadyl compound as a catalyst. Blowing air into a solution of these ingredients for five hours produces PPS with a yield of 90%.

Because the method does not produce NaCl, the product has reliable electrical characteristics, and is expected to expand applications for PPS.

Single technology produces three types of polyethylene

A production process which can synthesize low density polyethylene (LDPE), linear low density polyethylene (LLDPE) and high density polyethylene (HDPE) has been developed, claims Mitsui Petrochemical Industries Ltd of Tokyo.

The process uses a novel catalyst and different synthetic pathways. Conventionally, these three polyethylenes have been produced using different processes, at different plants, according to the company.

In the USA, Union Carbide Corp uses a method which can produce both LLDPE and HDPE.

Thin nylon film production method

A Langmuir-Blodgett method can be used to make films of nylon just 0.5 nm thick, says the Institute of Physical and Chemical Research (IPCR) in Tokyo.

When reacted with trifluoroacetic acid, hydrogen groups on the nylon are replaced with trifluoroacetyl. This makes the nylon soluble in organic solvents such as benzene and chloroform, the Institute reports.

To form the film, the nylon solution is placed in water. As it spreads out upon the surface, it evaporates and the trifluoroacetyl groups are replaced by hydrogen atoms. This causes the nylon to come out of solution, forming a monomolecular film.

Until now, nylon has proved difficult to form into films less than 200 nm thick because of its insolubility in conventional organic solvents.

Method for forming fine particles of polymers

Gamma radiation can be used to form fine particles of polymers, according to the Japan Atomic Energy Research Institute (JAERI) based in Tokyo.

JAERI's method involves irradiating a monomer solution with gamma rays which causes the precipitation of polymer particles with a diameter of 0.1–10 μm. The diameter of the particles, which can be easily isolated by filtration, is controlled by varying the intensity and the duration of the irradiation.

JAERI expects that the particles will find applications in drug delivery systems.

Polyimide superlattice made by deposition polymerization

A thin film of a polyimide superlattice has been formed by a research group, led by Associate Professor Okui, from the Tokyo Institute of Technology's Faculty of Engineering.

The researchers used a vacuum deposition polymerization technique. They placed a diamine and a carboxylic acid in the vacuum chamber, with an internal pressure of about one-ten-thousandth of an atmosphere. A film of polyimide, made of alternating diamine and carboxylic acid groups, was then deposited on a substrate at a rate of 2–6 nm a minute.

Growth of the film was observed to be temperature dependent. The polymer chains did not become vertically oriented at 25°C, but were arranged vertically when the substrate was held at 125–175°C. The temperature was then increased to 200°C in order to dehydrate the polyimide, forming a superlattice.

Plastic processing developed

A system for giving transparent plastics a hard coating has been developed by Nippon Dacro Shamrock (NDS) Co Ltd, a subsidiary of Nippon Oil & Fats Co Ltd of Tokyo, in association with the US company PPG Industries Inc.

The system uses NDS's technology for applying coatings to wide plastics parts with a 3–5 μm thick film of PPG's silicone-based agent. This gives the part resistance to chemical and mechanical damage, the developers claim.

NDS has established a subsidiary, Sarc Systems Inc, which will manufacture parts for supply to Asian countries. Sarc Systems has to invested about ¥1 billion in the construction of a facility at Nippon Oil & Fat's Totsuka factory in Kanagawa Prefecture. The facility, which was completed at the end of 1991, is expected to produce coating machinery worth ¥100 million in its first fiscal year of operation.

Method for recovering germanium and gallium

A method for recovering germanium, an important catalyst in the production of polyethylene terephthalate (PET), and gallium has been established by the Government Industrial Research Institute (GIRI) in Kyushu.

The Institute says that germanium in solution is selectively adsorbed by a methylglucamine-type chelate resin. In experiments, an ethyleneglycol solution was filtered through a membrane and then passed through a column filled with germanium-adsorbing fine particles of the chelate resin. The resin was then washed with dilute hydrochloric acid in order to recover the germanium.

Japan depends on imports for most of its germanium, about 60% of which is used as a catalyst for the production of PET. During PET production, 50–75% of the germanium ends up in a solution of ethyleneglycol, and until now it has been unrecoverable.

The Institute also says that zinc ferrocyanide and copper ferrocyanide selectively adsorb traces of gallium contained in an aluminium chloride solution. When 2 g of ferrocyanide was added to a 1.8 mole.l^{-1} of aluminium chloride solution containing 50 parts per million of gallium, 99% of the gallium was recovered. Copper ferrocyanide could recover 80% of the gallium.

Graft polymerization increases durability of polyethylene film

Graft polymerization can form a composite film of polyethylene (PE) and polyethylene terephthalate (PET) which is stronger than conventional PE films, claims Nisshin Steel Co Ltd of Tokyo.

In its process, the company stretches a PE film to six times its size and coats it with a thin layer of methacrylic acid. The coated film is laminated with a biaxially oriented PET film which is 25 μm thick and subjected to irradiation.

The film is then washed with water to remove residual monomer. A composite film results, with a graft polymerized structure on the inside and densely cross-linked PET on its surface.

Nisshin reports that scanning electron microscope images of the film's surface revealed none of the fine cracks which are characteristic of conventional PE stretch films.

Polyvinyl alcohol film production technique

A technique for producing polyvinyl alcohol (PVA) film has been devised by a research group from Ochanomizu University, led by Professor Shoji Ogawa.

The research group froze an 8% PVA solution at −20°C and then defrosted it at room temperature. This process was repeated five times in order to form a loosely cross-linked PVA gel film with a thickness of 1.5 mm.

The PVA gel film was then drawn in one direction in order to orient all of the molecules. The gel film was drawn to seven times its original length at a rate of 2 mm a minute. The stretched PVA film exhibited a tensile strength of 180 MPa and an elastic modulus of 80 MPa, according to the scientists.

High precision injection moulding machine

A prototype electric injection moulding machine is capable of producing parts with a surface roughness of 0.1 μm, according to its Tokyo-based developer.

Japan Steel Works Ltd has begun marketing a commercial version, the "J56ELD", which is targeted at firms manufacturing polygon mirrors for use in laser printers, lenses, and optical discs.

The machine features a mould flow analysis system which is used to maintain mould temperature and clamping force at optimum levels. Servo motors are used in the injection, mould opening and closing, and clamping mechanisms.

In tests using a machine with a clamping force of 55 tonnes, the firm claims that it was able to mould polygon mirrors with a surface roughness ranging from 0.05-0.2 μm. The mirrors were made of polycarbonate, measuring 44 mm in diameter and 6 mm in thickness.

Method for synthesizing highly polymerized polyphenylene sulphide resin

A bi-phasic method of synthesizing a highly polymerized polyphenylene sulphide (PPS) resin does not require the use of expensive catalysts, according to its developer, Kureha Chemical Industry Co Ltd of Tokyo.

Kureha Chemical found that polymerization of PPS resin is inhibited by residues of unreacted sodium sulphide which tend to cut the polymerizing chains. Based on this finding, Kureha developed the bi-phasic method in which concentrated PPS prepolymer in one phase polymerizes in the absence of residual sodium sulphide which is dissolved in the second phase:

dichlorobenzene and sodium sulphide. Conventional methods of increasing the degree of polymerization have usually involved the consumption of expensive catalysts such as lithium acetate.

Plastic recycling machine

A machine designed to recycle certain kinds of waste plastic, the "RE150/80", is now being sold by Hosokawa Micron Corp. Marketing will be targeted mainly at manufacturers of polyethylene film.

Scrap plastic is fed into the machine, melted, extruded into long thin strands and cut into pellets. The machine can process up to 120–150 kilograms of scrap plastic an hour.

The machine costs ¥2.2 million and the firm expects to sell ten of the machines each year.

Process for producing biodegradable plastics

A process for direct production of biodegradable plastics using carbon monoxide has been developed at the National Chemical Research Institute (NCRI) of the Agency of Industrial Science and Technology (AIST).

In the process, formaldehyde powder, which acts as a catalyst, is placed in a stainless steel vessel along with a solvent. Carbon monoxide gas is introduced into the vessel at a pressure of 15–20 MPa and, after heating to 150°C for two hours, polyglycolide powder, a polyester with a molecular weight of 10 000, is formed.

The polyglycolide exhibits a thermal cracking temperature of approximately 280°C and a melting point of approximately 200°C.

The researchers used the polyglycolide to mould a film 10 cm square and 0.25 mm thick. The researchers confirmed that the film had a biodegradability that is comparable to plastics produced by bacteria.

Resin compound for mould production

A heat-cured resin compound for producing plastic injection moulds has been developed by International Software Corp. "Rezomet" is based on an epoxy resin with a modulus of flexibility of 23.3 $kg.mm^{-2}$ to which metal powder is added (65–75% by weight) for better thermal conductivity.

The firm reports that injection moulds made with Rezomet can be used to mould parts made of ABS, polyolefin, polyacetal and polyphenylene ether resin. Tests have shown that the moulds can be used for up to 1000 shots. The firm claims that moulds made of Rezomet can also be used to form materials with high melting points, such as polyphenylene sulphide, without the need for cooling channels in the moulds.

Continuous surface modification system for plastic powder

A continuous system for surface modification of plastic powders has been developed by Fukuvi Chemical Industry Co Ltd, in collaboration with Shirayuri Women's University and the Research Development Corp of Japan.

The system uses two pipes (Coanda spiral units), one of which carries particles of plastic powder and another which carries the coating agent. The plastic powder is carried along in a stream of compressed air which forces the plastic particles to adopt a spiral flow along the curved surface of the pipe (Coanda effect). The plastic powder particles are sprayed with the coating agent resulting in their uniform encapsulation.

In experiments, in which silane was used to coat particles of methacrylic resin, the new process required less than one-third of the coating solution required by conventional methods and resulted in resin 10% stronger than is produced by conventional methods.

Thermoplastic produced from coffee bean husks

A method for producing a thermoplastic from coffee bean husks has been developed at the Industrial Products Research Institute in Tsukuba, scientists there claim.

The researchers subject the coffee bean husks, which contain large amounts of cellulose and lignin, to chemical processes which convert these compounds into a polymer. They say the polymer, when formed into a film, has a thermal decomposition temperature of 240°C, a glass transition temperature of 80°C, a tensile strength of 40 MPa, and an elastic modulus of 30 tonnes.

These values are comparable with those of general purpose plastics such as polystyrene.

Shape memory and intelligent materials

Shape memory resin from vinyl acetate

A shape memory resin based on vinyl acetate has been developed by the Agency of Industrial Science and Technology's Mechanical Engineering Laboratory in Ibaraki.

In experiments, a sample of the resin at 60°C was drawn out to a diameter of 0.8 mm, and cut into lengths of 9 cm. One of these samples was then cooled to 15°C and stretched to a length of 18 cm. When the stretched resin was immersed in water at 40°C, it shrank to 9 cm within five seconds, the Laboratory says.

The resin was again stretched at room temperature, to about 20 cm, and then re-immersed in 40°C water, once again restoring its length to the original 9 cm.

The Laboratory plans to investigate the mechanism of the material's memory with scanning electron microscopy and X-ray diffraction analysis.

Intelligent polymer

An intelligent polymer which expands and contracts with changes in temperature has been jointly developed by groups of researchers at Sophia University and Tokyo Women's Medical College.

The polymer takes the form of an agar-like gel and consists of acrylamide and acrylic acid, held together by a network of hydrogen bonds. At high temperatures, the hydrogen bonds break and the polymer expands. At low temperatures, the hydrogen bonds are restored and the polymer contracts, say the researchers, led by Professor Sakurai and Professor Ogata.

The scientists add that a functional chemical or enzyme may be placed inside the material's network structure, being released in a controlled manner when the temperature increases.

In experiments, a disc-formed sample of the polymer, 15 mm in diameter and 1 mm in thickness, was impregnated with an anti-tumour drug. The sample was then subjected to repeated cycles of heating and cooling, peaking at 30°C for periods of 30 minutes and dropping to 10°C for periods of 1.5 hours. When the temperature changed, expansion and contraction of the gel took place instantaneously, and release of the drug was predictable and controllable, they report.

Shrinkable polymer gel

A polymer gel which absorbs organic solvents and shrinks rapidly when an electric current is applied has been produced by a research group from Ibaraki University.

The material shrinks about ten times faster than conventional water-absorbing polymer gels and becomes electroconductive when irradiated with light, the scientists claim.

The material is a complex produced by dissolving a mixture of acrylamide-based polymer and a complexing agent, such as tetracyanobenzene, in an organic solvent. The material is then stacked in layers and electrified with a voltage of 10 V for several hours. This process produces a functional polymer gel consisting of regularly stacked benzene molecules.

In experiments, a 3 mm thick film of the gel was electrified, causing it to release the organic solvent and shrink. When irradiated with 400 nm light, it generated a current and displayed a conductivity of 0.01 $S.cm^{-1}$.

Temperature-dependent polymer particles

Functional polymer particles whose size and surface properties change with temperature have been developed by a research group from Keio University.

The particles are based on N-isopropylacrylamide, which has been mixed with between 1 and 0.1% methylene-bis-acrylamide, a cross-linking agent. At over 32°C, the particles have a diameter of 0.5 μm and a hydrophobic surface. At lower temperatures, the particles absorb water, acquire a hydrophilic surface, and swell to between two and four times their original size.

In general, proteins tend to adsorb onto hydrophobic surfaces, and experiments with human globulin indicated that while 2.4 ng of protein adsorbed to 1 g of the particles at 40°C, only 0.4 ng adsorbed to the particles at 25°C.

Shape memory material

A gel of polyvinyl alcohol (PVA) can memorize shapes, according to researchers at Shinshu University. Shape memory materials are generally used for pipe fittings and the tips of endoscopes. The PVA gel is expected to have applications in artificial blood vessels and organs, the scientists say.

To make the materials, the researchers freeze and then thaw a PVA solution to form the gel. They then mix this with glutaraldehyde to cross-link the polymer chains.

The cross-linked gel was formed and treated by dipping it in solvents, such as methanol. A different formation resulting from each treatment. The formation is destroyed upon boiling but restored by reintroducing the solvent.

The group demonstrated that a single gel could remember three different formations.

Business and financial information

Italian firm plans to enter Japanese market

Canon Group, an Italian firm located in Milan, is planning to enter the Japanese market with a polyurethane injection system designed for use in the production of insulating refrigerator parts.

The system injects polyurethane between the outer metal casing and the plastic interior of the refrigerator to create an insulating layer.

A major advantage of its system the Italian firm says is that it requires half the amount of chlorofluorocarbon compared with conventional units of this type.

The system will be imported and marketed by Canon Group's Japanese subsidiary Nippon Canon Inc which also plans to produce the system in Japan in the future.

Increased production of oxoalcohols

Kyowa Hakko Kogyo Co Ltd of Tokyo plans to invest ¥2.1 billion in establishing a plant for manufacturing propyleneglycol monomethylether acetate (PMA) and butyl acetate, at the Yokkaichi factory of its affiliate, Kyowa Yuka Co Ltd. The company says the move is in line with its plans to step up production of oxoalcohols.

The plant has a production capacity of 25 000 tonnes of PMA a year. In addition, the company has recently acquired Japan Oxocol Co Ltd, where Kyowa Hakko plans to step up production of high molecular weight alcohols and establish facilities for producing plasticizers for vinyl chloride.

Once the facilities are completed, Kyowa Hakko says it will be the third largest company in the oxoalcohol business; following the US-based Exxon Corp and the German group BASF AG.

Increased production of polymer microparticles

Sekisui Plastics Co Ltd of Osaka is investing ¥550 million in the construction of a plant for the production of its functional spherical polymer microparticle, "Tech-Polymer", at its factory in Shiga Prefecture.

The plant, which will have a production capacity of 250 tonnes a year, will be completed in 1992, and will double the Shiga factory's Tech-Polymer production capacity to 500 tonnes a year.

After completion of the factory, Sekisui Plastics expects sales of Tech-Polymer to increase to ¥1 billion, from the current ¥400 million.

Ethylene propylene rubber plant constructed

Japan Synthetic Rubber (JSR) Co Ltd of Tokyo plans to invest ¥8 billion in an ethylene propylene rubber (EPR) plant at its Kashima factory in Ibaraki Prefecture.

JSR says it will complete the 20 000 tonnes a year plant in late 1992. The company was producing 40 000 tonnes a year at its Yokkaichi factory in Mie Prefecture. JSR increased production at the Yokkaichi factory to 45 000 tonnes a year by 1992, giving the company a total EPR production capacity of 65 000 tonnes a year. JSR claims it will be the largest EPR producer in Japan.

Japan's EPR production capacity stands at 120 000 tonnes a year, worth ¥72 billion in 1990. The domestic demand for EPR increased by 13% in 1988, 12% in 1989, and 13% in 1990.

Sumitomo Heavy Industries to install computer integrated manufacturing system

Sumitomo Heavy Industries (SHI) Ltd of Tokyo is planning to install a computer integrated manufacturing (CIM) system at its plastic injection moulding machinery division.

The move is aimed at enhancing the division's product development capacity and making it more responsive to user demand, SHI says. The system is expected to help double the division's sales to ¥50 billion by the fiscal year 1994.

The CIM system will link the division's production control, sales management, quality control, and product development operations. When completed, the system will allow the firm's sales personnel to have estimates prepared automatically and will substantially reduce delivery times.

The system will also link SHI's own plants with those of the firm's affiliates. The system and installation costs are expected to total ¥1.5 billion.

SHI says it currently holds the fourth largest share of the domestic plastic injection moulding machine market. Once the new system is in operation, SHI aims to close the gap with its rivals Nissei Plastic Industrial Co Ltd and the Japan Steel Works Ltd.

Prices of ethylene and propylene raised

Mitsubishi Petrochemical Co Ltd of Tokyo and Nippon Petrochemicals Co Ltd have raised their charges for fixed expenses incurred during the production of ethylene and propylene for the first time in five years.

The two companies raised their charges for fixed expenses by ¥12–13 per kilogram in the first quarter of 1991, and planned to raise them by a further ¥5 per kilogram in the second quarter.

The prices charged for ethylene and propylene are calculated by adding the cost of fixed expenses, such as labour and equipment, to the varying price of the raw material, naphtha. The increases will result in a 5–10% increase in the prices of ethylene and propylene.

Mitsui Toatsu Chemicals to produce unsaturated polyester in Thailand

Mitsui Toatsu Chemicals Inc of Tokyo plans to establish a joint venture in Thailand for the production of unsaturated polyester resins.

The company, which already produces several industrial resins in Thailand — including plywood adhesives, polyvinyl chloride, and premixes — plans to construct an unsaturated polyester resin plant with a production capacity of 10 000–20 000 tonnes a year.

The company is reportedly planning to construct more resin production plants in Thailand in the near future.

Mitsubishi Kasei to produce polyester fibre raw material in Indonesia

Tokyo-based Mitsubishi Kasei Corp has established a joint venture in Indonesia for the production of pure terephthalic acid (PTA), a raw material used in polyester fibre production.

In association with Indonesian company PT Bakrie and Brothers, Mitsubishi Kasei has formed PT Bakrie Kasei Corp. The latter is capitalized at US$15 million, held 77.8% by Mitsubishi and 22.2% by Bakrie.

In late 1993, the joint venture will complete a 250 000 tonnes a year plant, the largest of its kind in Indonesia. The PTA will be supplied to customers in Indonesia and other south-east Asian countries.

Epoxy resin marketed

An epoxy resin for use in high performance composites is being marketed by Kokusai Chemical Co Ltd. The company expects the resin to be used in the construction of boats, surf-boards and sports materials.

The resin, "Kraft Resin Epoxy System Z-2/H-0.7", has additives which give it a bending strength of 200 MPa, three times that of conventional epoxies, Kokusai Chemical claims.

In addition, the hardened resin is clearer than conventional epoxies because it uses an amine-based hardener.

Seven kilograms of the resin cost ¥18 000 and the company aims to produce five to ten tonnes a month.

Tonen Chemical constructs production facilities

Tonen Chemical Co Ltd is constructing production facilities in Nishinasuno, Tochigi Prefecture. The company plans to base some of its affiliates' manufacturing activities at the new factory, in addition to its own operations.

The first section of the facility will produce double-layered high density polyethylene pipe. The company expects to manufacture the pipe at a rate of 3500 tonnes a year and achieve sales of ¥1.65 billion a year.

Phenolic resins for car and building interiors

Honen Corp of Tokyo plans to begin supplying phenolic resins for use in car interiors and as building materials.

The company says it has developed an adhesive for use in car interiors which is made from paper impregnated with one of its phenolic resins, "PR Resin". The adhesive is used to bond wood veneer to metal.

The company has also developed an expandable phenolic resin, "Honen Foam", which is designed for use with building materials. This product, which is already being supplied to two firms, has thermal insulation properties similar to polyurethane and polystyrene foams, both of which are widely used as heat-insulating building materials. The phenolic resin foam is resistant to fire.

Sekaicho Rubber establishes shoe rubber subsidiary

Nishimomiya-based Sekaicho Rubber Co Ltd has announced plans to establish a wholly-owned subsidiary for the production of shoe rubber.

The subsidiary, SEC Polymer, will be capitalized at ¥10 million and will have its head office and rubber processing factory at Sekaicho Rubber's Tokuyama factory in Yamaguchi Prefecture. The subsidiary will produce 6000 tonnes a year of rubber and it hopes to sell rubber worth ¥100 million in the first fiscal year of operation.

Vinyl acetate production facility completed

A ¥1 billion production facility for acetaldehyde, acetic acid and vinyl acetate has been completed by Showa Denko KK of Tokyo at its Oita factory in Kumamoto Prefecture.

The facility increases Showa Denko's acetaldehyde production capacity by 10 000 tonnes a year to 140 000 tonnes a year, its acetic acid production capacity by 20 000 tonnes a year to 150 000 tonnes a year, and its vinyl acetate production capacity by 40 000 tonnes a year to 120 000 tonnes a year.

Showa Denko has also signed an agreement whereby it will sell its excess production of vinyl acetate to Denki Kagaku Kogyo KK of Tokyo.

Exchange agreement for unsaturated polyester resins

Information on unsaturated polyester resins and vinyl ester resins will be exchanged as the result of an agreement between Takeda Chemical Industries Ltd of Osaka, the US firm Owens-Corning Fiberglas Corp and BASF AG of Germany.

The agreement provides for the exchange of technical information in areas such as product compositions, production methods and systems, raw materials, testing methods, and marketing information. The agreement compels the companies to fully disclose technical information to the other signatories who are then entitled to make full use of the information. Patent rights will be granted to the other signatories when necessary.

The agreement initially covers a period of five years, after which it will reviewed every year.

Research and development project on silica polymers

The Agency of Industrial Science and Technology (AIST) in Tokyo has started a ten-year research and development project on silica polymers.

Silica polymers are carbon-based polymers in which at least one carbon atom in each monomer has been replaced with silicon. Because silicon–silicon bonds are stronger than carbon–carbon bonds, silica polymers exhibit stronger bonding than carbon polymers. Silica polymers are also valued for their electrical properties.

AIST says the project will initially concentrate on developing monomer polymerization techniques.

Joint venture to market low flammability fibre

Osaka-based Teijin Ltd and Hoechst AG have signed an agreement establishing a joint venture for the production and marketing of the latter's general purpose low flammability polyester fibre, "Trevira CS".

The joint venture is based at Hoechst Japan's Tokyo office and capitalized with ¥500 million, 51% held by Hoechst Japan Ltd and 49% by Teijin. Operations began at the end of January 1991.

Teijin markets Trevira in Japan, and Hoechst Japan markets the fibre in East Asia and Oceania. The two companies were aiming to sell Trevira worth ¥600 million in 1991.

In addition, Teijin plans to invest about ¥1.2 billion in modifications to its Tokuyama factory for the production of 10 000 tonnes a year of Trevira. The modifications are scheduled to be completed by the middle of 1992.

High strength polyarylate fibre production increased

By the fiscal year 1994, Kuraray Co Ltd of Osaka plans to increase its capacity to produce the high strength polyarylate fibre "Vectran" to 800–1000 tonnes a year.

In March 1990, the company completed a 400 tonnes a year Vectran production facility at its Saijo Factory in Aichi Prefecture. In fiscal year 1991, the company planned to produce 120–130 t of Vectran, gradually increasing production until full capacity is reached in the fiscal year 1993. Further plant expansions will be carried out in order to meet the 1994 target.

Joint venture to sell polyethylene fibre

Toyobo Co Ltd of Osaka and the Dutch company DSM have announced the establishment of an equally-owned joint venture, Nippon Dainyma Co Ltd. The venture will produce "Dainyma", a high strength polyethylene fibre which was developed by the two companies.

The new company is based at Toyobo's head office and capitalized at ¥20 million. It will initially produce 200 tonnes a year of Dainyma at a plant in Toyobo's laboratory.

Dainyma is mainly used in:

- high strength ropes for marine use;
- protective work clothes and sportswear; and
- ski boards, helmets and speaker vibration plates.

Dainyma costs over ¥7000 per kilogram for the 1700 denier grade.

Mitsubishi Kasei to double polybutyrene terephthalate production capacity

Daia Foil Co Ltd, a subsidiary of Tokyo-based Mitsubishi Kasei Corp, plans to begin producing polybutyrene terephthalate (PBT).

Daia Foil plans to produce 5000 tonnes a year of PBT at its Yokkaichi factory in Mie Prefecture where it currently operates a 47 000 tonnes a year polyethylene terephthalate (PET) production facility. Mitsubishi Kasei currently produces 5000 tonnes a year of PBT at another location.

Mitsubishi Kasei has also announced it plans to begin importing polycarbonate from one of its Korean affiliates.

Sumitomo Chemical to increase polyvinyl chloride compound production

Plans to double production capacity of "Sumilit FLX", a polyvinyl chloride (PVC) moulding compound used for making items such as car-interior parts, have been announced by Sumitomo Chemical Co Ltd of Osaka.

The company plans to invest ¥1 billion in doubling Sumilit FLX production capacity at its Ehime factory in Ehime Prefecture from 5000 to 10 000 tonnes a year. It supplies 5000 tonnes a year to the car industry but is currently developing new applications for the product.

Sumilit FLX is a mixture of PVC powder, plasticizing agent, stabilizer and colouring. The powder is placed in a mould which is heated to 200–240°C.

Idemitsu to increase polystyrene production

¥5 billion will be invested in the construction of a 60 000 tonnes a year polystyrene factory in Johor, Malaysia, according to Petrochemicals (Malaysia) Sdn, an affiliate of Idemitsu Petrochemical Co Ltd of Tokyo. The facility is scheduled to begin operating in December 1992.

At present, 36 000 tonnes of polystyrene are produced each year in Malaysia, of which Petrochemicals (Malaysia) produces 30 000 tonnes. The factory will give the company a total production capacity of 90 000 tonnes a year.

Demand for polystyrene in Malaysia has risen sharply in recent years, due to the nation's increased output of electronic devices, and will reach 70 000 tonnes a year by 1993, Idemitsu predicts.

Kureha Chemical produces methacryl butadiene styrene resin in Singapore

Tokyo-based Kureha Chemical Industry Co Ltd has established a joint venture in Singapore which has begun operating a methacryl butadiene styrene (MBS) resin production factory.

The venture, Kureha Chemicals Singapore, is owned 67.5% by Kureha, 25% by Rohm and Haas Co of the USA, and 7.5% by Sumitomo Chemical Co Ltd from Osaka. The venture has invested about ¥4 billion in the construction of the 16 000 tonnes a year MBS resin production factory.

The venture's MBS resin will be marketed in Southeast Asia and Oceania, and the company hoped to achieve sales of ¥6–7 billion in the first fiscal year.

Polypropylene production

A venture for the production of polypropylene has been established by Ube Industries Ltd, in collaboration with Sumitomo Chemical Co Ltd of Osaka, Tokuyama Soda Co Ltd of Tokyo and Union Polymer Co Ltd.

The joint venture, Ube Polypro, is capitalized at ¥500 million, held 47.5% by Ube Industries, 29.69% by Sumitomo Chemical, 17.81% by Tokuyama Soda and 5% by Union Polymer.

Ube Polypro, based in Tokyo, has constructed a polypropylene production factory at Ube Industries's factory in Ube which has a polypropylene production capacity of 80 000 tonnes a year.

Increased marketing of high density polyethylene film

Tokyo-based Tonen Sekiyu Kagaku KK plans to increase its efforts in the marketing of its high density polyethylene film.

The double-axial drawn film exhibits high transparency and rigidity, and high resistance to shock and moisture, the company claims. It has recently constructed film production facilities at its Chidori Factory with a capacity of 4000 tonnes a year.

The company reports that it has received a large order for the film from a major US label manufacturer.

Production facility for polybutyrene terephthalate

A facility for the production of raw polybutyrene terephthalate (PBT) resin has been establishd by Dainippon Ink & Chemicals Inc of Tokyo.

The facility, located at the company's Kashima factory, has resin production capacity of 3000 tonnes a year and began operating in February 1991. Following completion of the facility, production of resin at the company's Amagasaki factory ceased.

The company also plans to construct facilities at its Kashima factory for the production of polythioether sulphone (PPES) and polyphenylene sulphide (PPS).

Anti-vibration rubber sales company set up in USA

In early 1991, Osaka-based Toyo Tyre & Rubber Co Ltd set up a joint venture in Michigan, USA, to handle sales of automotive vibration-damping rubbers.

The new firm, GT Automotive Co Ltd, is capitalized at US$100 000, held 40% by Toyo Tyre and the remainder by GenCorp Inc of the USA.

Using technology from Toyo Tyre, GenCorp has been producing automotive anti-vibration rubbers for sale through Toyo Tyre (USA) Corp (TTC) in California. GT Automotive will now take over anti-vibration rubber sales activities from TTC, which will then concentrate on tyre sales.

The new firm is expected to generate income of US$14 million a year after three years.

Rubber parts plant

In response to increased orders, automotive rubber parts maker Kinugawa Rubber Industrial Co Ltd of Chiba, an affiliate of Tokyo-based Nissan Motor Co Ltd, has built a plant in Fukushima Prefecture.

When operations began in December 1991, the ¥7.2 billion plant was producing sealing parts worth ¥7 billion a year. Once operating, current sealing parts production at its Chiba plant was discontinued and the latter now specializes in anti-vibration rubber parts production.

Joint venture to produce engineering plastics

Mitsui Toatsu Chemicals Inc of Tokyo and US chemical company ARCO Chemical Co have established a joint venture for the production and marketing of engineering plastics.

The joint venture, MTC ARCO, is capitalized at ¥200 million and based at ARCO's Japanese subsidiary, ARCO Chemical Japan Inc in Tokyo. In addition, beginning in May 1991, the joint venture invested about ¥800 million in the construction of a 5000 tonnes a year production facility for a styrene-based plastic, "Dylark", at Mitsui Toatsu's Osaka factory.

Dylark is used in the production of automobile parts, and was previously imported from the USA by ARCO Chemical Japan.

Adhesive resin production to be increased

Showa Denko KK of Tokyo plans to vastly expand its production of an adhesive resin which, it says, is becoming increasingly popular as a wrapping material for food and other goods. The resin, "Adote", shrinks at about 100°C, a lower temperature than conventional adhesive resins.

Showa Denko has been marketing about 300 tonnes a year of adhesive resin products on an experimental basis for about three years. Anticipating greater demand for such products, the company has been engaged in the development of new adhesive resins.

Adote will be available in a few dozen types, each for a different application. The company has started large-scale production of the resin and expects to eventually expand its sales to about 5000 tonnes a year.

Polyphenylene sulphide resin to be produced in Japan

The Japanese subsidiary of the US firm Phillips Petroleum, Phillips Petroleum International Ltd, plans to build a plant and a technical centre in Tokyo, for the production of polyphenylene sulphide (PPS) resins.

Phillips Petroleum, which was the original developer of PPS resin, already produces PPS-based polymers in Japan through Phillips Petroleum Toray Inc, a joint venture with Toray Industries Inc of Tokyo. Phillips Petroleum Toray Inc sells base polymer to Dainippon Ink & Chemicals which uses it to produce a range of PPS compounds.

Nippon Steel Chemical to increase resin compound production in USA

Nippon Steel Chemical Co Ltd plans to increase its production capacity of styrene resin, "Estyrene SE", at its US subsidiary, Thermofil Co.

The company, which acquired Thermofil in March 1989, plans to invest about ¥700 million in constructing a 1500 tonnes a year production line at one of Thermofil's three factories.

Estyrene SE has good dimensional stability and finishing characteristics, and is resistant to fire, light and heat. It is widely used to make housings for computer monitors.

Daicel Chemical to enter polystyrene business

Daicel Chemical Industries Ltd has signed an agreement with the US company Chevron Research and Technology Co under which Daicel will be allowed access to Chevron's polystyrene production technology.

Chevron's process technology, which Daicel will use in a polystyrene factory it plans to build, reportedly allows for concurrent continuous copolymerization of general purpose polystyrene, impact resistant polystyrene and acrylonitrile-styrene resin.

Daicel Chemical plans to invest about ¥13 billion in the construction of the 50 000 tonnes a year polystyrene production factory at its Amiboshi works in Hyogo Prefecture. Construction of the factory began in October 1991 and operations will begin in October 1992.

Ten–year project to develop biodegradable plastics from wood components

The Agriculture, Forestry and Fisheries Research Council of the Ministry of Agriculture, Forestry and Fisheries (MAFF), Tokyo, is planning a ten-year project to develop biodegradable plastics from modified cellulose and lignin extracted from wood. The project began in the fiscal year 1991.

Much of the research will be based on the results of the Ministry's ten-year "Biomass Conversion Project". One of the developments arising from that project is "lignocellophane", a plastic-like sheet developed by researchers at MAFF's Forestry and Forest Products Research Institute. The sheet was developed as a wrapping to prevent photo-oxidation of food.

The new biodegradable plastic project will take the results of these studies further as well as develop a new styrene-foam-like material from starch.

Sumitomo Chemical and Dow Chemical to produce polycarbonate resin

By 1993, Sumitomo Chemical Co Ltd and US company Dow Chemical Co plan to begin joint production of polycarbonate (PC) resins.

In 1988, Dow Chemical invested in a Sumitomo subsidiary, Sumitomo Naugatuck Co Ltd, and since then it has been researching the Japanese chemical market. Dow Chemical and Sumitomo had originally planned to begin production of PC resin in Japan in 1991. The companies have rescheduled their plans because of an expected glut of PC resin due to the large number of companies who are now building new PC production facilities.

Increased polystyrene production

Mitsubishi Kasei Polytec Co Ltd has constructed a 50 000 tonnes a year polystyrene (PS) production facility at its Yokkaichi factory in Mie Prefecture. The ¥5 billion facility was completed in late 1991. Mitsubishi Kasei Polytec already operates a PS facility with a production capacity of 160 000 tonnes a year.

Previously called Mitsubishi Monsanto Chemical, Mitsubishi Kasei Polytec was a joint venture of Mitsubishi Kasei Corp of Tokyo and Monsanto Co. Monsanto withdrew from the venture and the name was changed to Mitsubishi Kasei Polytec.

Market research for polyamide resin

BASF Engineering Plastic Co Ltd, a joint venture of BASF AG and Tokyo-based Mitsubishi Petrochemical Co Ltd, plans to research the viability of marketing BASF's new polyamide, "Ultramide T", in Japan.

Ultramide T is produced by copolymerization of caprolactam, hexamethylene diamine and terephthalic acid. It has high heat resistance (melting point: 295°C), high rigidity, displays chemical resistance and is easy to process, the company says. In addition, Ultramide T has high dimensional stability and a low rate of water absorption.

BASF Engineering Plastic plans to sell Ultramide T to manufacturers of electronics parts, at a price of ¥1500–1600 a kilogram.

Toray imports engineering plastics from USA

Toray Industries Ltd of Tokyo has begun importing engineering plastics from US company Eastman Kodak, which it will market in Japan.

The company plans to import 15 kinds of polyester-based plastics, including polycyclohexan dimethylene terephthalate (PCT) which is a high temperature resistant plastic only produced by Eastman Kodak.

Toray and Eastman Kodak eventually plan to establish a Japanese joint venture for the production of these plastics, probably when the Japanese market reaches 5000–10 000 tonnes a year.

Blow moulding technology development agreement

Blow moulding technology capable of producing large acrylonitrile butadiene styrene (ABS) parts will be jointly developed as a result of their agreement, according to Ube Industries Ltd and US company General Electric (GE) Co.

Blow moulding machines are half the cost of injection moulding machines, but are currently unable to handle large shapes because the resin is imperfectly transferred by the pressurized air. In addition, ABS resin has poor expansivity and breaks easily during the blow moulding process.

Under the plan, Ube Industries will develop a new blow moulding machine for large shapes and GE will develop high expansivity ABS resin. The two companies eventually aim to produce ABS mouldings for use in airplanes, automobiles and construction.

Drawn nylon film industry association established

The Drawn Nylon Film Industry Association of Japan has been established by five manufacturers.

The manufacturers are Unitika Ltd, Idemitsu Petrochemical Co Ltd, Kohjin Co Ltd, Toyobo Co Ltd and Mitsubishi Kasei Polytec Co Ltd. The association will conduct public relations activities and will work to promote demand for drawn nylon film.

Drawn nylon film was developed by Unitika in 1968, and demand is currently increasing at about 10% annually.

Increased production of flame-retardant film

Mitsubishi Plastics Industries Ltd plans to increase annual production of its flame-retardant polyester film by 150 000 m^2 to a total of 500 000 m^2.

The company is also planning to develop new product applications and is aiming to make the polyester film one of its main products within five years.

The film is electrically-insulative, heat resistant and has good processing properties. Demand for the film is increasing for use as electrical insulating material, electromagnetic shield material, flexible print substrate and as a sheet-heating element for use in motor-car mirrors.

Mass production of heat stretching polyester fibre

Teijin Ltd plans to mass produce a heat stretching polyester long fibre of its own development.

Conventional polyester long fibre shrinks when heated during treatments such as dyeing. The new fibre, however, shrinks when heated to temperatures below 110°C but stretches at temperatures over 110°C.

When the new fibre is mixed with conventional polyester long fibre, for use in fabrics and knits, the result is a soft-textured material. The company plans to sell the new fabrics for use in apparel and car-seat covers.

Nippon Petrochemicals constructs low-density polyethylene plant

Nippon Petrochemicals Co Ltd has invested about ¥10 billion in the construction of a 50 000 tonnes a year low-density polyethylene (LDPE) production plant at its Kawasaki Works in Kanagawa Prefecture. Operations started in October 1991.

Many Japanese and Korean companies, including Asahi Chemical Industry Co Ltd and Sumitomo Chemical Co Ltd, are either constructing or planning to construct their own LDPE production plants and it is predicted that there will be an oversupply of LDPE in the near future.

Silicone-modified imide resin marketed

Sumitomo Bakelite Co Ltd has acquired the right to produce and market a silicone-modified polyimide resin from US company Occidental Chemical Co.

The silicone-modified polyimide resin, which was developed by Occidental Chemical, has a low stiffness, low permittivity and high stability when compared with conventional polyimide resins. Occidental Chemical reportedly has a low capacity for production and limited resources for product development.

Sumitomo Bakelite has begun producing the resin at its Yokohama laboratory.

Kanegafuchi Chemical and ICI to cooperate in marketing vinylchloride additives

The UK-based company ICI and Kanegafuchi Chemical Industry Co Ltd of Osaka have signed a two-year agreement to cooperate in the marketing of their vinylchloride additives.

Both companies' products are reportedly bought by many of the same customers, so the move will save them money by using the same distribution channels.

Kanegafuchi Chemical currently produces 36 000 tonnes a year of the vinylchloride additive "Kaneka Ace B", through Kaneka Belgium NV. ICI produces the acryl-based vinylchloride additive "Diacon APA".

Daicel Chemical to transfer "Free Plastic" business to Daicel Kraft

Sakai-based Daicel Chemical Industries Ltd has announced plans to transfer its "Free Plastic" business activities to a subsidiary, Daicel Kraft Co Ltd.

Free Plastic is a craft material which becomes soft and workable when immersed in hot water at 60°C. Daicel Chemical has been marketing the plastic since 1987 and it says sales are now more than ¥100 million a year.

Resin plant construction suspended

Plans for the construction of a polycarbonate (PC) resin plant, which was to have been built in The Netherlands, have been suspended by Idemitsu Petrochemical Co Ltd of Tokyo and the Dutch chemical company DSM.

In 1988, the two companies established an equally-owned joint venture and were planning to build a 40 000 tonnes a year plant, to be constructed by late 1990. According to both companies, this plan was suspended owing to rising construction costs.

Idemitsu has developed its own PC production technology and is currently producing the resin in Brazil and Japan. The company also has plans to establish a facility in Taiwan.

Acrylonitrile butadiene styrene plant

A joint UK-based venture for the production and marketing of acrylonitrile butadiene styrene (ABS) resin has been established by Daicel Chemical Industries Ltd of Sakai and Hoechst AG.

The venture, located in the UK, will be a 3000 tonnes a year ABS production plant. In addition, Daicel's subsidiary in the USA, Dainepco, is constructing an ABS plant which will use Daicel's production technology.

CHAPTER NINE
TEXTILES

Fibres

Polyallylene fibre with higher elasticity

A polyallylene fibre which exhibits a higher elasticity than aramid types is under development at Kuraray Co Ltd of Osaka.

The company is currently marketing a polyallylene fibre, "Vectran", which exhibits an elasticity of 600 grams per denier. However, several aramid fibres show a higher elasticity than Vectran so the company is developing a polyallylene which will have greater elasticity and a wider range of applications.

Amorphous polyester fibre

A soft amorphous polyester fibre is being mass produced by Teijin Ltd at its Matsuyama works in Ehime Prefecture. The company began selling the fibre in late 1991, producing about 20 tonnes a year.

Fibre prevents adherence of algae and barnacles

Anti-fouling technology which prevents algae and barnacles from adhering to immersed surfaces such as ships' hulls has been developed at Toray Industries Inc in Tokyo.

The technology involves attaching a fluffy polyester cloth to the immersed surface. The cloth features densely packed 6 mm ultra-fine fibres which sway freely in the current.

In experiments, a concrete block covered in the cloth and another uncovered block were immersed in the sea off Shimane Prefecture. After 16 months, the uncovered block was completely overgrown with algae and barnacles, but the covered block was growth-free.

The production of the cloth involves two steps. Polyester is first blended with another polymer and spun into a thick yarn which is then woven into a fluffy cloth. The cloth is chemically treated to dissolve out the polymer material, leaving only ultra-fine polyester fibres in dense clumps.

Fibrous adsorbent for cobalt

A fibrous inorganic material that selectively adsorbs cobalt ions has been developed at the National Institute for Research in Inorganic Materials (NIRIM) in Ibaraki.

The fibrous material consists of tetratitanic acid. When packed in a column, the material adsorbs cobalt ions from a weak acid solution. The effect is temperature-dependant, with maximum adsorption occurring at about 100°C. At this temperature, the material adsorbs 99% of cobalt ions in a pH 4 solution.

The Institute expects the material to be used to adsorb radioactive cobalt ions from the cooling water of nuclear reactors.

Polyester microfibre filter water cleaning system

A water cleaning system that uses a polyester microfibre filter, "Torayrom RD", is being tested by Toray Industries Inc of Tokyo in the Biwako lake in Shiga Prefecture.

The microfibre, which is 1/20 the thickness of human hair, is woven into a dense fabric to form the filter. The standard system can filter 6000 tonnes of water a day, but larger systems can also be constructed, Toray says.

The system has been installed at Otsu City. In addition to the filtration unit, it is connected to a dehydration system, "Torayrom TM", which solidifies the filtered waste.

Results from the tests so far have shown that the distance of visibility in the lake water improved from 25 to about 52 cm after cleaning operations for one day. Toray plans to continue the tests, treating 250 tonnes of water a day.

Textiles

Deodorant material

A high performance, long life deodorant material based on porous iron, coated in a ferro-organic complex is the result of their joint studies, according to Nippon Steel Corp of Tokyo and Matsushita Electric Works Ltd of Osaka.

To form the deodorant material, a urethane foam is impregnated with an aqueous slurry of ferrous metal powder with a particle size of less than 10 μm. The foam is then fired to form a three-dimensional, iron-based porous material. The porous material is surface treated with an organic acid (ascorbic acid) to form a metal complex, and then chemically treated with alkali.

Odorous substances are adsorbed to the metal complex on the surface of the microstructured porous material. The metal-complex adsorption sites on the surface are supplied with electrons by the metallic iron and are thus able to retain their deodorizing function over a long time.

The material can physically and chemically deodorize almost all types of nitrogen- and sulphur-containing malodorous substances. Compared with activated charcoal, the material adsorbs ammonia 11.7 times faster and hydrogen sulphide 6.2 times faster and has 22 times the deodorization capacity, the companies claim.

Processing and equipment

Hard finishing of fabrics

The Research Institute for Polymers and Textiles in Tsukuba has developed a method for hard finishing fabrics.

Conventional methods usually require treating the fabric with formalin which is toxic and leaves residues in the fabric, according to the Institute. In the new process, fabric is placed in a solution of polyvinylalcohol (PVA), subjected to pressing and then irradiated.

The hardness can reportedly be controlled by altering the concentration of PVA. In experiments, the Institute processed cotton using a 1.5% PVA solution which resulted in levels of hardness 2–3 times that of the original cotton. When the fabric was washed 20 times, the hardness was only reduced by 20%.

Three-dimensionally crimped polyester long fibre

Suminoe Textile Co Ltd of Osaka says it has developed a continuous process for producing three-dimensionally crimped polyester long fibre on an automated production line. The company has also established an integrated production system for carpets using the long fibre.

"Sumitoron" is composed of fibres which are individually processed to form random three-dimensional curls. Compared with linear fibres, Sumitoron has improved bulk and elasticity, with reduced fluff, and displays chemical resistance and resistance to moisture, heat and light, the company says.

Suminoe Textile has also developed a dedicated dyeing system for the fibre. It has invested ¥2 billion in the construction of a 2000 tonnes a year integrated mass-production facility at its Shiga works. The company aims to achieve sales of ¥1 billion in the first year of marketing, and ¥3 billion in the second year.

Business and financial information

Fire-proof acrylic fibre facility

Asahi Chemical Industry Co Ltd of Tokyo plans to invest about ¥5 billion in the construction of a fire-proof acrylic fibre production facility at its Fuji factory in Shizuoka Prefecture.

The company's fire-proof acrylic fibre contains self-extinguishing polymers that contain halogen and metal groups. The fibre can be processed in the same fashion as conventional acrylic, is easy to dye, and can be made bacteria-proof and odour-proof through the use of additives.

By 1995, the company plans to have increased its production capacity from the current 700 tonnes a year to more than 1000 tonnes a year.

Toray Industries produces textiles in the UK

The UK subsidiary of Tokyo-based Toray Industries Inc, Toray Textiles Europe Ltd, plans to start producing fine textiles made from polyester long fibre.

The Japanese textile industry is exercising voluntary restrictions on its exports to Europe, so Toray has decided to begin local production in order to expand its sales. In October 1991, the UK company began producing fine textiles at a factory which Toray purchased from Courtaulds plc.

Toray plans to invest ¥15 billion in the construction of a factory which will carry out all textile production processes from weaving to dyeing. The plant will be partly operating by 1993 and full-scale production will begin in 1995.

Du Pont–Toray produces aramid fibre in Japan

Production of the aramid fibre Kevlar has begun at the Tokai plant, in Aichi Prefecture, of Du Pont-Toray Co Ltd — a joint venture between Toray Industries Inc of Tokyo and E.I. du Pont de Nemours and Co, based in Wilmington, Delaware, USA.

The Tokai plant will produce filament, stable, and chopped fibres at a total rate of 2500 tonnes a year, increasing to 5000 tonnes a year after three years. The output will be marketed in Asia and Oceania through Du Pont–Toray Kevlar Ltd, a joint venture sales company.

Du Pont-Toray is currently importing the aramid fibres as well as pulp-type fibre from the US firm's Richmond plant in Virginia. The company will continue to import the pulp-type fibre from the USA, but if demand increases it may begin producing in Japan.

Domestic production of aramid fibre by 1993

Production in Japan of high strength para-type aramid fibre is planned by 1993 by Nippon Aramid Co Ltd.

Nippon Aramid, which was established in 1987, has been marketing aramid fibre imported from Akzo. According to the company, demand for the material, which is useful as an asbestos replacement, is rapidly increasing and Akzo now plans to raise its production capacity from 5000 to 7000 tonnes a year.

Nippon Aramid will initially expand imports to meet the demand but expects to have constructed its factory by 1993 when it will begin replacing imports. Details, such as the production capacity and the location, are yet to be decided.

Nippon Aramid is a joint venture of Sumitomo Chemical Co Ltd of Osaka and the Dutch-based Enka BV, part of the Akzo NV Group.

Toray to expand polyester long fibre production

Toray Industries Inc plans to expand its production of polyester long fibre at its Indonesian affiliate, PT Indonesia Toray Synthetics (ITS).

ITS plans to invest ¥8 billion in constructing a new facility with a polyester long fibre production capacity of 840 tonnes a month. The new facility is scheduled to begin operations in March 1992.

ITS currently produces 1650 tonnes a month of polyester short fibre, 850 tonnes a month of nylon long fibre. The company has also started manufacturing 330 tonnes a month of polyester long fibre.

Mitsubishi Rayon to import non-woven fabrics from USA

A non-woven fabric used in the construction industry, "Colback", is to be imported by Mitsubishi Rayon Co Ltd of Tokyo from Akzo Co Ltd of the USA.

In 1995, Mitsubishi Rayon and Akzo plan to establish a joint venture for the production of Colback in Japan, and for its marketing in Asia and Oceania.

Colback, developed by Akzo, is made from composite fibres consisting of a polyester core surrounded by a sheath of nylon 6. The fabric has good two-dimensional expansion properties, and is resistant to heat and peeling.

Non-woven glass fabric factory to be built

The construction of a non-woven glass fabric factory in Nakatsu, Gifu Prefecture, will cost Honshu Paper Co Ltd about ¥2 billion, the company reports.

The facility will have a production capacity of 3600 tonnes a year and is scheduled to begin operating in mid-1992. Once operating, the company's existing factory in Nakatsu will be decommissioned.

The non-woven glass fabric to be produced has dimensional stability, water and weather resistance, and can be used as a substitute for asbestos, according to Honshu. The company currently produces 1200–1300 tonnes a year of the material, but increasing demand for asbestos substitutes has prompted the expansion.

Polyester long fibre spinning facility

In May 1991, Teijin Ltd of Tokyo began operating a 650 tonnes a month polyester long fibre spinning facility at its Matsuyama Minami factory in Ehime Prefecture.

The high-speed spinning facility produces more than 5000 metres of spun fibre a minute, five times faster than conventional spinning facilities, the company says.

In addition, Teijin plans to increase the polyester long fibre production capacity of its other factory. These changes will give the company a total polyester long fibre production capacity of 346.91 tonnes a day.

CHAPTER TEN
MEDICAL MATERIALS

Dental materials

Fluoride-releasing polymer for use as dental material

A fluoride-releasing polymer, "Teethmate F", can help to prevent the formation of caries, according to its inventor Kuraray Co Ltd of Osaka. The product is being marketed by Tokyo-based J. Morita Corp.

Kuraray produces Teethmate F by polymerizing a fluoride-containing monomer with a methacrylic acid ester. When the polymer comes into contact with water, it releases the fluoride which then reacts with the tooth enamel. This forms a stable layer of fluorapatite on the surface of the tooth, which protects it from the bacterial action that leads to decay.

The product is used by filling the small grooves on the surface of the molar teeth with a special applicator. The fluoride ions in the polymer are released slowly over three years, and remain at a concentration well within accepted safety levels, the company claims.

Teethmate F will be sold as a set priced at ¥13 100. Kuraray is predicting annual sales of ¥130 million in the first year.

Biocompatible prosthetic dental root

A prosthetic dental root manufactured from titanium and a titanium-based alloy has improved surface characteristics, according to its Tokyo-based developer.

Tokyo Bi-Tech Laboratories Inc manufactures its prosthetic dental root in nitrogen gas at 800°C. This causes a nitrogen compound film to form over the surface.

The film does not readily exfoliate and elution of ions is low. For this reason, the titanium alloy used in the tooth root reportedly has a higher level of biocompatibility and safety compared with conventional titanium alloys.

Stainless steel wire for orthodontic use

A stainless steel wire for use in the field of orthodontics is being manufactured by Suzuki Metal Industry Co Ltd of Tokyo.

According to the company, until now, all of the orthodontal stainless steel used in Japan has been imported, largely from the USA. Consequently, in 1988, Suzuki Metal Industry began development of an appropriate wire that has high strength, yet does not break when bent.

The wire has a tensile strength of 230 $kg.m^{-2}$, which is equivalent to that of the imported wires. Suzuki Metal Industry's product costs ¥30–200 for a 355 mm length of the wire.

The market for orthodontal stainless steel in Japan currently stands at around ¥1 billion. The company is predicting a 10% share of the domestic market in the first year of sales, and 30% in two to three years.

Prosthetic dental root marketed

A prosthetic dental root made from pure titanium, "Tiroot", is to be marketed by Toho Titanium Co Ltd of Tokyo through Yoshida Dental Trade Distribution Co Ltd, also based in Tokyo.

Toho Titanium says it began developing Tiroot in 1985 and was granted manufacturing approval in 1989. The company has been publicizing and marketing the prosthesis on a trial basis since then.

Tiroot will be sold as a set with peripheral apparatus, costing around ¥20 000 a root.

Toho Titanium describes itself as a titanium smelter which now wishes to expand into the medical field. The company hopes that annual sales of its medical materials will reach ¥500 million in 1992.

Photopolymerizing glass ionomer for filling teeth

3M Health Care Ltd has begun marketing "Vitrabond", a photo- polymerizing glass ionomer, as a dental filler for decayed teeth.

Glass ionomers can be used to adhere directly to the dentine of the tooth after removal of the decayed tissue. Moreover, they have very little irritative effect on the tooth. Vitrabond offers the further advantage of rapid hardening within 30 seconds when irradiated with light. The hardened material has a binding strength of 120 $kg.cm^{-2}$. Vitrabond is priced at ¥15 800.

Antibiotic hydroxyapatite for prosthetic bone and dental roots

Antibiotic hydroxyapatite powder for use in prosthetic bones and dental roots has been produced by Sangi Co.

Sangi produces the powder by combining hydroxyapatite with ions of silver, zinc and other heavy metals. The ions are adsorbed by replacing calcium in the surface of the hydroxyapatite. The compound has antibiotic effects against a wide range of bacteria and fungi, and is reportedly safe with a level of metal ion elution below 0.05 ppm. Moreover, it is thermostable up to 1000°C. It can also be used as an antibiotic additive in paper and synthetic resin.

Medical textiles

Toray Industries markets prosthetic vascular graft

A prosthetic vascular graft made from long general-purpose polyester fibres combined with polyester microfibres is being marketed by Tokyo-based Toray Industries Inc.

The vascular graft is designed for insertion within a vessel to cover damaged endothelial tissue for short periods to allow natural cellular regeneration. According to a Toray spokesperson, the polyester microfibres have a lower tendency to induce thrombosis than the regular polyester and fluororesin fibres used in conventional prosthetic vascular grafts.

Toray is already supplying the vascular grafts to the Baylor College of Medicine in Houston, Texas, USA, for use in treating patients with coronary occlusion.

The company estimates the market for prosthetic vascular grafts is several hundred million yen in Japan and several billion yen in the USA.

Chitosan produces antibiotic film for textiles

An antibiotic film made using chitosan has been developed by a research group at the Tochigi Prefectural Textile Industrial Laboratory. The group plans to use the film to develop antibiotic, deodorizing textiles.

Chitosan is formed by deacetylating chitin, which is a homopolysaccharide of N-acetyl-D-glucosamine that occurs naturally as a major component of the exoskeleton of crustaceans. The research group has been studying ways to use the antibiotic properties of chitosan in textiles.

The 0.1 mm film was produced by stretching chitosan in an aqueous acetic acid solution. The film was then chemically treated to prevent dissolution in water.

According to the results, more than 90% of the bacteria were destroyed. Other tests conducted so far show that the film is also effective against the trichophyte that causes dermatophytosis (athlete's foot).

The group is now trying to find ways to include the film in textiles without loss in its antibiotic properties.

Membrane eliminates viruses from blood

A hollow membrane made from regenerated fibre is capable of eliminating viruses from the blood, according to its developer Asahi Chemical Industry Co Ltd of Tokyo.

The membrane consists of cuprammonium rayon, produced by regenerating short cotton-seed linter fibres in a solution of cuprammonium hydroxide. The surface of the membrane is perforated by pores with a diameter of 30–100 nm.

The company claims the membrane can be manufactured with pores tailored to filter the hepatitis B virus (HBV), which has a 40 nm diameter, and the acquired immune deficiency syndrome (AIDS) virus (HIV), which has a 100 nm diameter. As cuprammonium rayon does not adsorb plasma proteins, the membrane eliminates the virus without altering the composition.

Asahi Chemical plans to start commercial production of the membrane in 1993 and has set an annual production target of 100 000 m^2.

Textile containing natural collagen marketed

A textile containing the natural protein collagen is being marketed by Sakai Textile Manufacturing Co Ltd of Fukui.

According to the company, "Miraface C" is biocompatible, non-allergic and does not irritate the skin. Moreover, it does not generate static electricity and exhibits moisture permeability, making it suitable for use in sportswear.

Collagen, a fibrous scleroprotein, is the major protein of connective tissue and the most abundant protein in higher animals.

Absorbable synthetic mesh for surgical use

A synthetic absorbable polyglycolic acid mesh for use in surgery is now being marketed by Lederle (Japan) Ltd.

"Dexon Mesh", manufactured by American Cyanamid Co Ltd, consists of woven polyglycolic acid fibres that degrade naturally within the body. The mesh is designed to cover damaged sites during and after surgery, thereby providing protection and support. Polyglycolic acid, which is used extensively in surgical sutures, is an inactive, non-antigenic, non-pyretic substance that does not cause side effects.

Dexon Mesh is priced at ¥74 400 for a box containing three packages.

Other materials

Synapse-like materials

Materials that function similarly to the natural synapses of optic nerves have been developed by researchers in Ibaraki. The scientists from the National Chemical Laboratory for Industry say the materials have photoelectric switching capacity, giving them potential for use as bio-elements.

The materials are Langmuir-Blodgett films formed from monomolecular layers of three organic compounds arranged on a quartz substrate. The compounds are tetracyano-p-quinodimethane (TCNQ), azobenzene, and a compound with a methylene chain that has either 6 or 14 alkyl groups.

The films are less than a micron thick. They exhibit rearrangements in their three-dimensional structure, leading to differences in electrical resistance when irradiated with ultraviolet (UV) or visible light.

According to the results of the group's experiments, the electrical resistance of one of the films increased by 40% when irradiated with UV light, then returned to the original level when irradiated with visible light, approximating the behaviour of an inhibitory synapse.

Electrical resistance of the second film increased in 6% increments when irradiated with UV light in the absence of visible light, similar to the function of a plastic synapse, which undergoes a gradual change in transmission properties in line with the specific type of stimulation.

The research group plans to develop the materials for use as bio-elements in neuro-computers.

Cultured cells recovered as organized tissue

A method to recover cultured animal cells as organized tissue has been developed by researchers in Tokyo.

Professor Yasuhisa Sakurai of the Institute of Medical Engineering at Tokyo Women's Medical College worked in collaboration with Kao Corp on a method for culturing cells on the company's culture plate.

The plate was coated with the polymer polyisopropylacrylamide which is hydrophobic above 32°C, and hydrophilic below. At 37°C, the cells adhere to the hydrophobic form of the polymer, forming tissue that slips off the plate when the temperature is lowered to less than 32°C.

In experiments using rat liver and vascular endothelial cells, Sakurai claims to have obtained a higher rate of cell proliferation on the polymer-coated culture plate than by conventional methods. Moreover, 100% of the cultured cells were recovered and more than 80% of their functional capacity retained.

Sakurai says the method may prove useful for providing cultured human liver, epithelial and endothelial tissue for drug toxicity and irritation testing.

Carrier for cell and microbe cultures

A carrier for cell and micro-organism cultures has been devised by Tokyo-based Toray Industries Inc.

The carrier is an electret consisting of a dielectric nonwoven polypropylene fabric in which a semi-permanent state of electric polarization has been set up.

Toray says pilot studies show that rat liver cells cultured on the carrier at 39°C can proliferate from 5000 cells/ml to 500 000 cells/ml in seven days. These figures suggest it can support the growth of around 50 times more cells than conventional carriers fabricated from plastic or glass.

A Toray research group is currently investigating optimal culture conditions for use with the carrier.

Prosthetic ventricular septum implanted in infant with single ventricle

The successful implant of a prosthetic ventricular septum in a one-month-old male infant born with a single ventricle has been achieved by Professor Yasuaki Naito of the Department of Thoracic Surgery at Wakayama Medical College's School of Medicine.

Naito's surgical team implanted a synthetic fibre membrane with a length of 10 cm and a width of 8 cm to fully separate the ventricle into two sections in a ten hour operation.

The single ventricle was discovered when the infant developed life-threatening feeding difficulties and cyanosis due to blood oxygen deficiency soon after birth. The infant has been discharged from hospital in a healthy condition after convalescing for 70 days.

Naito says he has implanted the prosthetic septum successfully in the ventricle of four children over the age of four years but, until now, has never attempted the procedure in a newborn infant.

Anti-bacterial materials

A glass which is anti-bacterial in nature has been developed and commercialized, according to claims by the Ishizuka Glass Co Ltd, based in Aichi Prefecture.

The glass contains metal ions such as silver or copper and these leach out over time killing any bacteria with which they come into contact. In addition, the glass prevents the build up of moulds and algae.

It can be used as a glass or combined with resins or papers to form composite materials.

Electrical stimuli contract and relax artificial muscle

A research group at the Agency of Industrial Science and Technology's (AIST) Mechanical Engineering Laboratory has used electrical stimuli to simulate contraction and relaxation of artificial muscle.

The group developed an artificial muscle consisting of a polyvinyl alcohol (PVA) hydrogel. The gel contracts when immersed in an acidic solution and elongates in an alkaline solution.

The gel was formed into a 15 mm long tube with an inner diameter of 3 mm. The group filled the gel tube with a hydrochloric acid solution, then placed it in a large plastic container containing a sodium hydroxide solution. A platinum electrode was inserted in each solution and a direct current of 40 mA was applied across the electrodes.

According to the results, a positive current flow through the hydrochloric acid within the gel tube caused it to elongate within 30 seconds, while a negative current flow caused it to contract.

The scientists say that electrical stimulation of artificial muscle allows better control of contraction and elongation than other methods used so far.

Processing and equipment

Technique for preparing two-dimensional crystals of bioactive polymers

A technique for preparing two-dimensional crystals could help in the analysis of the three-dimensional structure of bioactive polymers, according to the method's joint developers, the Protein Engineering Research Institute, the Research Development Corp of Japan (JRDC) and Iwate University.

The research group dissolves dehydro-abiethyleneamine (DHAA) in hexane and adds it to a solution of ribosome polymer. The DHAA floats on the surface of the polymer solution in a film 10–100 nm thick and promotes the crystallization of the ribosome into a two-dimensional crystal.

The two-dimensional crystal is then observed from various angles with an electron microscope, allowing elucidation of the three-dimensional structure by Fourier transform analysis.

Artificial respirator marketed

Baxter Ltd, the Japanese subsidiary of the US medical equipment manufacturer Baxter International Inc, has begun marketing a small volume, single-use artificial respirator.

The product is a membrane respirator that replaces the function of the lungs during cardiac surgery. By combining the sections for gas and heat exchange, Baxter reduced the circulatory volume of the device to 220 ml, which is about one-third the size of the company's conventional respirators. This reduces the amount of blood that must be circulated outside the body, without loss in gas exchange capacity.

The single-use artificial respirator will cost around ¥250 000. Baxter is hoping that sales of the product will raise the company's share of Japan's artificial respirator market over the next three years to 15% from the current level of 10%.

Sensitive high performance liquid chromatography

Shimadzu Corp of Kyoto has begun marketing a high performance liquid chromatography system, "LC-10A". The company says the system is three times more sensitive than its conventional counterparts.

LC-10A consists of a plunger pump which propels small quantities of the test sample through the system, an automatic reagent dispenser, an adsorption column, an ultraviolet to visible light (UV-VIS) detector for absorptiometry, a fraction collector, a controller, and a recorder.

The standard version of LC-10A costs ¥6.1 million. Shimadzu expects it to be used for biotechnological research and quality control in the pharmaceutical, food, medical, and chemical industries, and hopes for minimum annual sales of ¥18.3 billion.

Haemodialyser containing synthetic polymer dialysis membrane marketed

Haemodialysers containing a synthetic polymer hollow fibre membrane are being marketed by Tokyo-based Nikkiso Co.

The dialysis membrane consists of a polyester alloy formed by blending polyallylate and polyether sulphone. According to a Nikkiso spokesperson, the dialyser is biocompatible, capable of high-performance dialysis, and is not deformed by high-pressure steam sterilization.

In contrast to conventional units, "FL Series" dialysers reportedly eliminate the low molecular proteins that lead to amyloid deposition and related side effects.

The dialyser will cost ¥6800 and Nikkiso is predicting annual sales of some ¥800 million.

Business and financial information

Japan's biotechnology market trebled in 1990

Japan's market for biotechnology-based products trebled from ¥100 billion in 1989 to ¥295 billion in 1990, according to a survey by the Tokyo-based Ministry of Agriculture, Forestry and Fisheries (MAFF).

Drugs, detergents and equipment accounted for around 93% or ¥275 billion of the 1990 total. Non-processed foods accounted for 5% (¥10.4 billion) and tissue-cultured seedlings for 2% (¥6 billion).

The market for industrially processed food products grew, with oligosaccharides accounting for ¥7 billion, stevia sweeteners for ¥1.5 billion, and biotechnologically produced wines for ¥5 billion.

MAFF predicts that the entire bio-industry's research and development spending will be exceeded by product sales for the first time, although agricultural biotechnology still lags behind.

Rigid contact lens material supplied to Switzerland

Toray Industries Inc of Tokyo is supplying its material for rigid gas-permeable contact lenses, "Toray SR-310", to Ciba Vision AG of Switzerland.

Toray SR-310 is a fluorosilicon acrylate resin that has the highest oxygen permeability coefficient (Dk 150) of any contact lens material available worldwide, according to Toray. The company uses the material in its rigid contact lens, "Breath-O Hard CL", which was released in Japan in October 1990. The company expects the lens to achieve annual sales of ¥13 billion in 1993 in the domestic market alone.

Ciba Vision will market contact lenses fabricated from Toray SR-310 throughout Europe — sales in Germany have already begun.

Sumitomo Bakelite to import vascular grafts from USA

Sumitomo Bakelite Co Ltd of Tokyo has acquired sole import and marketing rights to vascular grafts manufactured by the US firm Corvita Corp.

The vascular grafts are manufactured from a material consisting of a special polymer combined with collagen and gelatin. The grafts reportedly have excellent biocompatibility and an elasticity equivalent to that of human blood vessels.

Sumitomo Bakelite expects to begin marketing the grafts in 1993 and is predicting that annual sales of the grafts and other Corvita products will amount to ¥1 billion in 1995 and ¥6 billion by 2000.

Under the agreement, Sumitomo Bakelite will also acquire the technical expertise necessary to allow a shift to licensed domestic manufacture of the grafts in the near future. The agreement was negotiated as part of the Sumitomo Bakelite's plan to achieve annual sales of cardiovascular products totalling ¥20 billion by 2000.

CHAPTER ELEVEN:
DIRECTORY

This chapter lists details of the commercial organizations which appear in the Source Book.

Each entry lists:

- The English form of the company name (entries are listed alphabetically according to this category);
- The Japanese form of the company name;
- The Tokyo Stock Exchange's (TSE) reference number for the company;
- A brief description of the company and its major activities;
- Contact details including full postal address, and telephone and facsimile numbers;
- Names of senior personnel;
- Details of the size of the company (capital and number of employees);
- Sales figures for recent years;
- Figures for research and development spending and capital investment;
- Sales breakdown by SIC code (SIC codes listed are those used in the USA, not Europe or UK). The breakdown figures are rounded up to nearest whole percent;
- A description of the company's major products;
- Research and development activity;
- Locations of factories and laboratory sites;
- Locations of subsidiaries (domestic and overseas);
- Supporting banks; and
- Shareholders.

At the end of the chapter, there is a product index which is organized according to the major SIC code subdivisions.

Advantest Corp.

Japanese Name	:	***Adobantesuto K.K.***
TSE Number	:	6857
Description	:	Manufacturer of electronic measuring instruments specializing in semiconductor testing devices. Advantest Corp. is an affiliate of Fujitsu Ltd..
Address	:	2-4-1, Nishi-Shinjuku, Shinjuku-ku, Tokyo 163
Tel & Fax	:	Tel: 03-3342-7500 Fax: 03-3342-7509
Statistics	:	Capital : 30092 million yen Employees : 1594
Officers	:	CEO : Mr. Hiroshi Ohura/President
Sales	:	March, 1988 - 54143 million yen March, 1989 - 70665 million yen March, 1990 - 74055 million yen March, 1991 - 81458 million yen
Capital Investment	:	7788 million yen
Sales Breakdown	:	see table below

SIC#	Product	Sales
3674	Semiconductor testing devices	80 %
382	Electronic measuring devices	20 %

Main Products	:	Electronic Equipment: Semiconductor testing devices, Electronic measuring instruments
Banks	:	Mitsui Taiyo Kobe Bank; Dai-Ichi Kangyo Bank; Long Term Credit Bank of Japan; Sanwa Bank; Bank of Tokyo; Kyowa Saitama Bank
Branches	:	Tokyo Sales Office (Nerima-ku, Tokyo)
Factories	:	Gunma Factory (Gunma-ken); Gunma Second Factory (Gunma-ken); Higashi Matsuyama Factory (Saitama-ken).

Aisin Seiki Co., Ltd.

Japanese Name	:	***Aisin Seiki K.K.***
TSE Number	:	7259
Description	:	Automobile parts manufacturer with bulk of production going to Toyota Motors.
Address	:	2-1, Asahi-cho, Kariya-shi, Aichi-ken 448
Tel & Fax	:	Tel: 0566-24-8329 Fax: 0566-24-8848
Statistics	:	Employees : 9727
Officers	:	CEO : Mr. Shigeo Aiki/President
Sales	:	March, 1988 - 370040 million yen March, 1989 - 406126 million yen March, 1990 - 458500 million yen March, 1991 - 525037 million yen
Sales Breakdown	:	SIC# Product 3465 Automobile parts
Main Products	:	Automobile Parts: Engine parts, Transmissions, Control and driving devices
Banks	:	Tokai Bank; Sanwa Bank; Mitsui Taiyo Kobe Bank; Chuo Trust and Banking
Branches	:	Overseas Offices: Los Angeles Office (California, U.S.A.); Brussels Office (Belgium); Singapore Office (Singapore)
Shareholders	:	Toyota Motor Corp [7203] - 22.0% ; Toyoda Automatic Loom Works [6201] - 4.9% ; Tokai Bank [8321] - 4.7% ; Mitsui Taiyo Kobe Bank [8314] - 4.2% ; Mitsui Trust and Banking [8401] - 4.0% ; Nippon Life Insurance - 3.4%

Ajinomoto Co., Inc.

Japanese Name		***Ajinomoto K.K.***
TSE Number	:	2802
Description	:	The company is Japan's largest integrated processor of foods, and a leader in advanced amino acid application technology. Exhibiting growth potential in pharmaceuticals and chemicals.
Address	:	1-5-8, Kyobashi, Chuo-ku, Tokyo 104
Tel & Fax	:	Tel: 03-3272-1111 Fax: 03-3297-8698
Statistics	:	Capital : 75400 million yen Employees : 5640
Officers	:	CEO : Ms. Kaoru Toba/President
Sales	:	March, 1988 - 432524 million yen March, 1989 - 460049 million yen March, 1990 - 477178 million yen March, 1991 - 504000 million yen
R&D Spending	:	19300 million yen Capital Investment: 24500 million yen
Sales Breakdown	:	SIC# Product 2099 Seasonings 2099 Processed foods 2834 Pharmaceuticals 28 Specialty chemicals
Main Products	:	Processed Foods; Food Additives: Ajinomoto (monosodium glutamate flavor enhancer), Emulsifiers, Seasonings; Biotech Products: Amino acids, Pharmaceuticals; Specialty Chemicals: N-Acrylamino acid and its salts, Adenine hydrochloride, Adenosine, Halogen-based flame retardant, Phosphoric acid ester plasticizers
Banks	:	Mitsubishi Trust and Banking; Mitsubishi Bank; Dai-Ichi Kangyo Bank; Bank of Tokyo
Branches	:	Tokyo Branch Office; Osaka Branch Office; Fukuoka Branch Office
Factories	:	Kawasaki Plant (Kawasaki-shi, Kanagawa-ken); Yokohama Plant (Yokohama-shi, Kanagawa-ken); Tokai Plant (Yokkaichi-shi); Kyushu Plant (Saga-ken); Kashi Plant (Kashi-shi)
Laboratories	:	Central Research Laboratories (Kawasaki-shi; Yokohama-shi); Food Products Development Laboratories (Kawasaki-shi); Technology & Engineering Center (Kawasaki-shi)
Subsidiaries	:	Knorr Foods Co., Ltd.; Ajinomoto General Foods, Inc.; Ajinomoto Frozen Foods Co., Ltd. (Kanto)
Overseas Subsidiaries	:	Ajinomoto U.S.A., Inc. (New Jersey, U.S.A.) Deutsche Ajinomoto GmbH (Hamburg, Germany)

Alithium Ltd.

TSE Number	:	Not listed
Description	:	Engaged in the development and production of aluminum-lithium alloys. Alithium Ltd. is a research venture initiated by the Japan Key Technology Center and seven major Japanese aluminum producers.
Address	:	Toranomon Takagi Bldg., 1-7-2, Nishi-Shinbashi, Minato-ku, Tokyo 105
Tel & Fax	:	Tel: 03-3595-1482 Fax: 03-3592-1285
Officers	:	CEO : Mr. Kenichiro Kitora/President Contact : Mr. Kenichiro Kitora/President
Sales Breakdown	:	SIC# Product 3334 Aluminum-lithium alloys
Main Products	:	Aluminum Alloys: 'AL-LI' aluminum-lithium alloy
Factories	:	Yokkaichi Plant (Mie-ken)
Laboratories	:	Yokkaichi Lab (Mie-ken)
R&D Areas	:	Aluminum-lithium alloys

Anelva Corp.

Japanese Name	:	***Nichiden Aneruba***
TSE Number	:	Not listed
Description	:	Manufacturer of vacuum equipment and components primarily for use in the production of semiconductors. Anelva Corp. is a wholly-owned subsidiary of NEC Corp..
Address	:	5-8-1, Yotsuya, Fuchu City, Tokyo 183
Tel & Fax	:	Tel: 0423-34-0211 Fax: 0423-60-2277
Statistics	:	Capital : 600 million yen Employees : 1200
Officers	:	CEO : Mr. Susumu Yasuda/President Contact : Mr. Masayuki Shimomura
Sales	:	March, 1988 - 25117 million yen March, 1989 - 30775 million yen March, 1990 - 37267 million yen March, 1991 - 44603 million yen
Sales Breakdown	:	

SIC#	Product	Sales
3559	Semiconductor manufacturing equipment	50 %
3559	Other equipment and components	50 %

Main Products	:	Industrial Machinery: Vacuum equipment and components for use in semiconductor manufacture
Banks	:	Sumitomo Bank; Sumitomo Trust and Banking; Bank of Tokyo; Kyowa Saitama Bank; Dai-Ichi Kangyo Bank; Bank of Yokohama; Sanwa Bank
Branches	:	Osaka Branch (Yodogawa-ku, Osaka-shi); Kyushu Branch (Fukuoka-shi); California Branch (California, U.S.A.)
Factories	:	Honsha Factory (Fuchu-shi, Tokyo); Fuji Factory (Yamanashi-ken)
Shareholders	:	NEC Corp. [6701] - 100%
Subsidiaries	:	Anelva Engineering Corp. (Fuchu-shi, Tokyo); Anelva Shizuoka Corp. (Fujinomiya-shi)

Araya Industrial Co., Ltd.

Japanese Name	:	***Araya Kogyo K.K.***
TSE Number	:	7305
Description	:	Major manufacturer of bicycle parts and steel pipes. Controlling over half of the Japanese market for bicycle rims, Araya Industrial is also active in robotics and precision pipes for office automation equipment.
Address	:	Araya Bldg., 2-12-12, Minami-Semba, Chuo-ku, Osaka 542
Tel & Fax	:	Tel: 06-253-0221 Fax: 06-253-0237
Statistics	:	Employees : 737
Officers	:	CEO : Mr. Nobuyuki Araya/President Contact : Mr. Kazuhiro Ikeda/Administrative Manager
Sales	:	March, 1989 - 25191 million yen March, 1990 - 27722 million yen March, 1991 - 31893 million yen

Sales Breakdown :

SIC#	Product	Sales
3312	Steel pipes	54 %
3751	Bicycle wheel rims	32 %
3751	Bicycles	10 %

Main Products	:	Steel Pipes: Precision pipes; Bicycles and parts: 'Tsubame' brand bicycles, Steel bicycle rims, Light alloy bicycle rims
Banks	:	Sanwa Bank; Fuji Bank; Hokkoku Bank; Daiwa Bank; Tokai Bank; Mitsui Taiyo Kobe Bank
Shareholders	:	Osaka Securities Finance [8512] - 7.8%; Sanwa Bank [8320] - 4.9%; Fuji Bank [8317] - 4.9%; Hokkoku Bank [8363] - 4.9%; Daiwa Bank [8319] - 4.6%; Nisshin Steel [5407] - 4.4%; Daido Life - 4.4%; Yasuda Trust and Banking [8404] - 3.5%
Overseas Subsidiaries	:	P.T. Pabrik Araya Indonesia (Indonesia)

Asahi Chemical Industry Co., Ltd.

Japanese Name : ***Asahi Kasei Kogyo K.K.***

TSE Number : 3407

Description : Asahi Chemical is a major chemical producer, with subsidiaries and affiliates worldwide. The company's strong R&D sections are continually developing new products such as coated fabrics, synthetic rubbers, factory automation systems, pharmaceuticals and prefabricated housing units. New biotechnology products include 'CDP Choline' Neuronal Agent, '7-ACA' for antibiotics, 'Sunfural' anticancer agent, 'Immusorba' immunoabsorbent column, and 'Sepacell' leukocyte removal filter in the area of medicines. The company plans to merge with Toyo Jozo and bolster production of pharmaceuticals and food industries.

Address : Hibiyamitsui Bldg., 1-1-2, Yaraku-cho, Chiyoda-ku, Tokyo 100

Tel & Fax : Tel: 03-3507-2730 Fax: 03-3507-2005

Statistics : Capital : 99115 million yen
Employees : 15331

Officers : CEO : Mr. Reiichi Yumikura/President

Sales : March, 1988 - 765483 million yen
March, 1989 - 817800 million yen
March, 1990 - 879300 million yen
March, 1991 - 965200 million yen

R&D Spending : 44000 million yen Capital Investment: 77600 million yen

Sales Breakdown :

SIC#	Product
2819	Acids
2821	Acrylic rubber
2823	Rayon fibers
2824	Polyester and vinyl fibers
286	Organic chemicals
2836	Bacterins
2024	Frozen yogurt
3841	Artificial kidney
2452	Prefabricated buildings

Main Products : Industrial Inorganic Chemicals: Hydrochloric acid, Ammonia, Soda; Specialty Chemicals: Acrylic esters, Acrylonitrile, Adenine hydrochloride (vitamin B4), Adenine sulphate, Adenosine and its salts, Amino acid, Ammonia cooling agent, Ethylene-vinyl acetate copolymer, Mosodium-L glutamate, Styrene monomer, Surface coating agent; Plastics Materials: Acrylic resins, Ion exchange resins, Styrene-chloroprene; Cellulose Man-made Fibers: Rayon fibers and products; Pharmaceutical Preparations: Veterinary preparations, Animal proteins; Biological Products: Bacterins, Blood derivatives; Ice Cream and Food Products: Frozen yogurt; Surgical Supplies: Plasma separator, Artificial kidney; Prefabricated Buildings: 'Hebel House' prefabricated buildings and construction materials

Banks : Sumitomo Bank; Dai-Ichi Kangyo Bank; Fuji Bank; Industrial Bank of Japan; Long Term Credit Bank of Japan

Branches : Tokyo Head Office (Chiyoda-ku); Osaka Head Office (Kita-ku); Nobeoka Office (Miyazaki-ken); Moriyama Office (Shiga-ken); Fuji Office (Shizuoka-ken)

Asahi Glass Co., Ltd.

Japanese Name	:	**Asahi Garasu K.K.**
TSE Number	:	5201
Description	:	Largest glass manufacturer in Japan, the company has diversified into many materials-related fields.
Address	:	2-1-2, Marunouchi, Chiyoda-ku, Tokyo 100
Tel & Fax	:	Tel: 03-3218-5555 Fax: 03-3214-5060
Statistics	:	Capital : 87245 million yen Employees : 10010
Officers	:	CEO : Mr. Jiro Furumoto/President Contact : Mr. Kazuo Kamiya/General Manager, International
Sales	:	December, 1988 - 834421 million yen December, 1989 - 925931 million yen December, 1990 - 1018090 million yen
R&D Spending	:	40000 million yen Capital Investment: 87800 million yen

Sales Breakdown :

SIC#	Product	Sales
28	Chemicals	36 %
3211	Glass products	31 %
32	Building materials	21 %
367	Electronic components and materials	5 %
3297	Ceramics and refractories	3 %
0000	Others	4 %

Main Products : Glass Products: Thin flat glass, Thick flat glass, Wired glass, Heat absorbent wired glass, Heat reflective glass, Compound glass, Reinforced glass, Mirror glass, Laminated glass, Glass bulbs for TVs and various displays; Building Materials: Lead glass tubing, Powdered glass, Glass tubing, Aluminum sashes, PVC sashes, ALC products, GRC-based products, Polycarbo-laminates, Sound-proofing products; Chemical Products: Soda ash, Caustic soda, Hydrochloric acid, Phosphoric acid, Liquid chlorine, Sodium hypochlorite, Propylene oxide, Propylene glycol, Polyol, Urethane-based water-proofing materials, Urethane-based flooring materials, Epichlor hydrine, Methyl chloride, Methylene chloride, Chloroform, Bromine, CFCs, Trichlor ethylene, Perchlor ethylene, PVC monomer, PVC polymer, 1.1.1 trichlor ethane, Chemical feeds, Potassium chloride, Caustic potash, Sodium chloride, Calcium chloride, Industrial-use ammonium chloride, Halon fire extinguishing agents, Magnesia clinkers, Zeolite, Magnesium carbonate, Lithium bromide, Fluorine-based surfactants, Fluororesins, Fluororubbers, Trichloro acetate, Fluororesin-based paints, Propylene hexafluoride, Ion exchange membranes, Testing reagents, Flame retardants, Pharmaceutical intermediates, Agrochemical intermediates, Foaming agents, Methyl chloride cooling agent; Refractories and Ceramics: Zirconite, Sillimanite bricks, Mullite bricks, Magnesia carbon bricks for steelmaking, Direct bond bricks for cement glass and steel production, Cement, Abrasion-resistant materials, Infra-red ceramic heater, Zirconium borate, Engineering ceramics products, Refractories; Electronics: Optical disks, Optical fiber applied products, Semiconductor ICs, Semi-custom ICs, Hybrid ICs, LCDs, Crystal applied products, PCBs; Other Products: Medical equipment, FRP precision molded products, Opthalmic products

Banks : Mitsubishi Bank; Dai-Ichi Kangyo Bank; Mitsubishi Trust and Banking; Fuji Bank

Laboratories : Research Center (Yokohama-shi); Advanced Glass R&D Center (Yokohama-shi)

Asahi Organic Chemicals Industry Co., Ltd.

Japanese Name	:	***Asahi Yukizai Kogyo K.K.***
TSE Number	:	4216
Description	:	Asahi Organic Chemicals manufactures and markets molded phenol resin products, engineering plastic resins, and plastic plumbing materials, and is diversifying into plumbing design and construction work.
Address	:	2-5955, Nakanose-cho, Nobeoka-shi, Miyazaki-ken 882
Tel & Fax	:	Tel: 0982-33-3311 Fax: 0982-21-8606
Statistics	:	Capital : 5000 million yen Employees : 942
Officers	:	CEO : Mr. Tadayoshi Kamimori/President Contact : Mr. Akio Taura/Manager, Export Division
Sales	:	March, 1988 - 40209 million yen March, 1989 - 40101 million yen March, 1990 - 42436 million yen March, 1991 - 45753 million yen
R & D Spending	:	400 million yen Capital Investment: 5000 million yen
Sales Breakdown	:	

SIC#	Product	Sales
2821	Plastic piping materials	68 %
2821	Industrial-use resins	24 %
3089	Molded materials	5 %

Main Products	:	Plastics Materials and Resins: Plastic piping materials, Industrial use resins, Molded phenol resin materials
Banks	:	Fuji Bank; Miyazaki Bank; Mitsui Taiyo Kobe Bank; Tokai Bank; Kagoshima Bank; Oita Bank
Factories	:	Nobeoka Factory (Nobeoka-shi); Tone Factory (Koga-shi); Aichi Factory (Niwa-gun); Shimonoseki Factory (Shimonoseki-shi); Tokiwa Factory (Iwaki-shi)
Laboratories	:	Nobeoka Factory (Nobeoka-shi)

Baxter Ltd.

TSE Number	:	Not listed
Address	:	4, Rokubancho, Chiyoda-ku, Tokyo 102
Tel & Fax	:	Tel: 03-3237-6611 Fax: 03-3237-6617
Statistics	:	Capital : 3930 million yen
		Employees : 1000
Officers	:	CEO : Mr. Jack L. McGinley/President
Sales	:	March, 1989 - 39144 million yen
Sales Breakdown	:	SIC# Product 2836 Blood component products 3841 Surgical supplies 3845 Dialysis hardware
Branches	:	Tokyo Branch (Minato-ku); Osaka Branch (Kita-ku); Sapporo; Sendai; Nagoya; Kanazawa; Hiroshima; Fukuoka; Kagoshima
Factories	:	Gifu Plant (Gifu-ken); Saitama Plant (Saitama-ken); Miyazaki Plant (Miyazaki-ken)
Laboratories	:	Tokyo

C. Itoh & Co., Ltd.

Japanese Name : ***Ito Chu Shoji K.K.***

TSE Number : 8001

Description : C.Itoh & Co. is one of Japan's three largest general trading firms. The company is diversifying into new fields, including energy production with uranium plant construction and development of aerospace equipment and supplies with the launching of Japan's Communications Satellite (JCSAT-I) in 1989.

Address : 2-5-1, Kita-Aoyama, Minato-ku, Tokyo 107-77

Tel & Fax : Tel: 03-3497-2121 Fax: 03-5474-7294

Statistics : Capital : 174711 million yen
Employees : 7273

Officers : CEO : Mr. Minoru Murofushi/President

Sales : March, 1988 - 14922000 million yen
March, 1989 - 15555900 million yen
March, 1990 - 20532700 million yen
March, 1991 - 20596000 million yen

Sales Breakdown :

SIC#	Product	Sales
505	Metal compounds	29 %
5084	General machinery and construction	26 %
516	Chemicals and energy	19 %
514	Foodstuffs and food products	11 %
5131	Textiles	11 %
5031	Timber products and commodities	5 %

Main Products : Chemicals and Allied Products: Metal compounds, Uranium mining; Industrial Machinery: Textile machines, Paper and pulp machines, Oil-refining machines, Power plant machines, Recording instruments, welding machinery; Construction and Mining Machines: Cranes, Dredges, Logging equipment, Shovels, Quarrying machinery; Plastics Materials: Synthetic resins; Petroleum Products; Coal, coke; Groceries and Related Products: Foodstuffs, beverages; Piece Goods and Notions: Textiles; Lumber, Plywood and Millwork: Lumber products

Banks : Dai-Ichi Kangyo Bank; Bank of Tokyo; Sumitomo Bank; Sumitomo Trust and Banking; Industrial Bank of Japan; Long Term Credit Bank of Japan

Branches : Osaka Head Office (Osaka-shi); Nagoya Branch (Nagoya-shi); Kyushu Branch (Fukuoka-shi); Hokkaido Branch (Sapporo-shi); Yokohama Branch (Yokohama-shi) and 18 other domestic and 90 overseas offices

Subsidiaries : C. Itoh Iron & Steel Co., Ltd.(Osaka-shi); C. Itoh Foods Co., Ltd.(Osaka-shi); C. Itoh Ceramics Corp. (Osaka-shi) and 101 other domestic subsidiaries

C.I. Kasei Co., Ltd.

Japanese Name	:	***Shiiai Kasei K.K.***
TSE Number	:	Not listed
Description	:	Manufacturer of PVC products.
Address	:	1-18-1, Kyobashi, Chuo-ku, Tokyo 104
Tel & Fax	:	Tel: 03-3535-4541 Fax: 03-3535-4545
Statistics	:	Capital : 2500 million yen Employees : 1022
Officers	:	CEO : Mr. Hiroshi Ishiya/President
Sales	:	September, 1988 - 37738 million yen September, 1989 - 40038 million yen March, 1990 - 21828 million yen
Sales Breakdown	:	SIC# Product 3081 PVC film 3084 PVC piping 3089 PVC sheeting
Main Products	:	PVC Products: PVC sheet, PVC piping, PVC film
Banks	:	Dai-Ichi Kangyo Bank; Mitsui Taiyo Kobe Bank; Saitama Bank; Mitsubishi Bank; Norinchukin Bank; Sumitomo Trust and Banking; Yasuda Trust and Banking
Branches	:	Osaka Branch (Osaka-shi); Nagoya Branch (Nagoya-shi); Fukuoka Branch (Fukuoka-shi); Tokyo Branch (Chuo-ku); Sendai Branch (Sendai-shi); Sapporo Branch (Sapporo-shi); Shizuoka Branch (Shizuoka-shi)
Factories	:	Shiga Factory (Koga-gun); Akashi Factory (Akashi-shi); Tochigi Factory (Haga-gun); Ageo Factory (Ageo-shi)
Laboratories	:	Engineering Lab. (Ageo-shi)
Shareholders	:	C. Itoh & Co., Ltd. [8001]
Subsidiaries	:	Hamapura Seisakusho K.K.; Yokobi K.K.; Bonpack K.K.

Canon Inc.

Japanese Name	:	***Kyanon K.K.***
TSE Number	:	7751
Description	:	Major manufacturer of copiers, cameras and computer peripherals. Known around the world for its cameras and copiers, Canon has expanded its product lines to become a comprehensive producer of OA and computer equipment. While enjoying expanding overseas sales across all product categories, the company has moved aggressively to increase its foreign production capacity with new plants in Asia and expanded capacity in the U.S. and Europe.
Address	:	Shinjuku Dai-ichi Seimei Bldg., 2-7-1, Nishi-Shinjuku, Shinjuku-ku, Tokyo 163
Tel & Fax	:	Tel: 03-3348-2121 Fax: 03-3349-8765
Statistics	:	Capital : 81408 million yen Employees : 18589
Officers	:	CEO : Dr. Keizo Yamaji/President
Sales	:	December, 1988 - 672277 million yen December, 1989 - 814226 million yen December, 1990 - 931296 million yen

Sales Breakdown :

SIC#	Product	Sales
3579	Copiers	32 %
3577	Computer peripherals	28 %
3861	Cameras and video equipment	19 %
366	Communications equipment	16 %
3827	Optical products and others	6 %

Main Products	:	Copiers and Copier Supplies: Copiers, Copier applications, Drums, Toners, BJ heads, BJ inks, Ink cartridges; Computer Peripherals: LBP, Information display systems, Printers; Image Systems: Facsimile transceivers, Telephone systems, Electronic filing systems, Micrographics equipment, Scanners, Magneto-optical disks; Computer and Information Systems: Personal publishing systems, Desktop publishing systems, Computers, Multi-media phone systems, Japanese word processors, Calculators, Electronic dictionaries, Electronic tools; Cameras and Video Equipment: Cameras, Still video system, Video equipment, Lenses; Optical Equipment: Semiconductor production equipment, Broadcasting equipment; Medical Equipment: X-ray mirror cameras, Non-mydriatic retinal cameras, IOL estimators, Auto ref-keratometers; Others: Audio visual aids, Electronic typewriters, NeXT computer, PBX equipment, Components, Production engineering
Banks	:	Fuji Bank; Mitsui Taiyo Kobe Bank; Bank of Tokyo; Dai-Ichi Kangyo Bank; Industrial Bank of Japan; Sumitomo Bank
Branches	:	Branch and Sales Offices Throughout Japan; Overseas Offices in: Beijing, Dalian, Shanghai, Guangzhou, Taipei, Seoul, Hong Kong
Factories	:	Toride Plant (Toride-shi); Ami Plant (Ami-shi, Ibaraki-ken); Fukushima Plant (Fukushima-shi); Utsunomiya Plant (Utsunomiya-shi); Tamagawa Plant (Kawasaki-shi); Kosugi Plant (Kawasaki-shi); Ueno Plant (Mie-ken); Nagahama Plant (Shiga-ken)
Laboratories	:	Canon Research Center (Atsugi-shi); Component Development Center (Hiratsuka-shi, Kanagawa-ken)
R&D Areas	:	Business and Communications Equipment: Various imaging and information technologies, Computer hardware and software, etc.; Optics and Optoelectronics; Energy: Solar cells, etc.;

Chubu Electric Power Co., Inc.

Japanese Name	:	***Chubu Denryoku K.K.***
TSE Number	:	9502
Description	:	The company supplies electric power to five prefectures: Aichi, Gifu, Mie, Nagano, and Shizuoka (west of the Fuji River).
Address	:	1, Toshin-cho, Higashi-ku, Nagoya-shi, Aichi-ken 461-91
Tel & Fax	:	Tel: 052-951-8211 Fax: 052-971-0729
Statistics	:	Capital : 369583 million yen Employees : 20191
Officers	:	CEO : Mr. Kohei Abe/President
Sales	:	March, 1988 - 1633700 million yen March, 1989 - 1660100 million yen March, 1990 - 1706700 million yen March, 1991 - 1829300 million yen
Sales Breakdown	:	SIC# Product Sales 4911 Electric power 100 %
Main Products	:	Electric power
Banks	:	Japan Development Bank; Industrial Bank of Japan; Long Term Credit Bank of Japan; Tokai Bank
Factories	:	Hamaoka Nuclear Power Station (Shizuoka-ken); Shin-Nagoya Thermal Power Station (Nagoya-shi, Aichi-ken); Yokkaichi Thermal Power Station (Mie-ken) and 186 other power stations
Laboratories	:	Electric Power Research & Development Center (Nagoya-shi); Electrotechnology Applications Research & Development Center (Nagoya-shi)
Shareholders	:	Meiji Mutual Life - 5.35% ; Nippon Life - 4.68% ; Daiichi Mutual Life - 3.68% ; Mitsui Taiyo Kobe Bank [8314] - 3.04% ; Tokai Bank [8321] - 2.93% ; Industrial Bank of Japan [8302] - 2.91%
Subsidiaries	:	Comres Corp. (Nagoya-shi, Aichi-ken); Chubu Telecommunications Co., Inc. (Nagoya-shi, Aichi-ken); Chubu Telemessage Co., Ltd. (Nagoya-shi, Aichi-ken); Toenec Corp. (Nagoya-shi, Aichi-ken)

Chugai Pharmaceutical Co., Ltd.

Japanese Name	:	***Chugai Seiyaku K.K.***
TSE Number	:	4519
Description	:	Major pharmaceutical manufacturer specializing in prescription drugs. The company was granted manufacturing approval for erythropoietin (EPO) in December, 1989. More recently it has applied for approval for granule-colony stimulating factor (G-CSF).
Address	:	2-1-9, Kyobashi, Chuo-ku, Tokyo 104
Tel & Fax	:	Tel: 03-3281-6611 Fax: 03-3281-2828
Statistics	:	Capital : 20762 million yen Employees : 3600
Officers	:	CEO : Mr. Hajime Sano/President
Sales	:	December, 1988 - 124492 million yen December, 1989 - 128595 million yen December, 1990 - 125504 million yen
R & D Spending	:	21500 million yen Capital Investment: 9000 million yen

Sales Breakdown :

SIC#	Product	Sales
2834	Metabolic agents	22 %
2834	Respiratory agents	18 %
2836	Hormones	15 %
2834	Digestive agents	10 %
283	Chemotherapy drugs	7 %
2834	Nervous system agents	5 %
2879	Insecticides	4 %
2834	Antibiotics	3 %
2835	Clinical reagents and equipment	2 %

Main Products	:	Pharmaceuticals: Antitumor and chemotherapy drugs, Metabolic drugs, Vitamins, Respiratory drugs, Antibiotics, Digestive drugs, Nervous system drugs; Chemical Products: Insecticides, Clinical reagents; Biological Products: Hormone preparations; Medical Equipment: Clinical equipment; Specialty Chemicals: Alfacalcidol, Diazinon, Dimethoate
Banks	:	Sumitomo Bank; Mitsui Taiyo Kobe Bank; Kyowa Saitama Bank; Saitama Bank; Norinchukin Bank; Mitsui Trust and Banking
Branches	:	Tokyo Branches (No.1 & No.2); Sendai Branch (Miyagi-ken); Sapporo Branch (Hokkaido); Nagoya Branch (Aichi-ken); Osaka Branch; Hiroshima Branch; Fukuoka Branch; Takamatsu Branch (Shikoku)
Factories	:	Ukima Factory (Kita-ku, Tokyo); Matsunaga Factory (Fukuyama-shi, Hiroshima-ken); Kyoseki Factory (Fukushima-ken); Fujieda Factory (Shizuoka-ken); Utsunomiya Factory (Tochigi-ken)
Laboratories	:	Screening Laboratory (Gotenba-shi); Development Laboratory (Toshima-ku); Shinshu Ina Laboratory (Nagano-ken)
Subsidiaries	:	Eiko Chemicals K.K. (Fukushima-ken)
Overseas Subsidiaries	:	Chugai Holding USA Inc. (New York, U.S.A.); Gen Probe Inc. (California, U.S.A)

Chugoku Kogyo Co., Ltd.

Japanese Name	:	***Chugoku Kogyo K.K.***
TSE Number	:	Not listed
Address	:	6-1-4, Sagisu, Fukushima-ku, Osaka 553
Tel & Fax	:	Tel: 06-454-1221 Fax: 06-454-0082
Statistics	:	Capital : 20 million yen Employees : 12
Officers	:	CEO : Mr. Katsumi Matsumoto/President Contact : Mr. Yoshitsugu Sawaki/Manager, Overseas Trade
Sales	:	February, 1988 - 737 million yen February, 1989 - 1130 million yen February, 1990 - 2053 million yen February, 1991 - 2150 million yen

Sales Breakdown :

SIC#	Product	Sales
2819	Inorganic chemicals	35 %
3341	Scraps and recoveries	30 %
5051	Non-ferrous metals (wholesale)	20 %
0000	Electronics and misc. materials	

Banks	:	Fuji Bank; Sumitomo Bank; Kyowa Saitama Bank; Fuji Bank; Mitsubishi Bank
Shareholders	:	Katsumi Matsumoto ; Keiko Matsumoto ; Fumie Kanto

Chukoh Chemical Industries, Ltd.

Japanese Name	:	***Chukoh Kasei Kogyo K.K.***
TSE Number	:	Not listed
Description	:	Manufacturer of synthetic resin products with emphasis upon the development and production of fluorocarbon products and applications.
Address	:	Toranomon Gojokai Bldg., 2-9-14, Toranomon, Minato-ku, Tokyo 105
Tel & Fax	:	Tel: 03-5251-7414 Fax: 03-5251-7413
Statistics	:	Capital : 110 million yen Employees : 336
Officers	:	CEO : Mr. Nobushige Kiso/President Contact : Mr. Nobushige Kiso/President
Sales	:	March, 1988 - 5300 million yen March, 1989 - 6067 million yen March, 1990 - 6394 million yen March, 1991 - 7530 million yen
Sales Breakdown	:	SIC# Product 2821 Synthetic resins and resin products 3089 Fluororesin coated glass fiber fabric 2821 Engineering plastics 3674 Integrated circuits and IC substrates 2891 Synthetic resin sealing materials
Main Products	:	Synthetic Resin Products: Fluororesin coated glass fiber fabrics, Carbon fiber reinforced resins, Engineering plastics, Polyphenylene sulfide coatings, Soft fluorocarbon resin (fluoronated rubber and fluorocarbon resin) sealing materials; Integrated Circuits: Fluororesin based copper clad IC laminates
Banks	:	Industrial Bank of Japan; Mitsubishi Bank; Mitsui Taiyo Kobe Bank; Sumitomo Bank; Fukuoka Bank; Juhachi Bank
Branches	:	Tokyo Branch (Minato-ku, Tokyo); Osaka Branch (Osaka-shi); Nagoya Branch (Nagoya-shi, Aichi-ken); Fukuoka Branch (Fukuoka-shi, Fukuoka-ken); Yokohama Sales Office (Yokohama-shi, Kanagawa-ken)
Factories	:	Matsuura Plant (Nagasaki-ken); Yokohama Plant (Kanagawa-ken)
Laboratories	:	Matsuura Laboratory (Nagasaki-ken); Yokohama Laboratory (Kanagawa-ken)
Shareholders	:	Chukoh Industries Ltd.; Nobushige Kiso

Cosmo Oil Co., Ltd., Group

Japanese Name	:	***Cosmo Sekiyu K.K.***
TSE Number	:	5007
Description	:	Japan's third largest oil distributor-wholesaler.
Address	:	1-1-1, Shibaura, Minato-ku, Tokyo 105
Tel & Fax	:	Tel: 03-3798-3211 Fax: 03-3798-3237
Statistics	:	Capital : 28332 million yen Employees : 3455
Officers	:	CEO : Mr. Hiroto Sumiyoshi/President Contact : Mr. Haruo Matsumoto/Senior Managing Director
Sales	:	March, 1988 - 1297590 million yen March, 1989 - 1209770 million yen March, 1990 - 1465310 million yen March, 1991 - 1713370 million yen
R & D Spending	:	5300 million yen Capital Investment: 31000 million yen
Sales Breakdown	:	

SIC#	Product	Sales
517	Gasoline	45 %
5172	Kerosene	23 %
5172	Fuel oil	20 %
5172	Lubricant oils	11 %
286	Organic chemicals	

Main Products	:	Fuel Products: Gasoline, Kerosene, Gas oil, Lubricants; Specialty Chemicals: Petroleum-based benzene, iso-Butylene, Naphtha, Rust preventative oil, Toluene
Banks	:	Industrial Bank of Japan; Sanwa Bank; Bank of Tokyo; Tokai Bank; Long Term Credit Bank of Japan; Tokai Bank; Toyo Trust and Banking; Nippon Credit Bank; Japan Development Bank; Sumitomo Bank; Mitsubishi Trust and Banking
Factories	:	Chiba Refinery (Ichihara-shi, Chiba-ken); Yokkaichi Refinery (Yokkaichi-shi); Sakai Refinery (Sakai-shi, Osaka); Sakaide Refinery (Sakaide-shi, Kagawa-ken)
Laboratories	:	Cosmo General Research Laboratories Co., Ltd.

Dai Nihon Glass Industries Ltd.

Japanese Name	:	***Dai Nihon Garasu Kogyo K.K.***
TSE Number	:	Not listed
Address	:	1-2-27, Miyashimo, Sagamihara-shi, Kanagawa-ken 229
Tel & Fax	:	Tel: 0427-72-1171 Fax: 0427-73-0167
Statistics	:	Capital : 300 million yen Employees : 247
Officers	:	CEO : Mr. Shohei Makino/President Contact : Mr. Shohei Makino/President
Sales	:	December, 1988 - 6779 million yen December, 1989 - 8502 million yen
Sales Breakdown	:	SIC# Product 3089 Fiber reinforced plastic (FRP) products 2821 Specialized resin products
Banks	:	Mitsubishi Bank; Dai-Ichi Kangyo Bank; Yokohama Bank; Mitsubishi Trust & Banking
Branches	:	Sales Office (Chiyoda-ku, Tokyo); Osaka Branch (Osaka-shi); Nagoya Office (Nagoya-shi, Aichi-ken); Fukuoka Office (Fukuoka-shi, Fukuoka-ken)
Factories	:	Factory (Sagamihara-shi, Kanagawa-ken)
Laboratories	:	Laboratory (Sagamihara-shi, Kanagawa-ken)
Shareholders	:	Asahi Glass Co., Ltd. [5201]

Dai-Ichi Kogyo Seiyaku Co., Ltd.

Japanese Name	:	***Dai-Ichi Kogyo Seiyaku K.K.***
TSE Number	:	4461
Description	:	Manufacturer of various surface active agents, with sidelines in fine and synthetic chemicals.
Address	:	Shin-Kyoto Center Bldg., Karasuna-Nishiiru, Shiokojidori, Shimogyo-ku, Kyoto 600
Tel & Fax	:	Tel: 075-343-6824 Fax: 075-343-1421
Statistics	:	Capital : 5577 million yen Employees : 1012
Officers	:	CEO : Mr. Tomoo Sugiyama/President Contact : Mr. Katsuhisa Kitano/Director
Sales	:	March, 1988 - 33695 million yen March, 1989 - 35280 million yen March, 1990 - 36526 million yen March, 1991 - 40110 million yen
R & D Spending	:	1678 million yen Capital Investment: 1594 million yen

Sales Breakdown :

SIC#	Product	Sales
2843	Surface active agents	47 %
2869	Water soluble polymers	25 %
2869	Urethane chemicals	14 %
2869	Resin additives	12 %
0000	Others	2 %

Main Products	:	Specialty Chemical Products: Alkyl salts, Alkyl esters, Water soluble polymers, Phenol anti-oxidant, Antistatic agents for fiber/plastics and rubber, Dispersion agents for food/sludge/photography/wax/pesticides/pulp and paper, Emulsifiers for pesticides/foodstuffs/oil and tar/asphalt/photography/wax, Stearic acid, Flocculants, Gelling agents, Higher alcohol for the distillation of sperm oil, Organic anti-oxidant, Polyoxylethylene alkyl ether, Sea surface pollution dispersant, Slip additive, Surface coating agent; Other Products: Toiletries, Construction, Water treatment systems, Energy, Textile goods
Banks	:	Fuji Bank; Yasuda Trust and Banking Company; Daiwa Bank; Kyoto Bank; Norinchukin Bank; Industrial Bank of Japan
Branches	:	Tokyo Branch (Chuo-ku); Nagoya Branch (Nagoya-shi); Kyushu Branch (Fukuoka-shi); Hokuriku Branch (Fukui-shi)
Factories	:	Kyoto Factory (Kyoto); Yokkaiichi Factory (Yokkaiichi-shi, Mie-ken); Ohgata Factory (Nakakubiki-gun, Niigata-ken)
Laboratories	:	Application Research Institute; New Material Research Institutes - 1 and 2 ; Process Development Institute; PPG Research Institute; PAAM Research Institute; PAO Research Institute; Chemical Analysis Institute
R&D Areas	:	Chemical Products: Organic synthesis, Surface chemistry technologies
Overseas Subsidiaries	:	Chin Yee Chemical Industries, Ltd.; Cho Il Synthesis Co., Ltd.

Daicel Chemical Industries, Ltd.

Japanese Name : ***Daisel Kagaku Kogyo K.K.***

TSE Number : 4202

Description : Large-scale manufacturer of acetic acid and acetic acid cellulose. A member of the Mitsui group of chemical companies, Daicel is enjoying steady increases in sales across all of its product categories due in part to plant innovations in the early 1980s. R&D stresses cellulosics, synthetic resins and synthetic resin films. The company is currently involved in a joint R&D venture with Onkyo Co., Ltd. to develop a cellulose audio speaker.

Address : 1, Teppo-cho, Sakai-shi, Osaka 590

Tel & Fax : Tel: 0722-27-3111 Fax: 0722-27-3000

Statistics : Capital : 36244 million yen
Employees : 3213

Officers : CEO : Mr. Akiro Kojima/President

Sales : March, 1988 - 134718 million yen
March, 1989 - 144712 million yen
March, 1990 - 154272 million yen
March, 1991 - 168195 million yen

R & S Spending : 8800 million yen Capital Investment: 24200 million yen

Sales Breakdown :

SIC#	Product	Sales
2869	Organic chemicals	45 %
2823	Cellulosics	30 %
2821	Plastics and films	18 %
376	Propellants and propulsion systems	5 %
2869	Functionals	2 %

Main Products : Industrial Organic Chemicals: Acetic acid, Acetic acid anhydride, Sorbic acid, Potassium sorbate, Chloracetyl chloride, Slip additives, Fatty acids, Lactic acid, Formic acid, Non-inflammable polymer flame retardant, Acrylic emulsion adhesives, Acrolien, Acetyl chloride, Synthetic resin emulsions; Cellulosics: Cellulose acetate, Nitrocellulose, Nitrocellulose OH-Less lacquer, CMC, HEC, CMHEC; Plastics Materials: AS resins, ABS resins, Coated OPP films, Coated PET films, Cellophane, Tinted films, Automobile air bags; Propellants and Propulsion Systems: Jet fighter propulsion systems and propellants, Helicopter propellants and fuels, Gun-powder and related chemicals for use in munitions, Rocket propellants

Banks : Mitsui Taiyo Kobe Bank; Mitsui Trust and Banking; Sanwa Bank

Branches : Nagoya Branch (Nagoya-shi, Aichi-ken); Fukuoka Branch (Fukuoka-shi, Fukuoka-ken)

Factories : Sakai Plant (Osaka); Kanzaki Plant (Hyogo-ken); Aboshi Plant (Hyogo-ken); Aboshi No. 2 Plant (Hyogo-ken); Harima Plant (Hyogo-ken); Ohtake Plant (Hiroshima-ken)

Laboratories : Tsukuba Research Center (Tsukuba-ken); Research Center Laboratory (Hyogo-ken); Filter Products R&D Laboratory (Osaka); Plastics Development Center (Osaka)

R&D Areas : Synthetic Resins: ABS resins, Polycarbonate resins, Polymer alloys; Cellulosic Materials: Pyrocellulose materials, Other cellulosic fibers and materials, Cellulose applications in audio speakers

Overseas Subsidiaries : Daicel (U.S.A.), Inc. (U.S.A.); Daicel (Europa) GmbH (Germany); Daicel Chemical (Asia) Pte. Ltd. (Singapore); Daicel Businex Inc. (U.S.A.); Daicel International Finance B.V. (Netherlands); Daicel Kisco Co., Ltd. (Taiwan); Dynepco Inc. (U.S.A.)

Daido Steel Co., Ltd.

Japanese Name	:	***Daido Tokushuko K.K.***
TSE Number	:	5471
Address	:	1-11-18, Nishiki, Naka-ku, Nagoya-shi, Aichi-ken 460
Tel & Fax	:	Tel: 052-201-5111 Fax: 052-221-9268
Statistics	:	Capital : 35898 million yen Employees : 7628
Officers	:	CEO : Mr. Toshio Kishida/President
Sales	:	March, 1989 - 287998 million yen March, 1990 - 302553 million yen March, 1991 - 326002 million yen
R & D Spending	:	5875 million yen Capital Investment: 26300 million yen
Sales Breakdown	:	SIC# Product 3312 Special and cast steel 3356 Titanium alloys 35 Machinery
Banks	:	Industrial Bank of Japan; Tokai Bank; Mitsubishi Trust and Banking; Dai-Ichi Kangyo Bank
Overseas Subsidiaries	:	Rotary Drilling Tools Manufacturing Pte., Ltd. (Singapore); Daido Steel (America) Inc. (Illinois, U.S.A.); Ohio Star Forge Co. (Ohio, U.S.A.); CSC Industries (Ohio, U.S.A.); NHK Cimebara Industria de Molas Lda. (Brazil)

Daido Steel Sheet Corp.

Japanese Name : ***Daido Kohan K.K.***

TSE Number	:	5454
Address	:	3-2-1, Kuise-Minamishinmachi, Amagasaki, Hyogo 660
Tel & Fax	:	Tel: 06-487-1717 Fax: 06-487-1830
Statistics	:	Capital : 5018 million yen Employees : 630
Officers	:	CEO : Mr. Tatsuo Nagano/President
Sales	:	March, 1988 - 45452 million yen March, 1989 - 50609 million yen March, 1990 - 54465 million yen
Sales Breakdown	:	SIC# Product Sales 3479 Galvanized steel sheets 81 %
Banks	:	Industrial Bank of Japan; Sumitomo Bank; Sanwa Bank; Mitsui Taiyo Kobe Bank; Fuji Bank; Mitsubishi Bank; Nippon Credit Bank
Branches	:	Tokyo Branch (Tokyo); Sapporo Branch (Hokkaido); Tohoku Branch (Sendai-shi, Miyagi-ken); Hokuriku Branch (Toyama-ken); Nagoya Branch (Nagoya-shi, Aichi-ken); Chugoku Branch (Hiroshima-ken); Kyushu Branch (Fukuoka-ken)
Factories	:	Amagasaki Plant (Amagasaki-shi, Hyogo-ken); Konan Works
Subsidiaries	:	Daido Kenzai Kogyo Co., Ltd.; Daido Tekki Corp.; Daido Sho-un Co., Ltd.; Toa Co., Ltd.; Kansai Kasei Kogyo Co., Ltd.

Daihen Corp.

Japanese Name	:	***Daihen K.K.***
TSE Number	:	6622
Description	:	Manufacturer and distributor of welding equipment, industrial robots and electric distribution equipment. Though electric distribution equipment still accounts for a majority of sales, Daihen Corp. has become a noted producer in the field of automated welding and handling robots, and R&D activity stresses these areas.
Address	:	2-1-11, Tagawa, Yodogawa-ku, Osaka 532
Tel & Fax	:	Tel: 06-301-1212 Fax: 06-308-6342
Statistics	:	Capital : 10581 million yen Employees : 1830
Officers	:	CEO : Mr. Keijiro Kobayashi/President Contact : Mr. Masanori Shiramizu/General Manager, International
Sales	:	March, 1988 - 51097 million yen March, 1989 - 56095 million yen March, 1990 - 65034 million yen March, 1991 - 78508 million yen
R & D Spending	:	1139 million yen Capital Investment: 3516 million yen
Sales Breakdown	:	SIC# Product Sales 361 Transformers and related products 59 % 3548 Welding equipment & industrial robots 42 %
Main Products	:	Transformers and Related Products: Transformers, Electric switches; Controlling Devices: Substations, Radio frequency power generators and automatic maching units, Supersonic wave generators; Welding and Cutting Equipment: Arc welding robots, Automatic MIG (metal electrode inert gas) welders, Plasma arc cutting machines, CO2 laser processing equipment; Industrial Robots: 'Almega V40' 6 axis vertical handling robots, Wafer conveyor robots, Clean room conveyance robots, 'OSACOM 870' handling robot controller; New Ceramics: Fluorine bonded ceramics
Banks	:	Sumitomo Bank; Sumitomo Trust and Banking; Hokaido Takushoku Bank; Mitsui Taiyo Kobe Bank; Mitsubishi Trust and Banking
Factories	:	Joso Plant (Yodogawa-ku, Osaka); Mie Plant (Mie-ken); Kanehira Plant (Fukushima-ku, Osaka); Chitose Plant (Chitose-shi, Hokkaido); Settsu Plant (Settsu-shi, Osaka)
Laboratories	:	Joso Technology Center (Yodogawa-ku, Osaka); Settsu Technology Center(Settsu-shi, Osaka)
Shareholders	:	Sumitomo Trust and Banking [8403] - 5.6% ; Kansai Denryoku K.K. [9503] - 5.28% ; Sumitomo Bank [8318] - 4.4% ; Mitsubishi Trust and Banking [8402] - 2.9%
Overseas Subsidiaries	:	OTC America Inc. (North Carolina, U.S.A.); Daihen Europe GmbH (Germany); Daihen Thailand (Bangkok, Thailand); Ekarat-Daihen Transformer Co., Ltd. (Bangkok, Thailand)

Daikin Industries, Ltd.

Japanese Name : ***Daikin Kogyo K.K.***

TSE Number	:	6367
Description	:	Leading manufacturer of commercial air-conditioning units. The company has diversified into fluorochemicals, oil hydraulic and lubrication equipment. High growth areas are in air-conditioning units but the company is advancing into computer graphics systems, medical equipment, cryogenics and robotics.
Address	:	Umeda Center Bldg., 2-4-12, Nakazaki-Nishi, Kita-ku, Osaka 530
Tel & Fax	:	Tel: 06-373-4351 Fax: 06-373-4388
Statistics	:	Capital : 27757 million yen Employees : 8200
Officers	:	CEO : Mr. Minoru Yamada/President
Sales	:	March, 1988 (4 months) - 76865 million yen March, 1989 - 275900 million yen March, 1990 - 293900 million yen March, 1991 - 344070 million yen
R & D Spending	:	14100 million yen Capital Investment: 35200 million yen

Sales Breakdown :

SIC#	Product	Sales
3585	Air-conditioning and refrigeration eqpt.	73 %
2821	Fluorochemicals	14 %
3569	Hydraulic and lubrication equipment	8 %
3559	Defense equipment	5 %

Main Products	:	Air-conditioning and Refrigeration Equipment: Residential Air-conditioners, Industrial Air-conditioning systems; Chemical Products: CFCs, Other fluorochemicals, CFC cooling agent, Slip additive; Industrial Machinery: Oil hydraulic and lubrication equipment; Computer Integrated Systems: CAE systems, Graphic systems, 'MOL-GRAPH', 'DYNA-ARC'; Special Industry Machinery: Defense ammunitions and missile warheads, Helium-based cryogenics equipment, Robotics, MRI equipment
Banks	:	Sumitomo Bank; Saitama Bank; Norinchukin Bank; Sumitomo Trust and Banking; Nippon Credit Bank; Industrial Bank of Japan; Long Term Credit Bank of Japan
Branches	:	Tokyo Branch (Shinjuku-ku); Nagoya Branch (Nagoya-shi); Hiroshima Branch (Hiroshima-shi); Fukuoka Branch (Fukuoka-shi)
Factories	:	Sakai Plant (Sakai-shi); Yodogawa Plant (Settsu-shi, Osaka); Shiga Plant (Kusatsu-shi, Shiga-ken); Kashima Plant (Kashima-gun, Ibaraki-ken)
Laboratories	:	Mechanical Engineering Laboratories (Sakai-shi); Electronic Engineering Laboratories (Kusatsu-shi); R&D Div. of Chemical Dept. (Settsu-shi); CAE Center (Sakai-shi); MEC Laboratories (Tsukuba-shi)
Subsidiaries	:	Daikin Plant Co., Ltd.; Tokyo Daikin Airconditioning and Engineering Co., Ltd. (Tokyo); Osaka Daikin Airconditioning and Engineering Co., Ltd. (Osaka)

Dainichiseika Color & Chemical Mfg. Co., Ltd.

Japanese Name : ***Dainichiseika Kogyo K.K.***

TSE Number : 4116

Description : Major producer of coloring agents for synthetic resins and fibers.

Address : 1-7-6, Nihonbashi-Bakurocho, Chuo-ku, Tokyo 103

Tel & Fax : Tel: 03-3662-7111 Fax: 03-3669-3924

Statistics : Capital : 5843 million yen
Employees : 1727

Officers : CEO : Mr. Osamu Takahashi/President
Contact : Mr. Yuzo Iwazaki/Overseas Dept. Manager

Sales : March, 1988 - 115364 million yen
March, 1989 - 114254 million yen
March, 1990 - 119744 million yen
March, 1991 - 130960 million yen

Capital Investment: 5143 million yen

Sales Breakdown :

SIC#	Product	Sales
285	Coloring agents	65 %
28	Chemical products	18 %
0000	Electronic materials and others	17 %

Main Products : Coloring Agents: Colorants for synthetic resins and others, Specialty photogravure ink, Lithographic inks, Brilliant carmine 6B, Lake pigment, Organic pigments; Chemical Products: Printing ink intermediates and pigments, Rubber intermediates and pigments, Synthetic resin intermediates and pigments, Paint intermediates and pigments, Fiber intermediates and pigments, Other fine chemicals; Electronics Materials: Magnetic iron oxide; Other Materials: Synthetic resins, Printing films, Printing machinery, Other printing materials

Banks : Mitsui Taiyo Kobe Bank; Sanwa Bank; Fuji Bank; Tokai Bank; Bank of Tokyo; Dai-Ichi Kangyo Bank; Ashikaga Bank; Yasuda Trust and Banking Company; Mitsui Trust and Banking Company; Industrial Bank of Japan; Nippon Credit Bank

Branches : Tokyo Branch (Chuo-ku, Tokyo); Osaka Branch (Osaka-shi); Nagoya Branch (Nagoya-shi, Aichi-ken)

Factories : Tokyo Factory (Tokyo); Tokai Factory (Iwata-shi, Shizuoka-ken); Kawaguchi Factory (Saitama-ken); Osaka Factory (Higashi Osaka-shi, Osaka)

Laboratories : Central Research Laboratory

Subsidiaries : Hokkaido Dainichiseika Kogyo K.K. (Sapporo); Oyodo Dainichiseika Kogyo K.K. (Oyodo); Kyushu Dainichiseika Kogyo K.K. (Hakata-shi)

Dainippon Ink & Chemicals, Inc.

Japanese Name : ***Dainippon Ink Kagaku Kogyo K.K.***

TSE Number : 4631

Description : Major manufacturer of printing supplies and equipment. Noted for its aggressive management policies, the company is currently diversifying into such areas as synthetic resins, electronic materials and biotechnical products. R&D stresses products within these fields which hold promise for applications in the information and electronics industries.

Address : DIC Bldg., 3-7-20, Nihonbashi, Chuo-ku, Tokyo 103

Tel & Fax : Tel: 03-3272-4511 Fax: 03-3278-8558

Statistics : Capital : 80843 million yen
Employees : 6693

Officers : CEO : Mr. Shigekuni Kawamura/President

Sales : March, 1988 - 417697 million yen
March, 1989 - 440658 million yen
March, 1990 - 469769 million yen
March, 1991 - 505209 million yen

R & D Spending : 22067 million yen Capital Investment: 28467 million yen

Sales Breakdown :

SIC#	Product	Sales
2821	Polymers related products	39 %
2893	Printing supplies and equipment	31 %
2821	Specialty plastics and compounds	16 %
0000	Others	14 %

Main Products : Plastics and Related Products: Synthetic resins, Resin coating products, High polymers, Plastic alloys, Plastics for electronics applications, Petrochemicals, Wrapping films and other packaging materials, Engineering plastics, Plastic compounds, Plastic colorants, Plastic molded products, Pressure sensitive adhesives, Plastic magnetic and optical recording medias, Misc. plastics and synthetic resin products, Plastics and resin raw materials; Printing Supplies and Equipment: Printing inks, Jet inks, Metal coating inks and chemicals, Toner, OPC drums, Special magnetic tapes, UV paints, Films for use in labels and stickers, Double-sided tapes; Pigments and Colors: Organic pigments, Brilliant carmine 6B, Acid dyes; Building Materials; Biotechnological Products: Agricultural chemicals, Pharmaceutical raw materials; Specialty Chemicals: Acephenone, Acrylic emulsion, Acrylic emulsion adhesives, Anti-foaming agent, Phenol anti-oxidant, Bactericides/Preservatives for foods/paints/trees/bamboo, Blowing agent, n-Butylene-1, iso-Butylene, Calcium stearate, Epoxy plasticizers, Epoxy resin curing agent, Emulsion ethylene-vinyl acetate copolymer, Organic phosphorous-based flame retardant, Organic compound flame retardant, Flocculants, Captive use formalin, Nicotine sulphate, Polyester plasticizers, Phthalic acid ester plasticizers, Alphatic acid plasticizers, Adipic acid plasticizers, Metallic soap-based polyvinyl chloride stabilizer, Rust inhibitor, Sizing agent, Slip additive, Surface coating agent

Banks : Fuji Bank; Mitsubishi Bank; Long Term Credit Bank of Japan; Mitsui Taiyo Kobe Bank; Kyowa Saitama Bank; Mitsubishi Trust and Banking; Yasuda Trust and Banking; Nippon Credit Bank

Branches : Osaka Branch; Nagoya Branch; Sapporo Branch; Shizuoka Branch; Kyushu Branch; Kyoto Branch; Sendai Branch; Shikoku Branch; Chugoku Branch; Hiratsuka Office; 8 Sales Offices throughout Japan

Laboratories : Research Laboratory (Sakura-shi, Chiba-ken)

Daiwabo Co., Ltd.

Japanese Name	:	***Daiwabo K.K.***
TSE Number	:	3107
Address	:	Higobashi Shimizu Bldg., 1-3-7, Tosabori, Nishi-ku, Osaka 550
Tel & Fax	:	Tel: 06-446-5111 Fax: 06-446-5392
Statistics	:	Capital : 10419 million yen Employees : 2336
Officers	:	CEO : Mr. Satoru Arinobe/President
Sales	:	March, 1988 (Adjusted term) - 82809 million yen March, 1989 - 80371 million yen March, 1990 - 72241 million yen

Sales Breakdown :

SIC#	Product	Sales
22	Textiles	73 %
0000	Industrial materials	26 %
25	Home furnishings	

Banks	:	Sanwa Bank; Dai-Ichi Kangyo Bank; Toyo Trust and Banking; Norinchukin Bank; Sumitomo Trust and Banking
Branches	:	Tokyo Branch (Chuo-ku, Tokyo)

Daiya Reform Corp.

Japanese Name	:	***Daiya Reforumu Kensetsu K.K.***
TSE Number	:	Not listed
Description	:	Engaged in the import, export and sale of chemical products, most notabley concrete waterproofing agents which the company exports from XPEX Company (Canada).
Address	:	3-4-1, Marunouchi, Chiyoda-ku, Tokyo 100
Tel & Fax	:	Tel: 03-3214-5575 Fax: 03-3214-5577
Sales Breakdown	:	SIC# Product 5169 Concrete waterproofing agents
Main Products	:	Chemical Products (Wholesale): Concrete waterproofing agents

Denki Kagaku Kogyo Co., Ltd.

Japanese Name	:	***Denki Kagaku Kogyo K.K.***
TSE Number	:	4061
Description	:	Major producer of petrochemicals and lime chemicals. The company started out in 1915 as a pioneer in the Japanese chemicals industry. Since that time, it has steadily expanded its business activities to become a comprehensive chemical manufacturer producing over 200 different products. Recently the company has diversified into such fields as specialty chemicals, fine chemicals, electronics, semiconductor processing and biotechnology.
Address	:	Sanshin Bldg., 1-4-1, Yuraku-cho, Chiyoda-ku, Tokyo 100
Tel & Fax	:	Tel: 03-3507-5032 Fax: 03-3508-2739
Statistics	:	Capital : 35283 million yen Employees : 4000
Officers	:	CEO : Mr. Bunichiro Shimura/President Contact : Mr. Takehiko Furuhashi/General Manager, Overseas
Sales	:	March, 1988 - 192519 million yen March, 1989 - 204548 million yen March, 1990 - 207819 million yen March, 1991 - 221820 million yen
R & D Spending	:	10212 million yen Capital Investment: 20261 million yen

Sales Breakdown :

SIC#	Product	Sales
286	Organic chemicals	65 %
3241	Cement	20 %
281	Inorganic chemicals	15 %

Banks	:	Dai-Ichi Kangyo Bank; Mitsui Trust and Banking
Factories	:	Omi Plant (Nishikubiki-gun); Omuta Plant (Omuta-shi, Fukuoka-ken); Chiba Plant (Ichihara-shi, Chiba-ken); Shibukawa Plant (Shibukawa-shi, Gumma-ken)
Laboratories	:	Research Center (Machida-shi, Tokyo)
Subsidiaries	:	Denka Seiken Co., Ltd. (Chuo-ku, Tokyo); Denka Seiyaku Co., Ltd. (Kawasaki-shi, Kanagawa-ken); Juzen Kagaku Co., Ltd. (Toyama-shi, Toyama-ken)

Denkoh Co., Ltd.

Japanese Name	:	***Denkoh K.K.***
TSE Number	:	Not listed
Description	:	Engaged in the manufacture and sale of industrial heating equipment, semiconductor manufacturing equipment and semiconductors. Denkoh Co., Ltd. maintains strong business ties with Mitsui & Co., Ltd. and its affiliates.
Address	:	6-1, Sakae-cho, Tachikawa-shi, Taito-ku, Tokyo 190
Tel	:	0425-37-1171
Statistics	:	Capital : 198 million yen Employees : 60
Sales	:	March, 1990 - 1231 million yen March, 1991 - 1391 million yen
Sales Breakdown	:	SIC# Product 3567 Infra-red heating devices 3567 Belt kilns 3674 Semicondutors and allied products 3559 Semiconductor manufacturing equipment
Main Products	:	Industrial Heating Equipment: Infra-red heating devices, Belt kilns; Semiconductors and Allied Products; Semiconductor Manufacturing Equipment
Banks	:	Tokyo Tomin Bank; Mitsui Taiyo Kobe Bank; Shoko-Chukin Bank; Fuji Bank; Mitsubishi Bank; Tokai Bank
Shareholders	:	Mitsui & Co., Ltd. [8031] - 1,250,000 Shares; Sachiko Yuri - 568,000 Shares; Mitsui Finance Service - 200,000 Shares

Dow Corning Toray Silicone Co., Ltd.

Japanese Name	:	***Tore Dau Koningu Shirikon K.K.***
TSE Number	:	Not listed
Description	:	Originally established in Tokyo as Toray Silicone Co., Ltd, the company changed its name to Toray Dow Corning Co., Ltd. in December of 1989. The company imports products from Dow Corning.
Address	:	Mitsui Bldg. No.6, 2-3-16, Nihonbashi-Marunouchi, Chuo-ku, Tokyo 103
Tel & Fax	:	Tel: 03-3246-1651 Fax: 03-3243-1577
Statistics	:	Capital : 3600 million yen Employees : 750
Officers	:	CEO : Mr. Masami Morinaka/President Contact : Mr. Masami Morinaka/President
Sales	:	November, 1988 - 28683 million yen November, 1989 - 31037 million yen November, 1990 - 34151 million yen
Sales Breakdown	:	SIC# Product 2869 Silicones
Main Products	:	Industrial Organic Chemicals: Silicone products, Organo silicone products
Banks	:	Mitsui Taiyo Kobe Bank; Dai-Ichi Kangyo Bank; Sumitomo Bank; Mitsubishi Bank; Long Term Credit Bank of Japan; Industrial Bank of Japan; Mitsui Trust and Banking; Citibank
Factories	:	Chiba Factory (Ichihara-shi); Fukui Factory (Sakai-gun, Fukui-ken)
Laboratories	:	Main Laboratory (Ichihara-shi)
Shareholders	:	Dow Corning Corp; Toray Co., Ltd.

Dowa Mining Co., Ltd.

Japanese Name	:	***Dowa Kogyo K.K.***
TSE Number	:	5714
Description	:	Dowa Mining is a comprehensive smelter of non-ferrous metals. Active from prospecting through mine management to smelting and the production of value-added metal products such as IC wafers, iron powder for tape coatings, rolled copper products, etc.
Address	:	Daiichi-Tekko Bldg., 1-8-2, Marunouchi, Chiyoda-ku, Tokyo 100
Tel & Fax	:	Tel: 03-3201-1062 Fax: 03-3201-7944
Statistics	:	Capital : 14965 million yen Employees : 1501
Officers	:	CEO : Mr. Sadaharu Niratani/President
Sales	:	March, 1988 - 138802 million yen March, 1989 - 168094 million yen March, 1990 - 190518 million yen

Sales Breakdown :

SIC#	Product	Sales
3569	Industrial machines	49 %
1031	Zinc	15 %
1041	Gold and silver	15 %
3351	Rolled copper	10 %
3669	Electronic materials	11 %

Main Products	:	General Industrial Machinery: Industrial machines; Metals: Lead, Zinc, Copper, Gold ores; Copper Rolling and Drawing: Rolled copper; Communications Equipment; Electronic Materials
Banks	:	Industrial Bank of Japan; Sanwa Bank; Kyowa Saitama Bank; Mitsui Trust and Banking; Toyo Trust and Banking

Du Pont Japan Ltd.

Japanese Name : ***Duponto Japan Rimiteddo***

TSE Number : Not listed

Description : Du Pont Japan is a subsidiary of US-based Du Pont and markets products imported from the parent company. The company is conducting R&D into manufacturing agrochemicals which are suitable for use in Asia.

Address : Shinnikko Bldg., Du Pont Tower, 2-10-1, Toranomon, Minato-ku, Tokyo 105

Tel : 03-3585-5511

Statistics : Capital : 164 million yen
Employees : 1100

Officers : CEO : Mr. M.J.R. Alan/President
Contact : Mr. M.J.R. Alan/President

Sales : March, 1989 - 92000 million yen

Sales Breakdown :

SIC#	Product
2879	Herbicides
3844	Mammography machines
2824	Acrylic fibers
2835	Diagnostic kits
3577	Electronic components
7374	Environmental-related software

Main Products : Agricultural Chemicals: Herbicides; X-ray Apparatus: Mammography machines; Organic Fibers: 'Dacron' polyester, 'Orlon' fibers; Diagnostic Substances: Diagnostic kits, DNA sequencers; Computer Peripherals: Electronic equipment; Data Processing Services: 'SAFER' computer-based software for identifying toxic vapor release

Branches : Osaka Branch; Sendai Branch; Sapporo Branch; Nagoya Branch; Fukuoka Branch

Factories : Utsunomiya Plant (Tochigi); Shimizu Plant (Shizuoka)

Laboratories : Du Pont Japan Central Research Laboratory (Yokohama-shi); Agricultural Science Laboratory (Tsukuba-shi); Du Pont Japan Imaging Technology Center (Kawasaki)

Electric Power Development Co., Ltd.

Japanese Name :	***Dengen Kaihatsu K.K.***
TSE Number :	Not listed
Description :	Quasi-public electric power company established by the Japanese government in conjunction with nine private electric utilities. Electric Power Development Co., Ltd. is currently involved in overseas technological cooperation projects in China and Turkey.
Address :	6-15-1, Ginza, Chuo-ku, Tokyo 104
Tel & Fax :	Tel: 03-3546-2211 Fax: 03-3544-0247
Statistics :	Employees : 2720
Officers :	CEO : Mr. Ichiro Fujiwara/President
Sales :	March, 1988 - 345336 million yen March, 1989 - 356910 million yen March, 1990 - 371430 million yen
Sales Breakdown :	SIC# Product Sales 4911 Electricity supply 90 %
Main Products :	Electric power generation and distribution
Banks :	Industrial Bank of Japan
Overseas Subsidiaries :	Energy Bot AS (Turkey); E.P.D.C. Pty., Ltd. (Australia)

Fuji Electric Co., Ltd.

Japanese Name	:	***Fuji Denki K.K.***
TSE Number	:	6504
Description	:	Leading producer of electronic machinery. A member of the Furukawa group of companies, Fuji Electric leads the Japanese market in sales of vending machines and is aggressively moving into the areas of factory automation and electronics.
Address	:	Shin Yuraku-cho Bldg., 1-12-1, Yuraku-cho, Chiyoda-ku, Tokyo 100
Tel & Fax	:	Tel: 03-3211-7111 Fax: 03-3215-8321
Statistics	:	Capital : 46769 million yen Employees : 14541
Officers	:	CEO : Mr. Takeshi Nakao/President Contact : Mr. Masatoshi Kato/Manager, Intl. Business Development
Sales	:	March, 1988 - 402301 million yen March, 1989 - 479183 million yen March, 1990 - 545232 million yen March, 1991 - 591964 million yen
R & D Spending	:	32000 million yen Capital Investment: 34200 million yen
Sales Breakdown	:	

SIC#	Product	Sales
3599	Heavy electrical equipment	43 %
367	Control/information systems & devices	42 %
36	Consumer electronics & other products	14 %

Main Products	:	Industrial Electrical Equipment: Power and energy systems, Electronic distribution equipment, Transformers, Water and sewage treatment systems, Process and factory automation equipment, Measuring equipment, Solar cells, Semiconductors and semiconductor production machinery, Vending machines; Consumer Electronics and Other Products: Vending machines, Fuel cells
Banks	:	Dai-Ichi Kangyo Bank; Industrial Bank of Tokyo; Kyowa Saitama Bank; Daiwa Bank; Yasuda Trust and Banking
Laboratories	:	Fuji Electric Corporate Research and Development Ltd. (Yokosuka-shi, Kanagawa-ken)
R&D Areas	:	Electronic Components

Fuji Electrochemical Co., Ltd.

Japanese Name	:	***Fuji Denki Kagaku K.K.***
TSE Number	:	6955
Description	:	Fuji Electrochemical is a primary producer of electronic parts including dry-cell batteries. The company is building production bases abroad and diversifying into optical communications equipment.
Address	:	5-36-11, Shinbashi, Minato-ku, Tokyo 105
Tel & Fax	:	Tel: 03-3434-1271 Fax: 03-3431-1375
Statistics	:	Capital : 8204 million yen Employees : 2112
Officers	:	CEO : Mr. Rinzo Iwai/President Contact : Mr. Fumiaki Nakamura/General Manager
Sales	:	March, 1988 - 61541 million yen March, 1989 - 65368 million yen March, 1990 - 72527 million yen March, 1991 - 76579 million yen
Capital Investment	:	3340 million yen

Sales Breakdown :

SIC#	Product	Sales
3679	Magnetic heads and supplies	46 %
3264	Ferrites, ceramics	28 %
3692	Dry-cell batteries	26 %

Banks	:	Tokai Bank; Dai-Ichi Kangyo Bank; Kyowa Saitama Bank; Daiwa Bank ; Industrial Bank of Japan
Branches	:	Sendai Branch (Miyagi-ken); Tokyo Branch (Tokyo); Yokohama Branch (Kanagawa-ken); Nagoya Branch (Aichi-ken); Osaka Branch (Osaka); Hiroshima Branch (Hiroshima-ken); Fukuoka Branch (Fukuoka-ken); Sapporo Branch
Factories	:	Fujimi Plant (Kosai-shi); Kosai Plant (Kosai-shi); Ohsuga Plant (Shizuoka); Komago Plant (Shizuoka); Sanyo Plant (Yamaguchi-ken); Washizu Plant (Kosai-shi
Laboratories	:	R&D Center (Kosai-shi, Shizuoka-ken)
Subsidiaries	:	Iwaki Electronics Co., Ltd. (Iwaki-shi); FDK Powertronics Co., Ltd. (Toyohashi-shi, Aichi-ken); FDK Mechatronics Co., Ltd. (Shizuoka-ken); FDK Engineering Co., Ltd. (Kosai-shi, Shizuoka-ken)

Fujitsu Laboratories Ltd.

Japanese Name	:	***Fujitsu Kenkyusho K.K.***
TSE Number	:	Not listed
Description	:	Engaged in a wide variety of R&D activities, the company is an R&D subsidiary of Fujitsu Ltd.
Address	:	1015, Kamiodanaka, Nakahara-ku, Kawasaki-shi, Kanagawa-ken 211
Tel & Fax	:	Tel: 044-777-1111 Fax: 044-751-7655
Statistics	:	Capital : 5000 million yen Employees : 1300
Officers	:	CEO : Mr. Mikio Yuki/President
Sales Breakdown	:	SIC# Product 8731 Information processing R&D 8731 Electronic systems R&D 8731 Personal systems R&D 8731 Telecommunications R&D 8731 Electronic devices R&D 8731 Materials R&D 8731 Space technology R&D
Main Products	:	Information Processing R&D: AI diagnostic shell, Neurochip, Financial neurocomputer, Conference machine translation system, Voice recognition system, 'IDATEN' pattern recognition system for production-line and driverless-car application, Highly parallel industrial computers, CAD systems, Ultrasonic medical diagnostic equipment, Biomagnetic measurement by SQUID superconducting magnetic sensor; Electronic Systems R&D: Holographic POS bar code reader, Real-time holographic fingerprint sensor, Hologram scanner for laser printer, Full color printer, Filter alternating color scanner, Still color picture restoration system, Perpendicular recording technology, Fixed magneto-optical disk drive, Component visual inspection system using 3-D imaging technology, LSI internal diagnosis, Intelligent assembly robots; Personal Systems R&D: Online hand written system, ISDN communication terminal, Computer graphics, Optical disk, Electronics packaging, Integrated Services Digital Network (ISDN), Optical shuttlebus multiplexing system, Digital audio processing system, High-speed multi-media communications system, Image encoding, Holonic network, Coherent lightwave communications, Optical switching system, Radio communications, Satellite communications; Electronic Devices R&D: ULSI, Microfabrication process technology, High electron mobility transistor, Infra-red devices, Integrated modulator/DFB laser light source for optical semiconductors, Wavelength-tunable narrow linewidth laser, Twin PIN photodiode, Bistable laser, Josephson 4-bit superconducting processor, Single chip SQUID, Resonant-tunneling hot electron transistor, Electron wave interference devices, Atomic layer aptiaxy, High-temperature superconductor, Crystal growth stimulation, Display devices; Materials R & D: Multilayer ceramic circuit board with copper conductor, Superconducting ceramics, PMSS thermal-resistant resin, Biosensors, Rewritable optical disk thin film material, Diamond film, Materials Engineering: Material characterization; Space Technology R & D: Radioastronomy, Space-biology, Remote sensing, High-accuracy positioning and distance measuring technology, Material processing, Space robotics
Laboratories	:	Kawasaki Laboratories (Kanagawa-ken); Atsugi Laboratories (Kanagawa-ken)
Shareholders	:	Fujitsu [6702]

Fujitsu Ltd.

Japanese Name	:	***Fujitsu K.K.***
TSE Number	:	6702
Description	:	Engaged in the manufacture and sale of computers, communications equipment and electronic components. Japan's leading computer manufacturer, Fujitsu also enjoys a strong presence in domestic semiconductor and telecommunications markets. R&D expenditure, which has exceeded 10% of total sales in recent years, continues to stress mainline products with 64 megabyte DRAMs and 64 kilobyte SRAMs utilizing HEMTs receiving special attention.
Address	:	Marunouchi Center Bldg., 1-6-1, Marunouchi, Chiyoda-ku, Tokyo 100
Tel & Fax	:	Tel: 03-3216-3211 Fax: 03-3213-7174
Statistics	:	Capital : 22141 million yen Employees : 50168
Officers	:	CEO : Mr. Tadashi Sekizawa/President Contact : Mr. Makoto Ogura/Manager, Systems Engineering
Sales	:	March, 1988 - 1714420 million yen March, 1989 - 2004600 million yen March, 1990 - 2125670 million yen March, 1991 - 2337780 million yen
R & D Spending	:	325000 million yen Capital Investment: 184900 million yen
Sales Breakdown	:	SIC# Product Sales 357 Computer equipment 72 % 366 Communications equipment 17 % 367 Electronic components 13 %
Main Products	:	Computer Equipment: Supercomputers, General use computers, Minicomputers, Office computers and workstations, Personal computers, POS systems, Computer communications related equipment, Medical computer systems, Disk drives and related equipment, Word processors, Fax machines, Software, System consulting, VAN services; Telephone and Telegraph Apparatus: Telephones, Telephone switchboards and related equipment, In-house business communication network systems, Ocean floor transcontinental communications system, Optical communication systems, Various wireless communications equipment; Electronic Components: Logic ICs, Analog ICs, Hybrid ICs, DRAM memory ICs, SRAMs memory ICs, ROM memory ICs, Relays, Connectors, Keyboards, Display units
Banks	:	Dai-Ichi Kangyo Bank; Industrial Bank of Japan; Mitsui Taiyo Kobe Bank; Kyowa Saitama Bank; Yokohama Bank; Mitsubishi Bank
Branches	:	98 domestic branches; 20 foreign branches
Factories	:	Kawasaki Plant (Kawasaki-shi, Kanagawa-ken); Oyama Plant (Oyama-shi, Tochigi-ken); Nagano Plant (Nagano-shi, Nagano-ken); 11 other factories throughout Japan
Laboratories	:	Information Systems Laboratory (Oda-ku, Tokyo); Kansai System Laboratory (Chuo-ku, Osaka); Kawasaki Laboratory (Kawasaki-shi, Kanagawa-ken), Atsugi Laboratory (Atsugi-shi, Kanagawa-ken)
R&D Areas	:	Electronics: Communications equipment, ATM technologies, Computer graphics systems, 64 megabyte DRAMs, 64 kilobyte SRAMs utilizing HEMTs
Subsidiaries	:	Fujitsu Business Systems Ltd. (Tokyo); Fujitsu Electrochemical Co., Ltd. (Tokyo); PFU Ltd. (Kanagawa-ken); Over 100 other subsidiaries throughout Japan
Overseas Subsidiaries	:	Fujitsu America, Inc. (U.S.A.); Fujitsu Korea Ltd. (R.O.K.); Fujitsu Europe Ltd. (U.K.); Fujitsu Australia Ltd. (Australia); 147 other subsidiaries throughout the world

Furukawa Co., Ltd.

Japanese Name	:	***Furukawa Kikai Kinzoku K.K.***
TSE Number	:	5715
Description	:	A manufacturer of machinery, metals, chemicals, electric materials, and fuels. Currently diversifying into aluminum foil for condensors, and truck cranes.
Address	:	2-6-1, Marunouchi, Chiyoda-ku, Tokyo 100
Tel & Fax	:	Tel: 03-3212-6561 Fax: 03-3287-0696
Statistics	:	Capital : 23497 million yen Employees : 2347
Officers	:	CEO : Mr. Yutaka Okumura/President
Sales	:	March, 1988 - 101538 million yen March, 1989 - 128613 million yen March, 1990 - 153432 million yen March, 1991 - 163981 million yen
Capital Investment	:	6328 million yen

Sales Breakdown :

SIC#	Product	Sales
35	Machinery	51 %
33	Metals	34 %
291	Fuel and energy	7 %
28	Chemicals	7 %
0000	Electronic materials	2 %

Main Products :

Construction Equipment: Wheel loaders, Power shovels, Crawler shovels, Bulldozers, Snowploughing dozers, Snow loaders, Forestry tractors, Vibration rollers, Road rollers, Pneumatic-tyre rollers, On-vehicle-type cranes, Truck cranes, Crane carriers; Rock Drills: Hydraulic/pneumatic rock drills, Hydraulic/pneumatic crawler drills, Hydraulic/pneumatic tunnel jumbos (wheel type, crawler type, etc.), Hydraulic breakers, Pneumatic hand breaker, Hydraulic rotary drills, Hydraulic mini pile drivers; Plant Equipment: Conveyors, Conveying facilities, Storage-release facilities, Crushing facilities, Screens, Feeders, Classifiers, Used tyre carbonization combustion facilities, Superhigh-pressure water jets, Copper smelting facilities; Steel Structures: Steel bridges, Gates, Steel frames; Parking Systems: Above and below ground parking systems; Environmental Protection Equipment: Electrostatic precipitators, Waste water treatment systems, Night soil treatment systems, Incineration plants; Pumps: Slurry pumps, Sewage pumps, Screw pumps, Furukawa-Sulzer centrifugal pumps and turbine pumps, Submersible sludge pumps, Specialized pumps, Pump plant equipment; Castings: High-manganese steel castings, High-chrome cast iron castings, Sabel reinforcement wear-resistant castings, Low alloy steel-type wear-resistant castings, Maxichrome crushing balls, Special wear-resistant linings; Bowling Equipment: Bowling systems; Metals: Copper, Gold, Silver, Crushed stone, Aluminum electrolytic foil for capacitors, Capacitor parts; Chemicals: Titanium dioxide, Cuprous oxide, Copper oxide, Sulfuric acid, Arsenic acid, Polymer iron sulfate, Chemical plaster, Titanium carbide; Electronic Materials: High-purity metallic arsenic, High-purity phosphorous, High-purity sulphur, Gallium arsenide polycrystal, High-purity copper oxide, High-purity bismuth oxide, Calcogenide glass, Infra-red ray transmission glass, Electrolytic iron, Exhaust gas treatment unit for semiconductors, Medical sticking contact needles; Fuels: Fuel oil, Gasoline, Light oil, Kerosene, Lubricating oil, Liquid petroleum gas, Coal, Coke

Banks	:	Dai-Ichi Kangyo Bank; Industrial Bank of Japan; Long Term Credit Bank of Japan
Branches	:	Osaka Branch (Osaka-shi); Chugoku Local Office (Okayama-shi); Kyushu Branch (Fukuoka-shi); Sapporo Branch (Sapporo-shi); Nagoya Branch (Nagoya-shi); Hokuriku Local Office (Toyama-shi); Sendai Branch (Sendai-shi, Miyagi-ken)
Laboratories	:	Hino Research Laboratory (Hino-shi, Tokyo)
Subsidiaries	:	Furukawa Real Estate Co., Ltd. (Chuo-ku, Tokyo); Furukawa Rock Drill Sales Co., Ltd. (Chiyoda-ku, Tokyo); Furukawa UNIC Corp. 9 machinery sales companies; 12 UNIC crane sales companies
Overseas Subsidiaries	:	Furukawa GmbH (Germany); Gougler Industries Inc. (Ohio, US); Furukawa Distribution S.A. (France); Furukawa Equipement S.A. (France); Furukawa Finance Europe B.V. (Holland); Southern Copper Ltd. (Australia)

Furukawa Electric Co., Ltd. ; The

Japanese Name	:	***Furukawa Denki Kogyo K.K.***
TSE Number	:	5801
Description	:	One of Japan's two largest electric wire and cable manufacturers. Holds patents for optical fiber coating technology. Pioneer in commercialization of fluoric glass fibers. Leader in production of shape memory alloys and superconductive wire.
Address	:	2-6-1, Marunouchi, Chiyoda-ku, Tokyo 100
Tel & Fax	:	Tel: 03-3286-3001 Fax: 03-3286-3694
Statistics	:	Capital : 55832 million yen Employees : 6969
Officers	:	CEO : Mr. Kengo Tomomatsu/President
Sales	:	March, 1988 - 470068 million yen March, 1989 - 525694 million yen March, 1990 - 577759 million yen March, 1991 - 628917 million yen
R & D Spending	:	15769 million yen Capital Investment: 34890 million yen
Sales Breakdown	:	SIC# Product 3643 Wires and cables 3356 Non-ferrous products 3351 Rolled copper products
Main Products	:	Current carrying Wire Devices: Wires and cables, Superconductive wire; Copper products: Rolled copper products; Light Metal Products: Shape memory alloys; Coating materials: Optical fiber coatings, Fire-retardant electric cable coatings for use in atomic power plants; Glass Fiber products: Fluoric glass fibers
Banks	:	Dai-Ichi Kangyo Bank; Long Term Credit Bank of Japan; Kyowa Bank; Yokohama Bank; Joyo Bank
Laboratories	:	Yokohama Laboratory (Yokohama-shi); Nikko Laboratory (Nikko-shi); Hiratsuka Laboratory (Hiratsuka-shi); Chiba Laboratory (Ichihara-shi); Information and Electronics Laboratory (Hiratsuka-shi); Optical Technology Laboratory (Ichihara-shi)

Hayashi Chemical Industries Co., Ltd.

TSE Number	:	Not listed
Description	:	Engaged in the manufacture and sale of dyestuffs, agricultural chemicals and ceramics. The company is also active in R&D in the area of new materials and superconductivity.
Address	:	31, Dogo, Nichimachi, Ishihara, Kisshoin, Minami-ku, Kyoto 601
Tel	:	075-681-8331
Statistics	:	Capital : 30 million yen Employees : 77
Sales	:	October, 1989 - 2652 million yen October, 1990 - 2705 million yen
Sales Breakdown	:	SIC# Product 2865 Dyestuffs 2869 Synthetic dyes 287 Soil ameliorants and fertilizers 3299 New ceramics
Main Products	:	Dyes and Dyestuffs: Organic and synthetic dyes, Dye-related chemicals; Agricultural Chemicals: Soil ameliorants, Fertilizers; New Ceramics: Superconductive ceramic powders
Banks	:	Sumitomo Bank; Bank of Kyoto; Sanwa Bank
Factories	:	Kusatsu Plant (Shiga-ken)
R&D Areas	:	Advanced Materials: Superconductive materials
Shareholders	:	Noboru Hayashi - 21,500 Shares

Hitachi Cable, Ltd.

Japanese Name	:	***Hitachi Densen K.K.***
TSE Number	:	5812
Address	:	2-1-2, Marunouchi, Chiyoda-ku, Tokyo 100
Tel & Fax	:	Tel: 03-3216-1611 Fax: 03-3214-5779
Statistics	:	Capital : 25478 million yen Employees : 5859
Officers	:	CEO : Mr. Hiroji Hashimoto/President
Sales	:	March, 1988 - 232971 million yen March, 1989 - 270786 million yen March, 1990 - 303641 million yen March, 1991 - 348585 million yen
R & D Spending	:	11433 million yen Capital Investment: 23098 million yen

Sales Breakdown :

SIC#	Product	Sales
3357	Optical fibers and wires	53 %
3677	Copper products and coils	17 %
3559	Electronics and leadframe products	17 %
3559	Semiconductor and rubber products	5 %

Banks : Sanwa Bank; Dai-Ichi Kangyo Bank; Fuji Bank; Joyo Bank; Industrial Bank of Japan; Toyo Trust and Banking

Hitachi Chemical Co., Ltd.

Japanese Name	:	***Hitachi Kasei Kogyo K.K.***
TSE Number	:	4217
Description	:	General processor of synthetic resins, emphasizing products for use in electrical/electronic equipment. Also produces housing components and molded automobile parts.
Address	:	2-1-1, Nishi-Shinjuku, Shinjuku-ku, Tokyo 160
Tel & Fax	:	Tel: 03-3346-3111 Fax: 03-3346-2977
Statistics	:	Capital : 15019 million yen Employees : 5403
Officers	:	CEO : Mr. Ryoji Yokoyama/President Contact : Mr. Kiyoshi Nakagawa/Director, Overseas Div.
Sales	:	March, 1988 - 249753 million yen March, 1989 - 276431 million yen March, 1990 - 283028 million yen March, 1991 - 314658 million yen
R & D Spending	:	13600 million yen Capital Investment: 21900 million yen
Sales Breakdown	:	

SIC#	Product	Sales
367	Electronics materials	41 %
286	Organic materials and resins	27 %
2821	Housing, construction materials	23 %
281e	Inorganic materials	7 %
283	Pharmaceuticals	2 %

Main Products	:	Electronics Materials: Materials and instruments for printed wiring boards, Printed wiring boards, Epoxy molding compounds for semiconductor encapsulation; Specialty Chemicals: Halogen-based flame-retardant, Organic phosphorous-based flame retardant, Vulcanizing agent, Phthalic acid ester plasticizers, Aliphatic acids plasticizers; Other Products: Housing materials, Molded automobile parts
Banks	:	Sanwa Bank; Fuji Bank; Dai-Ichi Kangyo Bank
Branches	:	Kansai Sales Office (Osaka); Chubu Sales Office (Nagoya-shi, Aichi-ken); 4 other sales offices; 19 branch offices
Factories	:	Yamazaki Works (Hitachi-shi, Ibaraki-ken); Shimodate Works (Shimodate-shi, Ibaraki-ken); 7 other factories
Laboratories	:	Tsukuba Research Laboratory (Tsukuba-shi, Ibaraki-ken); 2 other laboratories
Overseas Subsidiaries	:	Advanced Interconnection Technology, Inc. (New York, U.S.A.); Hitachi Chemical Electro-Products, Inc. (Georgia, U.S.A.); Hitachi Chemical (Hong Kong) Ltd. (Hong Kong); Hitachi Chemical Europe GmbH (Germany); etc.

Hitachi Maxell, Ltd.

Japanese Name	:	***Hitachi Maxell K.K.***
TSE Number	:	6810
Description	:	Manufacturer of magnetic and optical storage mediums and electronics-related products. A subsidiary of Hitachi, Ltd., Hitachi Maxell is Japan's largest producer of magnetic tape and one of the largest producers of floppy disks in the world. The company is currently emphasizing the development of optical disks and other information processing and storage products as well as electronic hardware and consumer electronics products. Close to half of Hitachi Maxell's production is sold overseas.
Address	:	1-1-88, Ushitora, Ibaraki-shi, Osaka 567
Tel & Fax	:	Tel: 0726-23-8101 Fax: 0726-23-8133
Statistics	:	Capital : 9134 million yen Employees : 2528
Officers	:	CEO : Mr. Hiroshi Watanabe/President Contact : Mr. Isamu Arai/Managing Director, Sales Department
Sales	:	March, 1988 - 132414 million yen March, 1989 - 133861 million yen March, 1990 - 137467 million yen March, 1991 - 139765 million yen
R & D Spending	:	8551 million yen Capital Investment: 9900 million yen
Sales Breakdown	:	

SIC#	Product	Sales
3695	Magnetic recording tape	61 %
363	Electronic appliances	21 %
36	Electronic equipmert and batteries	16 %

Main Products	:	Magnetic Tapes: Video tapes, Audio tapes, Computer tapes, Professional-use tapes, Magnetic tickets, Prepaid cards, Ink-rolls for thermal printers; Electronics: Floppy disks, Video floppy disks, Optical disks, Memory cards, IC cards, Cash cards, IC card reader/writer, IC card terminals, Corporate banking systems, Memory card reader/writer, Software, CD-ROMs, Image filing systems; Batteries: Dry cell batteries, Button type batteries, Lithium batteries, Rechargeable batteries; Electrical Appliances and Equipment: Water purifiers, Satellite dishes, Accessories for audio and video, Battery production plant
Banks	:	Sanwa Bank; Dai-Ichi Kangyo Bank; Fuji Bank; Bank of Tokyo; Tokai Bank; Toyo Trust and Banking
Branches	:	Sales Head Office (Chuo-ku, Tokyo); Tokyo Branch (Chuo-ku, Tokyo); Osaka Branch (Kita-ku, Osaka); Nagoya Branch (Nagoya-shi, Aichi-ken)
Factories	:	Kyoto Factory (Otokuni-gun, Kyoto); Fukuchiyama Factory (Fukuchiyama-shi, Kyoto); Tsukuba Factory (Mitsukaido-shi, Ibaraki-ken); Osaka Factory (Ibaraki-shi, Osaka)
Laboratories	:	Kyoto Laboratory (Otokuni-gun, Kyoto); Tsukuba Laboratory (Mitsukaido-shi, Ibaraki-ken)
R&D Areas	:	Kyoto Laboratory: 9-hour recording video tape; Tsukuba Laboratory: Perpendicular recording-type floppy disks using barium ferrite magnetic material, 7 gigabyte capacity optical disk, Metal tape with high oxidization-resistant qualities
Subsidiaries	:	Kyushu Hitachi Maxell, Ltd. (Fukuoka-ken); Maxell Seiki Ltd. (Otokuni-gun, Kyoto); Maxell Hokuriku Seiki Ltd. (Toyama-ken)

Hitachi Metals Co., Ltd.

Japanese Name : ***Hitachi Kinzoku K.K.***

TSE Number : 5486

Address : Chiyoda Bldg., 2-1-2, Marunouchi, Chiyoda-ku, Tokyo 100

Tel & Fax : Tel: 03-3284-4511 Fax: 03-3287-1956

Statistics : Capital : 19879 million yen
Employees : 8546

Officers : CEO : Mr. Koji Matsuno/President

Sales : March, 1988 - 251720 million yen
March, 1989 - 288711 million yen
March, 1990 - 323035 million yen
March, 1991 - 350440 million yen

R & D Spending : 8200 million yen Capital Investment: 28951 million yen

Sales Breakdown :

SIC#	Product	Sales
3714	Automobile parts	23 %
3499	Magnetic and electronic materials	22 %
3312	Specialty steels	22 %
346	Machine and building frames	14 %
3317	Steel pipe and fittings	13 %
35	Plant and equipment	6 %

Banks : Sanwa Bank; Fuji Bank; Industrial Bank of Japan; Toyo Trust and Banking; Dai-Ichi Kangyo Bank; Bank of Tokyo

Branches : Kansai Branch (Chuo-ku, Osaka); Kyushu Branch (Fukuoka-ken); Chubu Branch (Aichi-ken); Chugoku Branch (Hiroshima-ken); Tokai Branch (Shizuoka-ken); Kita-Kanto Branch (Saitama-ken); Yokohama Branch (Kanagawa-ken)

Factories : Kyushu Works (Fukuoka-ken); Moka Works (Tochigi-ken); Kuwana Works (Mie-ken); Wakamatsu Works (Fukuoka-ken); Yasugi Works (Shimane-ken); Kumagaya Plant (Saitama-ken); Electronic Products Works (Tochigi-ken)

Hitachi, Ltd.

Japanese Name	:	***Hitachi Seisakusho K.K.***
TSE Number	:	6501
Description	:	Nation's largest comprehensive manufacturer of heavy electric plant and equipment. The company has many promising subsidiaries in varied fields, including chemicals, wire and cable manufacturing, and shipbuilding. The company is also expanding into the aerospace field.
Address	:	4-6, Kanda-Surugadai, Chiyoda-ku, Tokyo 101
Tel & Fax	:	Tel: 03-3258-1111 Fax: 03-3253-2186
Statistics	:	Capital : 36332 million yen Employees : 80677
Officers	:	CEO : Mr. Katsushige Mita/President
Sales	:	March, 1989 - 3232000 million yen March, 1990 - 3525250 million yen March, 1991 - 3788810 million yen
R & D Spending	:	391898 million yen Capital Investment: 306500 million yen
Sales Breakdown	:	SIC# Product 3559 Special Industrial machinery 3569 General Industrial machinery 3579 Office automation equipment 3599 Industrial machinery, except electrical 5063 Power generation plants
Main Products	:	Heavy electric plant; Consumer electronics; Communications and electronic equipment; Industrial machinery and plants; Traffic equipment; Aerospace
Banks	:	Industrial Bank of Japan; Dai-Ichi Kangyo Bank; Sanwa Bank; Fuji Bank; Tokai Bank
Laboratories	:	Basic Laboratory. (Hiki-gun); Central Research Laboratory (Kokubunji-shi, Tokyo)
Overseas Subsidiaries	:	Hitachi America Ltd. (New York, U.S.A.); Hitachi Far East Ltd. (Hong Kong); Hitachi Europe Ltd. (London, UK)

Hoechst Gosei Co., Ltd.

Japanese Name : ***Hekisuto Gosei K.K.***

TSE Number : Not listed

Description : The company produces and sells synthetic resin emulsions using the technology of its two parent companies, Nippon Synthetic Chemical Industry and Hoechst Japan.

Address : Hoechst Bldg., 4-10-33, Akasaka, Minato-ku, Tokyo 107

Tel & Fax : Tel: 03-3582-3921 Fax: 03-3584-5918

Statistics : Capital : 540 million yen
Employees : 185

Officers : CEO : Mr. Hiroaki Horiuchi/President
Contact : Mr. Akiyoshi Takano/Manager, Sales Department

Sales Breakdown :

SIC#	Product	Sales
2821	Synthetic resins	95 %
286	Specialty chemicals	5 %

Main Products : Plastics Materials: Synthetic resin emulsions; Specialty Chemicals: Sea surface pollution precipitant

Banks : Bank of Tokyo; Mitsubishi Bank; Dai-Ichi Kangyo Bank; Daiwa Bank; Ogaki Kyoritsu Bank; Tokai Bank; Mitsui Taiyo Kobe Bank; Sanwa Bank; Sumitomo Bank

Branches : Tobu Branch (Minato-ku, Tokyo); Seibu Branch (Osaka); Chubu Branch (Nagoya)

Factories : Ogaki Factory (Ogaki-shi, Gifu-ken)

Laboratories : Shizuoka Laboratory (Ogasa-gun)

Shareholders : Hoechst Japan Ltd.; The Nippon Synthetic Chemical Industry Co., Ltd.

Overseas Subsidiaries : Hoechst Chemical Industries Ltd. (Bangkok, Thailand); P.T. Pulosynthetics Co., Ltd. (Indonesia)

Hoechst Mitsubishi Kasei Co., Ltd.

Japanese Name	:	***Mitsubishi Kasei Hekisuto K.K.***
TSE Number	:	Not listed
Description	:	Produces natural dyestuffs.
Address	:	4-10-33, Akasaka, Minato-ku, Tokyo 107
Tel & Fax	:	Tel: 03-3582-8452 Fax: 03-3582-2375
Statistics	:	Capital : 3600 million yen Employees : 340
Officers	:	CEO : Mr. Hiroshi Watanabe/President Contact : Mr. Hiroshi Watanabe/President
Sales Breakdown	:	SIC# Product 2861 Natural dye products
Main Products	:	Dyestuffs
Branches	:	Osaka Branch (Osaka-shi); Tokyo Branch (Minato-ku); Nagoya Branch (Nagoya-shi)
Factories	:	Kitakyushu Works (Kitakyushu-shi); Oh-hama Factory (Ogasa-gun)
Laboratories	:	Osaka Technical Center (Osaka-shi); Kitakyushu Technical Center (Kitakyushu-shi); Laboratory (Kitakyushu-shi)
Shareholders	:	Mitsubishi Kasei Corp. [4010]; Hoechst Japan Ltd.

Honda Engineering

TSE Number	:	Not listed
Description	:	Engaged in the development and manufacture of machine tools and dies. A subsidiary of Honda Motor Co., Honda Engineering was established in order to develop and supply machine tools for its parent company. Sales have been declining rapidly in recent years.
Address	:	1-10-1, Shinsayama, Sayama-shi, Saitama-ken 350-13
Tel	:	0429-53-0111
Statistics	:	Capital : 1200 million yen Employees : 2450
Sales	:	February, 1989 - 60645 million yen February, 1990 - 51697 million yen February, 1991 - 32600 million yen
Sales Breakdown	:	SIC# Product 354 Machine tools and dies
Main Products	:	Machine Tools and Dies: Titanium alloy grinding equipment
Banks	:	Mitsubishi Bank; Kyowa Saitama Bank; Hachijuni Bank
Shareholders	:	Honda Motor Co., Ltd. [7267] - 100%

Honda Motor Co., Ltd.

Japanese Name	:	***Honda Giken Kogyo K.K.***
TSE Number	:	7267
Description	:	World's largest manufacturer of motorcycles and Japan's third largest producer of automobiles. Honda sells well over half of its products overseas, most notably in the U.S. where the company maintains its own design and production facilities. In addition to transportation equipment, the company also produces a variety of powered equipment for farm, marine and home use. R&D is conducted primarily through the company's subsidiaries, Honda R&D Co., Ltd. and Honda Engineering Co., Ltd., and stresses automobile safety and other automotive-related technologies.
Address	:	2-1-1, Minami-Aoyama, Minato-ku, Tokyo 107
Tel & Fax	:	Tel: 03-3423-1111 Fax: 03-3243-8947
Statistics	:	Capital : 85289 million yen Employees : 31635
Officers	:	CEO : Mr. Nobuhiko Kawamoto/President
Sales	:	March, 1988 (Adjusted term) - 1249740 million yen March, 1989 - 2636770 million yen March, 1990 - 2748860 million yen March, 1991 - 2800200 million yen
R & D Spending	:	189199 million yen Capital Investment: 115800 million yen
Sales Breakdown	:	

SIC#	Product	Sales
3711	Automobiles	78 %
3751	Motorcycles	10 %
37	Parts and other	9 %
35	Farm and misc. combustion powered products	3 %

Main Products	:	Automobiles and Automobile Parts; Motorcycles and Motorcycle Parts; Other Combustion Powered Products: Lawn mowers, Marine motors, Water pumps, Portable electric generators, Snow removal equipment, Four wheel ATVs; Farm Machinery: Soil cultivators, Harvesters, Binders
Banks	:	Mitsubishi Bank; Mitsubishi Trust and Banking; Tokai Bank; Bank of Tokyo; Industrial Bank of Japan; Kyowa Saitama Bank; Sanwa Bank
R&D Areas	:	Automobile Technologies: Air bag systems, Seat belt tension adjustment systems

Honen Corp.

Japanese Name : ***K.K. Honen Kooporeeshon***

TSE Number	:	2601
Description	:	Top ranking producer of soya bean oil and second in edible oils. Has recently expanded into soya bean-related health foods and specialty chemicals.
Address	:	Mitsui-Seimei-Bldg., 1-2-3, Otemachi, Chiyoda-ku, Tokyo 100
Tel & Fax	:	Tel: 03-3211-6511 Fax: 03-3211-6507
Statistics	:	Capital : 8552 million yen Employees : 855
Officers	:	CEO : Mr. Masaji Shima/President Contact : Mr. Masaji Shima/President
Sales	:	March, 1989 (Adjusted term) - 47442 million yen March, 1990 - 88600 million yen March, 1991 - 90968 million yen
R & D Spending	:	600 million yen Capital Investment: 2600 million yen
Sales Breakdown	:	SIC# Product 2076 Vegetable oils 2046 Corn starch 286 Specialty chemicals
Main Products	:	Vegetable Oil and Fat products; Processed Vegetable Products: Corn starch; Adhesives ; Specialty Chemicals: Emulsifiers for foodstuffs
Subsidiaries	:	Sugiyama Chemical Laboratories Foundation (Yokohama-shi, Kanagawa-ken); Rakuyo Shokuhin Co., Ltd. (Yokohama-shi, Kanagawa-ken); Honen Finance Co., Ltd. (Chiyoda-ku, Tokyo); Honen Shoji K.K. (Chiyoda-ku, Tokyo)

Honshu Paper Co., Ltd.

Japanese Name	:	***Honshi Seishi K.K.***
TSE Number	:	3862
Address	:	5-12-8, Ginza, Chuo-ku, Tokyo 104
Tel & Fax	:	Tel: 03-3543-1862 Fax: 03-3545-9035
Statistics	:	Capital : 38976 million yen Employees : 6261
Officers	:	CEO : Mr. Yoshinobu Yonezawa/President
Sales	:	March, 1988 - 366974 million yen March, 1989 - 387173 million yen March, 1990 - 389278 million yen March, 1991 - 412973 million yen
R & D Spending	:	1950 million yen Capital Investment: 32231 million yen
Sales Breakdown	:	SIC# Product Sales 2621 Paper and paperboard products 50 % 2671 Processed paper products 48 % 28 Chemical products 3 %
Banks	:	Dai-Ichi Kangyo Bank; Mitsui Taiyo Kobe Bank; Long Term Credit Bank of Japan; Mitsui Trust and Banking; Norinchukin Bank
Branches	:	Sapporo Office (Sapporo-shi, Hokkaido); Sendai Office (Sendai-shi, Miyagi-ken); Nagoya Office (Nagoya-shi, Aichi-ken); Osaka Office (Osaka-shi); Hiroshima Office (Hiroshima-shi, Hiroshima-ken); Kyushu Office (Fukuoka-shi, Fukuoka-ken)
Factories	:	Kushiro Mill (Hokkaido); Edogawa Mill (Tokyo); Fuji Mill (Shizuoka-ken); Iwabuchi Mill; Matsumoto Mill (Nagano-ken); Nakatsu Mill (Oita-ken); Shiga Mill (Shiga-ken); Saga Mill (Saga-ken); Oita Mill (Oita-ken)
Laboratories	:	R&D Laboratory. (Edogawa-ku, Tokyo)
R&D Areas	:	Paper Related Technologies: Recycling technologies, Magnetic recording papers, Ink-jet printing papers, Thermal papers, Synthetic papers, Paper coating technologies; New Materials: Inorganic fibers
Overseas Subsidiaries	:	JANT Pty., Ltd. (Papua, New Guinea); Kogol Zorin Co. Pty., Ltd. (Papua, New Guinea); Toraiuoru Pte., Ltd. (Singapore); Eften Inc. (Illinois, U.S.A.); Eften Europe B.V. (Holland)

Horiba Ltd.

Japanese Name	:	***Horiba Seisakusho K.K.***
TSE Number	:	6856
Description	:	Leading manufacturer of measuring equipment specializing in pollution and exhaust emissions analyzers. An affiliate of Hitachi, Ltd., Horiba Ltd. maintains a strong financial posture by virtue of its commanding presence in the growing automobile exhaust analyzer market. The company stresses R&D activities and is involved in an R&D partnership with ABX S.A.R.L. (France). Semiconductor analyzers and optoelectronics are among the product categories Horiba has targeted for future development.
Address	:	2, Miyanohigashi-machi, Kisshoin, Minami-ku, Kyoto 601
Tel & Fax	:	Tel: 075-313-8121 Fax: 075-321-8312
Statistics	:	Capital : 4585 million yen Employees : 970
Officers	:	CEO : Mr. Masahiro Ohura/President Contact : Mr. Masahiro Ohura/President
Sales	:	March, 1988 - 16775 million yen March, 1989 - 17889 million yen March, 1990 - 19792 million yen March, 1991 - 23461 million yen
R & D Spending	:	2108 million yen Capital Investment: 1081 million yen
Sales Breakdown	:	see table below

SIC#	Product	Sales
3829	Automotive emissions analyzers	41 %
3826	Scientific analyzers	25 %
384	Medical analyzers	21 %
36	Electronic equipment	13 %

Main Products	:	Measuring Equipment: Automotive emissions analyzers, Air pollution measuring devices, pH meters, Liquid particle counters, Particle size distribution analyzers, Blood electrolyte analyzers; Electronic Equipment: Information processing equipment
Banks	:	Tokai Bank; Bank of Kyoto; Sanwa Bank; Dai-Ichi Kangyo Bank; Mitsui Taiyo Kobe Bank; Mitsubishi Bank; Sumitomo Bank; Daiwa Bank; Bank of Tokyo; Industrial Bank of Japan
Branches	:	Tokyo Branch (Chiyoda-ku, Tokyo); Tokyo Sales Office (Chiyoda-ku, Tokyo); Osaka Sales Office (Yodogawa-ku, Osaka); Nagoya Sales Office (Nagoya-shi, Aichi-ken)
Factories	:	Kyoto Plant (Minami-ku, Kyoto); Kessho Factory (Minami-ku, Kyoto)
R&D Areas	:	Scientific Instrumentation: Fourier transform infra-red spectrophotometer, High performance industrial pH meter
Subsidiaries	:	S.G. Instruments Co., Ltd. (Chiyoda-ku, Tokyo); K.K. Kos (Minami-ku, Kyoto); H.W. Co., Ltd. (Minami-ku, Tokyo)
Overseas Subsidiaries	:	Horiba Korea Ltd. (Republic of Korea); Horiba Instruments Ltd. (England); Horiba Europe GmbH (Germany); Contec Engineering (Italy); Horiba Instruments Inc. (California, U.S.A.)

Hosiden Corp.

Japanese Name :	***Hoshiden K.K.***
TSE Number :	6804
Description :	Manufacturer of consumer electronics parts and components. Hosiden is increasingly emphasizing information and communications equipment and components, with LC display elements figuring prominently. The company maintains several overseas production facilities in Asia and the UK, and has been devoting energy to strengthening its sales networks in Asia, the U.S.A. and Europe.
Address :	1-4-33, Kitakyuhoji, Yao-shi, Osaka 581
Tel & Fax :	Tel: 0729-93-1010 Fax: 0729-94-5101
Statistics :	Capital : 5198 million yen Employees : 1375
Officers :	CEO : Mr. Kenji Furuhashi/President
Sales :	March, 1988 - 47788 million yen March, 1989 - 53412 million yen March, 1990 - 60426 million yen March, 1991 - 72480 million yen

Sales Breakdown :

SIC#	Product	Sales
36	Electronic machine components	63 %
36	Composite parts and other	17 %
3651	Audio parts and components	15 %
36	Electromagnetic parts	5 %

Main Products :	Electronic Parts and Components: Electronic jacks, Electronic sockets, Color LC display elements, Computer keyboards, Audio equipment parts, Video equipment parts, Information processing equipment parts
Banks :	Fuji Bank; Sanwa Bank; Sumitomo Bank; Yasuda Trust and Banking; Mitsui Taiyo Kobe Bank
Branches :	Hong Kong Office (Hong Kong)
Factories :	Taiwan; Republic of Korea; Malaysia; United Kingdom
Laboratories :	R&D Center (Kobe-shi, Hyogo-ken)
Shareholders :	Osaka Securities Finance [8512] - 6.7% ; Fuji Bank [8317] - 5.0% ; Nippon Life - 4.0% ; Nichido Fire and Marine [8760] - 4.0% ; Yasuda Trust and Banking [8404] - 3.9% ; Sumitomo Mutual Life - 3.4%
Subsidiaries :	Hosiden Seiko Corp.; Hosiden Tohoku Corp.; Hosiden Kyushu Corp.; Hosiden Niigata Corp.; Hosiden F.D. Corp.; Hosiden Plastics Corp.; Hosiden Wakayama Corp.; Hosiden Shoji Corp.; Satolex Corp.

Hoya Corp.

Japanese Name	:	***Hoya K.K.***
TSE Number	:	7741
Description	:	Major maker of optical glass and glass for semiconductor manufacturing. The company has a large domestic market share in eyeglasses and crystal glass and has started production of magnetic disks in the U.S. Profits dropped in 1989 resulting from being cited for a violation of the Drugs, Cosmetics and Medical Instruments Law. High value-added products such as semiconductor photomasks and medical instruments will support the rebound of profits in the coming fiscal year.
Address	:	2-7-5, Nakaochiai, Shinjuku-ku, Tokyo 161
Tel & Fax	:	Tel: 03-3952-1151 Fax: 03-3952-1314
Statistics	:	Capital : 5736 million yen Employees : 2973
Officers	:	CEO : Mr. Tetsuo Suzuki/President Contact : Mr. Seiichi Toda/Deputy General Manager, International
Sales	:	March, 1989 (6 months) - 48600 million yen March, 1990 - 103000 million yen March, 1991 - 110000 million yen
R & D Spending	:	4798 million yen Capital Investment: 9928 million yen
Sales Breakdown	:	SIC# Product 3851 Eyecare products 3264 Molded porcelain parts for electronics 3229 Crystal and glassware 3695 Disks and tapes 3674 High-density ICs 3825 Laser oscillators
Main Products	:	Ophthalmic Goods: Eyeglasses, Contacts, Special lenses, Binoculars; Porcelain electrical Supplies: Molded porcelain parts; Pressed and Blown Glass: Crystal and glassware; Magnetic and Optical Recording: Disks and tapes, photosensitive glass; Semiconductors and Related Equipment: High-density ICs, Computer components; Electrical Signal Measuring Equipment: Laser oscillators, Optical isolators, Acousto-optic modulators, Deflectors, Frequency shifters
Banks	:	Sanwa Bank; Saitama Bank; Kyowa Bank
Branches	:	Electric Div., Kansai Sales Office (Kyoto-shi); Medical Div., Osaka Sales Office (Osaka-shi); Glass Div., Sendai Marketing Center (Sendai-shi); 18 other centers; Crystal Div, Nagoya Branch (Nagoya-shi) and 20 other branches
Factories	:	Hachioji Factory (Hachioji-shi); Nagasaka Factory (Kitakuma-gun, Yamanashi-ken); Itsukaichi Factory (Nishitama-gun, Tokyo); Hamura Factory (Nishitama-gun); Musashi Factory (Iruma-shi, Saitama-ken)
Laboratories	:	R&D Center (Akishima-shi); Kodama Research Laboratory (Kodama-gun, Saitama-ken)
Subsidiaries	:	Hoya Credit Co., Ltd. (Shinjuku-ku); Hoya Lens Co., Ltd. (Koga-gun, Shiga-ken); 10 other domestic subsidiaries
Overseas Subsidiaries	:	Hoya Europe B.V. (Netherlands); Hoya Corp. U.S.A. (California, U.S.A.); 21 other overseas subsidiaries

Idemitsu Kosan Co., Ltd.

Japanese Name	:	***Idemitsu Kosan K.K.***
TSE Number	:	Not listed
Description	:	Comprehensive energy supplier engaged primarily in the production and supply of petroleum fuels. In addition to its energy-related activities, Idemitsu Kosan Co. also manufactures fine chemicals and agrochemicals and is active in the biotechnical field where it is working on the development of monoclonal antibodies.
Address	:	3-1-1, Marunouchi, Chiyoda-ku, Tokyo 100
Tel & Fax	:	Tel: 03-3213-9332 Fax: 03-3201-7677
Statistics	:	Capital : 1000 million yen Employees : 5497
Officers	:	CEO : Mr. Shosuke Idemitsu/President
Sales	:	March, 1989 - 1488470 million yen
Sales Breakdown	:	SIC# Product 2911 Petroleum products 2836 Monoclonal antibodies 28 Fine chemicals 287 Agricultural chemicals
Main Products	:	Petroleum Products: Petrol; Biological Products: Monoclonal antibodies; Chemical Products: Fine chemicals, Agricultural chemicals; Advanced Materials: Super engineering thermoplastic resin polyethernitrile 'ID300', Functional thin film, Ultrafine particle titanium dioxide
Laboratories	:	Central Research Laboratory (Kimitsu-shi, Chiba-ken)

Idemitsu Petrochemical Co., Ltd.

Japanese Name	:	***Idemitsu Sekiyu Kagaku K.K.***
TSE Number	:	Not listed
Description	:	Manufacturer of synthetic resins and related products. An affiliate of Idemitsu Kosan Co., the company specializes in the production of special synthetic resin products for use by the automaking and other industries. In addition to synthetic resins, Idemitsu Petrochemical Co. has recently begun production of optical disks, and the company is also active in the area of biotechnology, where it has developed a mass production technique for gamma-linolenic acid.
Address	:	3-1-1, Marunouchi, Chiyoda-ku, Tokyo 100
Tel & Fax	:	Tel: 03-3213-9361 Fax: 03-3214-7337
Statistics	:	Capital : 2000 million yen Employees : 2400
Officers	:	CEO : Mr. Mutsumi Hongo/President
Sales	:	March, 1988 - 245779 million yen March, 1989 - 272285 million yen March, 1990 - 320598 million yen March, 1991 - 361121 million yen
Sales Breakdown	:	SIC# Product 2821 Synthetic resins and plastics 3081 Functional films 3572 Optical disks 2836 Biological products
Main Products	:	Synthetic Resins and Plastics: Polyolefin resins, Engineering plastics for use by the automobile industry, Polyphosphazene resins, Protein plastics, Protein rubber compound and inks, Functional films; Specialty Chemicals: Petroleum-based benzene, Ethylene, Propylene, Slip additive, Styrene monomer, Toluene, Xylene; New Material: Synthesized diamond thin film; Recording media: Optical disks; Biotechnical Products: Gamma-linolenic acid mass produced by fungi culture technique

Ishikawajima Precision Casting

Japanese Name	:	***Ishikawajima Seimitsu Chuzo K.K.***
TSE Number	:	Not listed
Description	:	Manufacturer of precision castings.
Address	:	3-5-1, Mukaidai-cho, Tanashi-shi, Tokyo 188
Tel	:	0424-66-1382
Statistics	:	Capital : 450 million yen Employees : 79
Sales	:	March, 1990 - 4000 million yen March, 1991 - 4010 million yen
Sales Breakdown	:	SIC# Product 3369 Precision castings
Main Products	:	Precision castings
Banks	:	Dai-Ichi Kangyo Bank; Mitsui Taiyo Kobe Bank; Kyowa Saitama Bank
Branches	:	Ishikawajima Harima Heavy Industries Co., Ltd. [7013] - 630,000 Shares; Ishikawajima Shibaura Machine - 270,000 Shares
Factories	:	Nishi-Tama Plant (Nishi-Tama-gun, Tokyo)

Ishikawajima-Harima Heavy Industries Co., Ltd

Japanese Name	:	***Ishikawajima Harima Jukogyo K.K.***
TSE Number	:	7013
Description	:	Manufacturer of industrial machinery, aerospace products, ships and ocean structures.
Address	:	Shin Otemachi Bldg., 2-2-1, Otemachi, Chiyoda-ku, Tokyo 100
Tel & Fax	:	Tel: 03-3244-5111 Fax: 03-3244-5131
Statistics	:	Capital : 64925 million yen Employees : 15280
Officers	:	CEO : Mr. Kosaku Inaba/President
Sales	:	March, 1988 - 714714 million yen March, 1989 - 615489 million yen March, 1990 - 632810 million yen March, 1991 - 731335 million yen

Sales Breakdown :

SIC#	Product	Sales
35	Industrial machinery and plant	67 %
3728	Aerospace	18 %
3731	Shipbuilding and ocean structures	15 %

Main Products	:	Machinery and Plant: Vacuum sintering furnace for metals, Advanced composite processing equipment, Multi-purpose coating system; Aerospace: Aircraft engines, Space-related equipment; Shipbuilding: Cargo ships, Passenger ships, Ocean structures
Banks	:	Dai-Ichi Kangyo Bank; Mitsui Taiyo Kobe Bank; Tokai Bank; Industrial Bank of Japan; Long Term Credit Bank of Japan
Branches	:	Overseas Offices: London Rep. Office (London); Paris Rep. Office (Paris); Kuala Lumpur Rep. Office (Malaysia)

Isuzu Motors Ltd.

Japanese Name	:	***Isuzu Jidosha K.K.***
TSE Number	:	7202
Description	:	Mid-sized motor vehicle manufacturer specializing in light trucks. An affiliate of General Motors Corp. (U.S.), Isuzu Motors is also well-known as a producer of quality diesel engines as well as light trucks. The company is currently losing money due in part to losses experienced by U.S. joint venture. Also affecting profits is a decline in exports, which account for nearly half of the company's gross sales.
Address	:	Bellport A-kan, 6-26-1, Minami-Oi, Shinagawa-ku, Tokyo 140
Tel & Fax	:	Tel: 03-5471-1111 Fax: 03-5471-1036
Statistics	:	Capital : 633470 million yen Employees : 13427
Officers	:	CEO : Mr. Kazuo Tobiyama/President
Sales	:	October, 1988 - 1023300 million yen October, 1989 - 1092190 million yen October, 1990 - 1195870 million yen
Sales Breakdown	:	see below

SIC#	Product	Sales
3713	Light trucks and buses	34 %
3714	Engines, parts and other	28 %
3713	Heavy trucks and buses	23 %
3711	Passenger cars and vans	15 %

Main Products	:	Motor Vehicles: Heavy trucks, Buses, Light trucks, Mini-buses, Passenger vehicles, Vans, Engines, Motor vehicle parts
Banks	:	Dai-Ichi Kangyo Bank; Industrial Bank of Japan; Sanwa Bank; Toyo Trust and Banking; Mitsubishi Trust and Banking; Mitsubishi Bank
Branches	:	Parts Center (Tsurumi-ku, Yokohama-shi, Kanagawa-ken)
Factories	:	Kawasaki Plant (Kawasaki-shi, Kanagawa-ken); Fujisawa Plant (Fujisawa-shi, Kanagawa-ken); Tochigi Plant (Tochigi-ken); Hokkaido Plant (Tomakomai-shi, Hokkaido)
Laboratories	:	Hokkaido Testing Site (Hokkaido); Research Centers (Fujisawa, Los Angeles, Detroit)
Subsidiaries	:	Zexel Co., Ltd. (Tokyo); Prees Kogyo Co., Ltd. (Kawasaki-shi, Kanagawa-ken) and 59 other domestic subsidiaries

Iwamoto Seisakusho Co., Ltd.

TSE Number	:	Not listed
Description	:	Manufacturer of precision machinery.
Address	:	354, Hazukashi Furukawa-cho, Fushimi-ku, Kyoto 612
Tel	:	075-933-4841
Statistics	:	Capital : 8 million yen Employees : 88
Sales	:	October, 1989 - 950 million yen October, 1990 - 950 million yen
Sales Breakdown	:	SIC# Product 3599 Precision machinery
Main Products	:	Precision machinery
Factories	:	Kyoto Plant (Kyoto)
Shareholders	:	Shinsuke Iwamoto

Iwate Iron Co., Ltd.

Japanese Name	:	***Iwate Seitetsu K.K.***
TSE Number	:	Not listed
Description	:	Manufacturer of ferrous metal products. Iwate Iron is also active in the development of composite metal materials for molding and other applications.
Address	:	1-96-2, Tsutsumigaoka, Kitakami-shi, Iwate-ken 024
Tel & Fax	:	Tel: 0197-63-4582 Fax: 0197-63-4582
Statistics	:	Capital : 160 million yen Employees : 220
Officers	:	CEO : Mr. Yasushi Obara/President Contact : Mr. Gyozo Kawahara/Managing Director, R&D Division
Sales	:	March, 1989 - 3700 million yen March, 1990 - 4400 million yen March, 1991 - 5000 million yen
Sales Breakdown	:	SIC# Product 331 Iron and steel products 3399 Composite metal molding materials
Main Products	:	Ferrous Metal Products: Pig iron, Castings; Metal Composite Materials: Nickel polymer molding materials
Banks	:	Shokochukin Bank
Factories	:	Main Factory (Kitakami-shi, Iwate-ken)
Laboratories	:	Laboratory and New Material R&D Center (Kitakami-shi, Iwate-ken)
R&D Areas	:	Metal Products: Composite metal materials, Boriding process surface treatment system, Metal polymer compounds for molding
Shareholders	:	Nippon Steel Corp. [5401] ; Mitsui and Co. [8031]

J. Morita Corp.

TSE Number	:	Not listed
Description	:	Engaged in the development and manufacture of medical equipment and techniques.
Sales Breakdown	:	SIC# Product 3841 Medical instruments
Main Products	:	Medical Instruments: Microwave lithotripsy system, Bone surgery laser systems

Japan Aircraft Manufacturing Co., Ltd.

Japanese Name	:	***Nippon Hikoki K.K.***
TSE Number	:	7407
Description	:	Aircraft manufacturer. An affiliate of Kawasaki Heavy Industries, Japan Aircraft Manufacturing is also engaged in the production of boats and various industrial machines and components.
Address	:	3175, Showa-machi, Kanazawa-ku, Yokohama-shi, Kanagawa-ken 236
Tel & Fax	:	Tel: 045-773-5100 Fax: 045-773-5102
Statistics	:	Capital : 6017 million yen Employees : 1734
Officers	:	CEO : Mr. Teruaki Yamada/President
Sales	:	March, 1988 - 23504 million yen March, 1989 - 24885 million yen March, 1990 - 27098 million yen March, 1991 - 29043 million yen
R & D Spending	:	419 million yen Capital Investment: 2771 million yen
Sales Breakdown	:	

SIC#	Product	Sales
372	Aircraft and parts	56 %
372	Aircraft repairs	29 %
35	Industrial machinery and equipment	3 %
3732	Motor boats	3 %

Main Products	:	Aerospace: Aircraft, Aircraft parts, Aircraft repair; Other Machinery: Boats, Industrial equipment
Banks	:	Dai-Ichi Kangyo Bank; Chuo Trust and Banking; Kyowa Saitama Bank; Bank of Yokohama; Daiwa Bank; Industrial Bank of Japan; Mitsui Taiyo Kobe Bank
Branches	:	Tokyo Office (Chiyoda-ku, Tokyo); Gifu Office (Gifu-ken)
Factories	:	Yokohama Plant (Kanazawa-ku, Yokohama-shi, Kanagawa-ken); Atsugi Plant (Yamato-shi, Kanagawa-ken)
Subsidiaries	:	Nikki Kosan K.K. (Kanazawa-ku, Yokohama-shi, Kanagawa-ken); Nikki Electronic Equipment K.K. (Yokohama-shi, Kanagawa-ken); Nikki Seiki K.K. (Yokohama-shi, Kanagawa-ken)

Japan Metals & Chemicals Co., Ltd.

Japanese Name : ***Nippon Jukagaku Kogyo K.K.***

TSE Number	:	5562
Address	:	8-4, Koami-cho, Nihonbashi, Chuo-ku, Tokyo 103
Tel & Fax	:	Tel: 03-3667-1331 Fax: 03-3668-6902
Statistics	:	Capital : 6800 million yen Employees : 1070
Officers	:	CEO : Mr. Tadaichi Tomioka/President
Sales	:	December, 1988 - 71405 million yen December, 1989 - 78802 million yen December, 1990 - 78258 million yen
R & D Spending	:	654 million yen Capital Investment: 2875 million yen

Sales Breakdown :

SIC#	Product	Sales
3313	Ferroalloys	65 %
3313	Electrometallurgical materials	15 %
28	Chemicals and related products	12 %
4911	Geothermal electricity generation	5 %

Banks	:	Dai-Ichi Kangyo Bank; Industrial Bank of Japan; Long Term Credit Bank of Japan; Hokuriku Bank; Norinchukin Bank
Laboratories	:	Chuo Laboratory (Yamada-gun, Gumma-ken); Tsukuba Laboratory (Tsukuba-shi, Ibaraki-ken)

Japan Organo Co., Ltd.

Japanese Name	:	***Organo K.K.***
TSE Number	:	6368
Description	:	Major supplier of water engineering systems.
Address	:	5-5-16, Hongo, Bunkyo-ku, Tokyo 113
Tel & Fax	:	Tel: 03-5689-7103 Fax: 03-3815-0483
Statistics	:	Capital : 5005 million yen Employees : 887
Officers	:	CEO : Mr. Hirokatsu Maeda/President Contact : Mr. Yoshihiko Aosaka/Manager, Overseas Bus. Development
Sales	:	September, 1988 - 43983 million yen March, 1989 (6 months) - 25986 million yen March, 1990 - 50160 million yen March, 1991 - 64211 million yen

Sales Breakdown :

SIC#	Product	Sales
3559	Customized plant	69 %
28	Resin, carbon and chemicals	17 %
3569	General plant	11 %
0000	Others	3 %

Main Products	:	Water System Engineering: Municipal water supply systems, Municipal sewage treatment systems, Industrial water treatment systems
Banks	:	Fuji Bank; Industrial Bank of Japan
Branches	:	Osaka Branch Office (Osaka-shi); Sapporo Office (Sapporo-shi); Sendai Office (Sendai-shi); Nagoya Office (Nagoya-shi); Hiroshima Office (Hiroshima-shi); Fukuoka Office (Fukuoka-shi); Seoul Office
Factories	:	Kashiwa Plant; Tsukuba Plant; Satte Plant
Laboratories	:	Central Research Laboratory (Toda-shi)
Overseas Subsidiaries	:	Organo (America) Inc. (Georgia, U.S.A.); Organo (Europe) Ltd. (U.K).; Organo (Thailand) Co., Ltd. (Bangkok); Organo (Asia) Sdn. Bhd. (Malaysia)

Japan REC Co., Ltd.

Japanese Name : **Nippon REC K.K.**

TSE Number	:	Not listed
Description	:	Engaged in the development and manufacture of non-metallic synthetic resin-based composite materials for use in printed circuit boards, ICs, LSIs and hybrid ICs.
Address	:	3-5-1, Duo-cho, Takatsuki-shi, Osaka 569
Tel & Fax	:	Tel: 0726-69-5231 Fax: 0726-69-5230
Statistics	:	Capital : 30 million yen Employees : 11
Officers	:	CEO : Mr. Hiroshi Hirano/President Contact : Mr. Atsushi Okuno/Engineering Director
Sales	:	March, 1988 - 70 million yen March, 1989 - 80 million yen March, 1990 - 100 million yen March, 1991 - 120 million yen

Sales Breakdown :

SIC#	Product	Sales
2821	Synthetic resin composite materials	70 %
3699	Non-metal electrical materials	30 %

Main Products	:	Synthetic Resin Materials: 'TI-01' epoxy resin impregnated aramide fiber PCB (printed circuit board) material, Epoxy resin-based corrosion-resistant solder resist, COB epoxy resins
Banks	:	Kyowa Saitama Bank
Branches	:	Tokyo Branch (Kyobashi-ku, Tokyo)
Factories	:	Takatsuki Factory (Takastuki-shi, Osaka)
Laboratories	:	Research Center (Takatsuki-shi, Osaka)
Shareholders	:	Sanyu Kogyo K.K. - 100%
Subsidiaries	:	Sanyu Resin K.K.; Sanyu Giko K.K.
Overseas Subsidiaries	:	Sanyu U.S.A. (Texas, U.S.A.)

Japan Steel Works, Ltd.

Japanese Name	:	***Nihon Seikosho K.K.***
TSE Number	:	5631
Address	:	Hibiya-Mitsui Bldg., 1-1-2, Yuraku-cho, Chiyoda-ku, Tokyo 100
Tel	:	03-3501-6111
Statistics	:	Capital : 19694 million yen Employees : 3964
Officers	:	CEO : Mr. Naohiko Yagi/President
Sales	:	March, 1988 - 102008 million yen March, 1989 - 108691 million yen March, 1990 - 118744 million yen March, 1991 - 133275 million yen
R & D Spending	:	4104 million yen Capital Investment: 18933 million yen
Sales Breakdown	:	

SIC#	Product	Sales
3559	Plastic processing machinery	37 %
3542	Molding machines and compressors	32 %
3462	Generator components and engines	16 %
3577	Circuit board machines	8 %
3443	Pressure vessels and reactors	7 %

Banks	:	Mitsui Taiyo Kobe Bank; Takugin; Mitsui Trust and Banking; Sumitomo Trust and Banking; Bank of Yokohama
Branches	:	Osaka Branch; Fukuoka Branch; Nagoya Branch; Hiroshima Branch; Sapporo Branch; Niigata Branch; Sendai Branch; Overseas Branches: New York, U.S.A.; California, U.S.A.; Dusseldorf, Germany; Singapore; Teheran, Iran; Peking, China
Factories	:	Muroran Plant (Hokkaido); Hiroshima Plant; Yokohama Plant; Kashima Plant (Ibaraki)
Laboratories	:	Research Center for Advanced Technologies (Chiba-ken)
Overseas Subsidiaries	:	Japan Steel Works America, Inc. (New York, Los Angeles, Houston); TSW (Pte) Ltd. (Singapore)

Japan Synthetic Rubber Co., Ltd.

Japanese Name	:	***Nihon Gosei Gomu K.K.***
TSE Number	:	4185
Description	:	Among the top producers of ABS resins and synthetic rubber. Currently diversifying into the biosynthetics field and functional materials.
Address	:	2-11-24, Tsukiji, Chuo-ku, Tokyo 104
Tel & Fax	:	Tel: 03-5565-6500 Fax: 03-5565-6630
Statistics	:	Capital : 23143 million yen Employees : 2899
Officers	:	CEO : Mr. Tatsuo Asakura/President
Sales	:	March, 1989 - 192495 million yen March, 1990 - 194298 million yen March, 1991 - 203409 million yen
Sales Breakdown	:	SIC# Product 2822 SBR, BR, IR, NBR, PDM 2821 ABS, AS AES plastics
Main Products	:	Synthetic Rubbers: SBR, BR, IR, NBR, PDM; Emulsions: Paper coating latex (PCL), Acrylic emulsion; Plastics: ABS, AS, AES; Specialty Chemicals: Formic acid, Surface coating agent
Banks	:	Industrial Bank of Japan
Branches	:	Osaka Branch (Chuo-ku, Osaka); Nagoya Branch (Naka-ku, Nagoya); Hiroshima Branch (Naka-ku, Hiroshima); London Branch (Middlesex, UK)
Factories	:	Yokkaichi Plant (Yokkaichi-shi); Chiba Plant (Ichihara-shi); Kashima Plant (Kashima-gun)
Laboratories	:	Electronics Research Laboratory (Kawasaki-shi); Research Center (Tsukuba-shi); Development Center (Yokkaichi-shi); Technical Center (Yokkaichi-shi)
Subsidiaries	:	Japan Elastomix Co., Ltd (Tokyo); Japan Latex Compounding Co., Ltd. (Yokkaichi-shi); Japan Coloring Co., Ltd. (Yokkaichi-shi); Japan Optical Disk Corp (Kohfu-shi); Nichigo Shoji Co., Ltd. (Tokyo)
Overseas Subsidiaries	:	Korea Latex Compounding Co., Ltd. (Korea); Kumho E.P. Rubber Co., Ltd. (Korea); UCB-JSR Electronics S.A. (Belgium)

JEOL Ltd.

Japanese Name	:	***Nihon Denshi K.K.***
TSE Number	:	6951
Description	:	Major manufacturer of optoelectronic products. A member of the Mitsubishi group of companies, JEOL is noted for its technological expertise in the field of optoelectronics and is Japan's leading producer of electron microscopes. 1991 R&D expenditures approached 10% of gross sales and were focused primarily upon 8 inch wafer semiconductor testing systems.
Address	:	3-1-2, Musashino, Akishima-shi, Tokyo 196
Tel & Fax	:	Tel: 0425-43-1111 Fax: 0425-46-3353
Statistics	:	Capital : 3240 million yen Employees : 1509
Officers	:	CEO : Mr. Takashi Takeuchi/President Contact : Mr. Shuichi Murakami/General Manager, Overseas Development
Sales	:	March, 1988 - 44037 million yen March, 1989 - 48022 million yen March, 1990 - 50569 million yen March, 1991 - 47860 million yen
R & D Spending	:	4506 million yen Capital Investment: 1282 million yen

Sales Breakdown :

SIC#	Product	Sales
3829	Electron optical devices	51 %
3826	Analytical instruments	29 %
35	Industrial equipment	13 %
3845	Electromedical equipment	8 %

Main Products	:	Analytical and Control Instruments: Electron microscopes, Scanning microscopes, Electron probe X-ray microanalyzers, FT NMR spectrometers and mass spectrometers; Optical Instruments: Electron guns/power supplies; Electromedical Equipment: Fully-automated amino acid analyzers, Other automated biochemical analyzers; Industrial Equipment: Semiconductor testing equipment, Semiconductor production equipment, Electron beam lithography systems, Other electron beam systems, Ion beam lithography systems, Other ion beam equipment, Ion plating equipment, Plasma generating equipment, Transistor inverters
Banks	:	Mitsubishi Bank; Nippon Credit Bank; Bank of Tokyo.
Factories	:	Akishima Factory (Akishima-shi, Tokyo)
Laboratories	:	Kansai Application Research Center
R&D Areas	:	Optoelectronics; Analytical Equipment: Digital analyzers, NMR technologies; Industrial Machinery: Medical Equipment: Semiconductor Testing Equipment: 8 inch wafer testing systems

JGC Corp.

Japanese Name	:	***Nikki K.K.***
TSE Number	:	1963
Description	:	Leading plant engineering firm enjoying a worldwide reputation in oil refining, petrochemical and LPG plants. The company is moving into the nuclear fuel reprocessing field.
Address	:	2-2-1, Otemachi, Chiyoda-ku, Toyko 100
Tel & Fax	:	Tel: 03-3279-5441 Fax: 03-3273-8047
Statistics	:	Capital : 7140 million yen Employees : 2344
Officers	:	CEO : Mr. Eiji Watanabe/President
Sales	:	March, 1989 - 133898 million yen March, 1990 - 185863 million yen March, 1991 - 278983 million yen
R & D Spending	:	2224 million yen Capital Investment: 3652 million yen
Sales Breakdown	:	SIC# Product 2911 Petrochemicals 1541 Chemical plants 351 Gas-related equipment
Main Products	:	Petroleum Refining: Petroleum, Chemicals, Gas treatment; Plant Construction and Engineering: General manufacturing facilities, Nuclear power plants
Banks	:	Mitsui Taiyo Kobe Bank; Sanwa Bank; Fuji Bank; Sumitomo Bank; Dai-Ichi Kangyo Bank
Branches	:	London Office (London); Singapore Office (Singapore); Nuclear Rep. Office (Washington D.C.)

JSP Corp.

Japanese Name	:	***JSP K.K.***
TSE Number	:	7942
Description	:	Manufacturer of foamed plastic products. A subsidiary of Mitsubishi Gas Chemical Co. Inc., JSP produces a wide range of products including food containers, cushioning materials, agricultural materials, display materials, architectural insulating materials, automobile parts and sports equipment. The company is involved in overseas production in the U.S. and plans to set up production facilities in S.E. Asia and Europe. In fiscal 1991 fixed capital investment stood at over 20% of gross sales, primarily as a result of expenditures necessary for the completion of the company's new Yokkaichi plant. R&D continues to stress mainline plastic foam products.
Address	:	2-1-1, Uchisaiwai-cho, Chiyoda-ku, Tokyo 100
Tel & Fax	:	Tel: 03-3503-6988 Fax: 03-3581-2476
Statistics	:	Capital : 6513 million yen Employees : 662
Officers	:	CEO : Mr. Masayo Uchiyama/President Contact : Mr. Masayo Uchiyama/President
Sales	:	March, 1988 - 27749 million yen March, 1989 - 31414 million yen March, 1990 - 31638 million yen March, 1991 - 34617 million yen
R & D Spending	:	1125 million yen Capital Investment: 7954 million yen
Sales Breakdown	:	

SIC#	Product	Sales
3086	Polystyrene foam sheets	45 %
3086	Polyolefin foams	25 %
3086	Polystyrene foam boards	21 %

Main Products	:	Foamed Plastic: Foamed polystyrene sheets and boards, Foamed polyolefin materials; Foamed Plastic Products: Shock-proof packaging materials, Food containers, Cushioning materials, Materials for use by the automobile industry, Building insulation materials, Housing interior materials, Commercial display materials, Materials for sporting goods, etc.
Banks	:	Long Term Credit Bank of Japan; Sumitomo Bank; Mitsubishi Trust & Banking Corp.; Mitsubishi Bank; Yokohama Bank; Ashikaga Bank; Tokyo Sowa Bank; Norinchukin Bank
Laboratories	:	Kanuma Research Center (Kanuma-shi, Tochigi-ken)
R&D Areas	:	Polystyrene Foam Products: Plystyrene foam sheets and boards
Subsidiaries	:	Nihon Xanpak K.K. (Minato-ku, Tokyo); Seihoku Package K.K. (Kashiwa-shi, Chiba-ken)
Overseas Subsidiaries	:	JSP America Inc. (New York, U.S.A); J&V Foam Products Inc. (California, U.S.A)

Kajima Corp.

Japanese Name	:	***Kajima Kensetsu K.K.***
TSE Number	:	1812
Description	:	Kajima Corp. is one of Japan's major general contractors and is rated number three worldwide, in terms of construction order bookings. The company has developed software applications for a variety of structures including clean rooms, warehouses, skyscrapers and domes. Kajima has completed 146 dam construction projects and has achieved success with its large-calibre shield tunnel technology, which was applied in the construction of the Teito Rapid Transit Subway System. Kajima is developing its vibration-control technology and incorporated this technology in the AMD, recently completed in Tokyo. The company has diversified into real estate, hotel management, health foods and is utilizing R&D efforts to automate the building process, computerize building maintenance and introduce labor-saving techniques.
Address	:	1-2-7, Motoakasaka, Minato-ku, Tokyo 107
Tel & Fax	:	Tel: 03-3404-3311 Fax: 03-3470-1444
Statistics	:	Capital : 621480 million yen Employees : 12850
Officers	:	CEO : Mr. Akira Miyazaki/President Contact : Mr. Shoichi Ohori/Managing Director
Sales	:	March, 1988 (6 months) - 440623 million yen March, 1989 - 1241370 million yen March, 1990 - 1419800 million yen March, 1991 - 1701660 million yen

Sales Breakdown :

SIC#	Product	Sales
1522	Dormitories, hotel construction	40 %
1622	Civil engineering	20 %
1542	Domes, commercial construction	20 %
1541	Warehouses and clean rooms	10 %
6552	Real estate development	9 %
8712	Architectural design and engineering	1 %

Main Products	:	Residential Construction: Company dormitories, Condominiums, Hotel construction; Civil Engineering Projects: Bridges, Tunnels and Others; Non-residential Construction: Domes (Parasol Dome), Special structures; Commercial Construction; Industrial Buildings: Various types of warehousing structures, Clean rooms; Real Estate Development: Subdividing, Developing; Architectural Services: Design and Engineering
Banks	:	Sumitomo Bank; Mitsui Taiyo Kobe Bank; Kyowa Saitama Bank; Mitsui Trust and Banking; Sumitomo Trust and Banking
Laboratories	:	Kajima Technical Research Institute (Chofu-shi, Tokyo)

Kanebo, Ltd.

Japanese Name	:	***Kanebo K.K.***
TSE Number	:	3102
Description	:	Major spinning firm and the second largest cosmetics producer in Japan. Currently diversifying into a broad range of products including foods, advanced plastics and computer software. R&D expenditures likewise cover a number of different areas with plastics and synthetic materials figuring prominently.
Address	:	1-5-90, Tomobuchi-cho, Miyakojima-ku, Osaka 534
Tel & Fax	:	Tel: 06-922-8125 Fax: 06-922-8128
Statistics	:	Capital : 29762 million yen Employees : 9918
Officers	:	CEO : Mr. Junji Itoh/Chairman Contact : Mr. Nobuo Takehara/Executive Manager, International
Sales	:	March, 1988 (Adjusted term) - 381819 million yen March, 1989 - 481558 million yen March, 1990 - 501950 million yen March, 1991 - 527771 million yen
ending	:	11400 million yen Capital Investment: 14000 million yen

Sales Breakdown :

SIC#	Product	Sales
2844	Cosmetics	28 %
2221	Synthetic fibers	18 %
23	Clothing	17 %
2211	Natural fibers	16 %
20	Foods	9 %
3299	New materials	5 %
2834	Pharmaceuticals	4 %
737	Computer software	3 %

Main Products	:	Natural Fibers: Cotton, Wool, Silk; Synthetic Fibers: Nylon, Polyester, Acrylics; Fashion goods: Men's and women's apparel, Artificial leather, Artificial chamois, Stockings; Cosmetics and Toiletries; Foods: Chewing gum, Noodles, Canned beverages, Candy, Ice cream; Synthetic Resins: PET resins, 'Agri-Sheet' plastic films, Various resins associated with synthetic fiber production; Other Materials: Abrasive materials for use in grinding wheels; Pharmaceuticals; Computer Software
Banks	:	Mitsui Taiyo Kobe Bank; Sumitomo Trust and Banking; Fuji Bank; Tokai Bank
Branches	:	New York Office (U.S.A); Zurich Office (Switzerland); Paris Office (France); London Office (UK); Honk Kong Office (HK)
Factories	:	Nagahama Plant; Hokuriku Plant; Suzuka Plant (Suzuka-shi); Ogaki Plant (Ogaki-shi); Tottori Plant (Tohaku-cho); Odawara Plant (Odawara-shi); Gosei Hikaku Plant (Hofu-shi, Yamaguchi-ken)
Laboratories	:	Osaka Research Center (Miyakojima-ku, Osaka)
R&D Areas	:	Electronics: Biotechnology: New Materials: Organic Chemicals: High polymers, PET resins, Large-scale PC separators

Kanebo-NSC Ltd.

TSE Number	:	Not listed
Description	:	Manufacturer of special adhesives. Kanebo-NSC is a joint venture of National Starch and Chemical Corp. and Kanebo, Ltd..
Address	:	Kanebo Senri Bldg., 1-6-5, Nishi-Senba, Mono-shi, Osaka 562
Tel & Fax	:	Tel: 0727-28-4701 Fax: 0727-28-3333
Officers	:	CEO : Mr. Kenji Takahashi/President Contact : Mr. Kenji Takahashi/President
Sales Breakdown	:	SIC# Product 2891 Adhesives
Main Products	:	Adhesives: Heat-resistant adhesives, Hot melt adhesives, Acrylic emulsion adhesives
Branches	:	Tokyo Sales Office (Chiyoda-ku, Tokyo)
Shareholders	:	National Starch & Chemical Corp. ; Kanebo, Ltd. [3102]

Kaneka Corp

Japanese Name	:	***Kanegafuchi Kagaku Kogyo K.K.***
TSE Number	:	4118
Description	:	Manufacturer of chemical products, foodstuffs and electronics materials with emphasis on PVC resins and soda. Kanegafuchi Chemical Industry is widely seen as a leader in PVC resin and other high polymer technologies, and the company produces a wide range of advanced resin products as well as various other chemical products including pharmaceuticals, an area in which sales have been exhibiting double digit growth in recent years. The company continues to emphasize PVC resins and has earmarked additional funds for the improvement of PVC production facilities.
Address	:	Asahi Shimbun Bldg. 9F, 3-2-4, Nakanoshima, Kita-ku, Osaka 530
Tel & Fax	:	Tel: 06-226-5050 Fax: 06-226-5177
Statistics	:	Capital : 29750 million yen Employees : 3594
Officers	:	CEO : Mr. Tadasu Tachi/President
Sales	:	March, 1988 - 183544 million yen March, 1989 - 198629 million yen March, 1990 - 210505 million yen March, 1991 - 232578 million yen
R & D Spending	:	11500 million yen Capital Investment: 25800 million yen
Sales Breakdown	:	

SIC#	Product	Sales
2821	PVCs and soda	21 %
2821	Special synthetic resins	21 %
2821	Synthetic resin products	19 %
2833	Pharmaceuticals and foodstuffs	18 %
3699	Electronics materials	12 %
2824	Synthetic fibers	9 %

Main Products	:	Synthetic Resins: PVC resin, PVC modifier, PVC compound resins, Solution type PVC, Heat resistant ABC family resins, Flame retardant ABC resins, Modified PET resins for injection molding, Liquid polymer for moisture curing system sealants; Synthetic Resin Products: Polystyrene foams and papers, Polyethylene foam, Weather resistant acrylic films, Polymide film, PVC foamed board; Industrial Chemicals: Caustic soda, Sodium sulfite, Hydrochloric acid, Acetone, Stearic acid, Gelling agent, Vinyl chloride monomer; Foodstuffs: Shortening, Magarine, Other edible oils and fats; Electronics Materials: Magnetic wires, Copper clad laminates, Amorphous silicon carbide solar cells; Pharmaceuticals and Related Products: Glutathione, Ubidecarenone (CoQ10), Ergostrol, Plasma separators, LDL absorption column, Plasmapherhesis unit
Banks	:	Mitsui Taiyo Kobe Bank; Tokai Bank; Daiwa Bank; Nippon Credit Bank
Branches	:	Tokyo Branch (Minato-ku, Tokyo); Nagoya Sales Office (Nagoya, Aichi-ken); Hokkaido Sales Office (Hokkaido); Kyushu Sales Office (Fukuoka-ken); Overseas Offices: Brussels Office (Belgium)
Factories	:	Takasago Plant (Takasago-shi, Hyogo-ken); Osaka Plant (Setsutsu-shi, Osaka); Shiga Plant (Otsu-shi, Shiga-ken); Kashima Plant (Kashima-gun, Ibaraki-ken)

Kankyo Sozai Kenkyusho K.K.

Japanese Name	:	***Kankyō Sozai Kenkyusho K.K.***
TSE Number	:	Not listed
Description	:	Manufacturer of solvents and organic composite materials for removing organic compounds and other pollution control functions.
Address	:	2-102-7, Yagoto-cho, Kasugai-shi, Aichi 486
Tel	:	0568-82-9798
Sales Breakdown	:	SIC# Product 2899 Pollution control materials 2869 Organic solvents
Main Products	:	Pollution Control Materials: Organic solvents, Resin coated carbon material for removing organic compounds

Kansai Electric Power Co., Inc.

Japanese Name	:	***Kansai Denryoku K.K.***
TSE Number	:	9503
Description	:	Japan's second largest electric utility. Supplying the Kansai area (Osaka, Kyoto and Kobe), Kansai Electric Power Co. is widely recognized and respected as a technological and managerial innovator within the power industry. The company has developed a diversified power generation base and is currently stressing hydroelectric generation through dam construction.
Address	:	3-3-22, Nakonoshima, Kita-ku, Osaka 530
Tel & Fax	:	Tel: 06-441-8821 Fax: 06-443-0233
Statistics	:	Employees : 24737
Officers	:	CEO : Mr. Yoshihisa Akiyama/President
Sales	:	March, 1988 - 2012290 million yen March, 1989 - 2019200 million yen March, 1990 - 2075300 million yen March, 1991 - 2245010 million yen

Sales Breakdown :

SIC#	Product	Sales
4911	Commercial and industrial use	65 %
4911	Residential use	35 %

Main Products	:	Electric Generation: Petroleum, Nuclear, Hydroelectric, LNG; Electricity Supply Services: Residential, Commercial, Industrial
Banks	:	Industrial Bank of Japan; Sumitomo Bank; Sanwa Bank
Laboratories	:	Research Laboratory (Amagazaki-shi, Hyogo-ken)
Shareholders	:	Osaka City - 8.5% ; Nippon Life - 4.9% ; Kobe City - 2.7%

Kansai Paint Co., Ltd.

Japanese Name	:	***Kansai Peinto K.K.***
TSE Number	:	4613
Description	:	Engaged in the production of paints, thinners and related products.
Address	:	4-3-6, Fushimi-machi, Chuo-ku, Osaka 541
Tel & Fax	:	Tel: 06-203-5531 Fax: 06-203-5018
Statistics	:	Capital : 25658 million yen Employees : 3048
Officers	:	CEO : Mr. Masuo Tanaka/President Contact : Mr. Yuzo Yamauchi/General Manager, Overseas Dept
Sales	:	March, 1988 - 135214 million yen March, 1989 - 143638 million yen March, 1990 - 151760 million yen March, 1991 - 166521 million yen
R & D Spending	:	4108 million yen Capital Investment: 8554 million yen

Sales Breakdown :

SIC#	Product	Sales
2851	Synthetic resin-based paints	82 %
3081	Marking film, others	10 %
2851	Thinners	6 %
2851	Lacquers	1 %
2851	Oil paints	

Main Products	:	Paint Products: Automotive paints, Construction paints and coatings, Metal paints, Marine paints and coatings, Anticorrosive coating, Epoxy resin enamels and varnishes; Related Products: Marking films, Electro-deposition ultraviolet resists for PCBs
Banks	:	Sanwa Bank; Dai-Ichi Kangyo Bank; Fuji Bank
Factories	:	Amagasaki Plant (Amagasaki-shi); Nagoya Plant (Nishikamo-gun); Tokyo Plant (Ota-ku); Hiratsuka Plant (Hiratsuka-shi); Kanuma Plant (Kanuma-shi)
Laboratories	:	Laboratory (Hiratsuka-shi)
Subsidiaries	:	Kansai Paint Marine Coatings Co., Ltd. (Tokyo, Osaka, and other locations); KP Trading Co., Ltd. (Tokyo, Osaka, Amagasaki, Hiratsuka); KP Wing Co., Ltd. (Tokyo, Osaka); KP Finance Ltd. (Osaka); Kanape Ad. (Tokyo, Osaka)

Kao Corp.

Japanese Name	:	***Kao K.K.***
TSE Number	:	4452
Description	:	Kao was originally a soap producer, but has invested heavily in R&D and marketing and now produces detergents, shampoos, cosmetics and is moving into plastics, agrichemicals and information equipment. The company registered an impressive sales growth rate over the last five years and is advancing abroad with production plants in Thailand and Taiwan. The company has over eight laboratories in Japan and opened up its own training institute, the Kasumigaura Creativity Center, in 1989.
Address	:	1-14-10, Nihonbashi, Kayaba-cho, Chuo-ku, Tokyo 103
Tel & Fax	:	Tel: 03-3660-7111 Fax: 03-3660-7044
Statistics	:	Capital : 61570 million yen Employees : 6768
Officers	:	CEO : Mr. Fumikatsu Tokiwa/President
Sales	:	March, 1988 - 490019 million yen March, 1989 - 521347 million yen March, 1990 - 545432 million yen March, 1991 - 570161 million yen
R & D Spending	:	28200 million yen Capital Investment: 61200 million yen

Sales Breakdown :

SIC#	Product	Sales
2841	Detergents	40 %
2844	Shampoos, bath agents	20 %
2676	Disposable diapers	10 %
2843	Sulfonated oils	5 %
2821	Polyurethanes	5 %
2869	Plasticizers	5 %
2899	Fatty acids	5 %
3652	Floppy disks	5 %
2079	Cooking oil	5 %

Main Products	:	Soap and Detergents: Household detergents, Bleaches, 'Moya' and 'Famiri' brands; Toilet Preparations: 'Essential' shampoos, Rinses, Scrubs, 'Sofina' line of cosmetics, 'Biore' body cleanser; Sanitary Paper Products: Disposable diapers; Surface Active Agents: Sulfonated oils and greases; Plastics Materials and Resins: Polyurethanes, Polyacrylates, Polyesters; Industrial Organic Chemicals: Super plasticizers, Fatty acids, Oleic acid, Lauric acid, Myristic acid, Palmitic acid, Stearic acid, Gel agent, Higher alcohols for the reduction of coconut oil, Polyoxyethylene alkyl ether, Slip additives, Alkyl chlorides, Glycerin, Antifoaming agents, Antistatic agents for fibers and plastics/rubbers, Bactericides/preservatives for foods and paints, Dispersion agents for pulp and paper/pesticides/photography, Emulsifiers for foodstuffs/oil and tar/asphalt/wax/photography; Magnetic Recording Media: Floppy disks; Edible Fats and Oils: 'Econa' cooking oils
Banks	:	Mitsui Taiyo Kobe Bank; Kyowa Saitama Bank; Sanwa Bank; Sumitomo Bank; Industrial Bank of Japan
Branches	:	Osaka Branch (Osaka-shi)
Factories	:	Wakayama Plant (Wakayama-shi); Tokyo Plant; Kawasaki Plant (Kanagawa); Sakata Plant (Yamagata-shi); Kyushu Plant (Fukuoka-shi); Tochigi Plant (Tochigi-shi); Kashima Plant (Ibaraki-shi); Toyohashi Plant (Toyohashi-shi)
Laboratories	:	Wakayama Laboratory (Wakayama-ken); Tokyo Laboratory (Tokyo); Tochigi Laboratory (Tochigi-ken); Kashima Laboratory (Saga-ken)

Kawasaki Heavy Industries, Ltd.

Japanese Name :	***Kawasaki Jukogyo K.K.***
TSE Number :	7012
Description :	The company originally focused its efforts on shipbuilding, but its gradual diversification into the heavy machinery industry has led it to become a comprehensive heavy machinery and engine maker, maintaining a leading position in the aerospace, shipbuilding and heavy vehicle production industries. Recent joint ventures include projects with Harland and Wolff, the UK's largest shipbuilder, for the production of ship engines, and BREL, a British Rail spinoff, to provide 1,400 subway car chassis for the London Underground, and negotiations are underway with Rolls Royce and British Aerospace for non-defense-related business projects.
Address :	2-1-18, Nakamachi-dori, Chuo-ku, Kobe-shi, Hyogo-ken 650
Tel & Fax :	Tel: 078-371-9530 Fax: 078-371-9568
Statistics :	Capital : 66552 million yen Employees : 16600
Officers :	CEO : Mr. Hiroshi Oba/President
Sales :	March, 1988 - 579731 million yen March, 1989 - 767068 million yen March, 1990 - 812812 million yen March, 1991 - 891624 million yen
R & D Spending :	16200 million yen Capital Investment: 35100 million yen

Sales Breakdown :

SIC#	Product	Sales
372	Aerospace	26 %
35	Machinery	23 %
3751	Consumer products and components	18 %
1629	Plant and steel structure engineering	18 %
3731	Shipbuilding	9 %
374	Rolling stock	7 %

Main Products : Shipbuilding: Ships, Small boats, Naval vessels, Jet foils and other high speed craft, Ocean development equipment, Maritime parts and equipment; Carriages: Rail cars, Monorails, Cable cars, Ropeway gondolas, Self-propelled cranes; Aerospace: Aircraft, Missiles, Electronic equipment, Space-related equipment-machinery and components, Jet engines, Gas turbine engines; Industrial Machinery: Crushers, Pulverizers, Cast products, Precision cast products, Solid refuse processing facilities, Industrial waste treatment facilities, Industrial boilers, Nuclear power generation equipment, Power generators, Municipal garbage incinerators, Ash treatment facilities, High pressure containers, Hydraulic equipment and machinery, Steam turbines, Gas turbines, Diesel engines, Ventilation equipment, Gas compression modules, Variable pitch propellors, Side thrusters, Gas turbine generators, Road sealing equipment, Mining equipment; Production Equipment: Steelmaking plant, Cement plant, Sugar plant, Water plant, Textile plant, Chemical plant, Food processing plant, Construction machinery, Water treatment plant, Industrial waste treatment plant, Automatic metal ore processing equipment, Laser processing equipment, Industrial robots; Steel Structures and Products: Bridges, Reservoir tanks, Water pipes, Flood gates, Airport facilities, Rocket launching facilities, Steel pipe structures; Consumer Products: Motorcycles, Water craft, Air cooled gasoline engines, Automotive transmissions

Banks	:	Dai-Ichi Kangyo Bank; Mitsui Taiyo Kobe Bank; Kyowa Saitama Bank; Daiwa Bank; Sumitomo Bank; Sanwa Bank; Bank of Tokyo; Industrial Bank of Japan; Long Term Credit Bank of Japan; Norinchukin Bank
Laboratories	:	Akashi Technical Laboratories (Akashi-shi); Gifu Technical Laboratories (Kakumuhara-shi, Gifu-ken)
R&D Areas	:	Iodine Laser Technology; Mid-sized High Efficiency Gas Turbine; Coal Partial Combustion Technology
Shareholders	:	Mitsui Trust and Banking [8401] - 5.39% ; Nippon Life - 4.56% ; Dai-Ichi Kangyo Bank [8311] - 4.54% ; Tokio Marine and Fire [8751] - 3.84% ; Kawasaki Steel [5403] - 2.63%

Kawatetsu Magnecs Co., Ltd.

TSE Number	:	Not listed
Description	:	Manufacturer of solvents and organic composite materials for removing organic compounds and other pollution control functions.
Address	:	1-5, Mihara-dai, Chonan-machi, Chosei-gun, Chiba 297-01
Tel	:	0475-46-2211
		2821

Kinugawa Rubber Industrial Co., Ltd.

Japanese Name	:	***Kinugawa Gomu Kogyo K.K.***
TSE Number	:	5196
Address	:	330, Naganuma-cho, Chiba, Chiba-ken 281
Tel & Fax	:	Tel: 0472-59-3111 Fax: 0472-50-1829
Statistics	:	Capital : 5307 million yen Employees : 2898
Officers	:	CEO : Mr. Takashi Matsuura/President Contact : Mr. Shin Shimizu/Manager, Sales Dept.
Sales	:	March, 1988 - 53625 million yen March, 1989 - 55224 million yen March, 1990 - 65605 million yen March, 1991 - 74102 million yen
R & D Spending	:	2015 million yen Capital Investment: 7200 million yen
Sales Breakdown	:	see table below

SIC#	Product	Sales
3714	Rubber weather strips	35 %
3714	Rubber vibration insulators	27 %
3714	Rubber interior parts and fittings	13 %
3714	Rubber hoses	8 %
3714	Rubber brake parts	6 %

Banks	:	Industrial Bank of Japan; Fuji Bank; Daiwa Bank; Chiba Bank; Yokohama Bank
Branches	:	Tokyo Branch (Chuo-ku); Nagoya Branch (Chikusa-ku); Fujisawa Billing Center (Fujisawa, Kanagawa-ken)
Factories	:	Chiba Plant (Chiba-shi); Mooka Plant (Tochigi-ken); Kyushu Plant (Nakatsu-shi)
Laboratories	:	Technology Development Center (Chiba-shi)
R&D Areas	:	Automobile Parts: Rubber automotive parts and components
Subsidiaries	:	Keishi Butsuryu K.K. (Chiba-shi, Chiba-ken); Kinugawa Kyoei K.K. (Chiba-shi, Chiba-ken)

Koa Oil Co., Ltd.

Japanese Name :	***Koa Sekiyu K.K.***
TSE Number :	5006
Description :	Oil refiner. Koa Oil is a member of the Caltex group of companies.
Address :	Nippon Bldg. 12F, 2-6-2, Otemachi, Chiyoda-ku, Tokyo
Tel & Fax :	Tel: 03-3241-8611 Fax: 03-3270-6928
Statistics :	Capital : 7260 million yen Employees : 1086
Officers :	CEO : Mr. Teruo Noguchi/President
Sales :	March, 1988 - 183335 million yen March, 1989 - 165176 million yen March, 1990 - 218867 million yen March, 1991 - 292655 million yen
R & D Spending :	1263 million yen Capital Investment: 27400 million yen

Sales Breakdown :

SIC#	Product	Sales
2911	Fuel oil	30 %
2911	Gasoline	25 %
2911	Kerosene	25 %

Main Products :	Petroleum Products: Gasolene, Fuel oils, Kerosene, Naptha, LPG
Banks :	Mitsui Taiyo Kobe Bank; Kyowa Bank; Mitsubishi Bank; Daiwa Bank; Sumitomo Trust and Banking; Japan Development Bank; Mitsubishi Trust and Banking Corp.; Mitsui Trust and Banking Company; Bank of Tokyo
Laboratories :	Osaka Laboratory (Takaishi-shi)
Shareholders :	Caltex Petroleum - 50.0% ; Nippon Oil [5001] - 6.9%

Kobe Steel, Ltd.

Japanese Name :	***Kobe Seikosho K.K.***
TSE Number :	5406
Description :	Among Japan's leading steelmakers, the company is actively developing its biotechnology division. The company manufactures rolled aluminum and rolled copper products.
Address :	1-3-18, Wakihama-cho, Chuo-ku, Kobe-shi, Hyogo-ken 651
Tel & Fax :	Tel: 078-261-5111 Fax: 078-261-4123
Statistics :	Capital : 212709 million yen Employees : 21303
Officers :	CEO : Mr. Sokichi Kametaka/President
Sales :	March, 1988 - 975932 million yen March, 1989 - 1176780 million yen March, 1990 - 1221780 million yen March, 1991 - 1321410 million yen
R & D Spending :	39700 million yen Capital Investment: 125000 million yen
Sales Breakdown :	SIC# Product 331 Iron and steel 3356 Welding rods 3369 Non-ferrous metals
Main Products :	Iron and Steel Products: Iron, Steel, Welding rods; Non-ferrous Metal Products: Rolled aluminium products, Rolled copper products; Machinery and Engineering; Information Electronics Products; Advanced Functional Materials; Biotechnology Products
Banks :	Dai-Ichi Kangyo Bank; Sanwa Bank; Mitsui Taiyo Kobe Bank; Industrial Bank of Japan; Yasuda Trust and Banking
Branches :	Tokyo Head Office (Chiyoda-ku); Osaka Branch (Osaka-shi); Nagoya Branch (Nagoya-shi); Hokkaido Office (Sapporo-shi); Shikoku Office (Takamatsu-shi); Kyushu Office (Fukuoka-shi)
Factories :	Kobe Works (Kobe-shi); Kakogawa Works (Kakogawa-shi, Hyogo-ken); Fujisawa Factory (Fujisawa-shi); Fukuchiyama Factory (Fukuchiyama-shi, Kyoto)
Laboratories :	Material Research Laboratories (Kobe-shi); Machinery Research Laboratories (Kobe-shi); Chemical Laboratories (Kobe-shi); Polymer Development Center (Kobe-shi)
Shareholders :	Nippon Life Insurance - 5.7% ; Sanwa Bank [8320] - 3.6% ; Dai-Ichi Kangyo Bank [8311] - 3.6% ; Yasuda Trust and Banking [8404] - 3.2%

Kokusai Chemical K.K.

Japanese Name	:	***Kokusai Chemical K.K.***
TSE Number	:	Not listed
Description	:	Synthetic resins manufacturer specializing in epoxy resins.
Address	:	2280, Yabe-cho, Totsuka-ku, Yokohama, Kanagawa-ken 244
Tel	:	045-811-1028
Statistics	:	Capital : 10 million yen Employees : 32
Sales	:	May, 1989 - 1037 million yen May, 1990 - 1042 million yen
Sales Breakdown	:	SIC# Product 2821 Synthetic resins
Banks	:	Kyowa Saitama Bank; Bank of Yokohama
Shareholders	:	Hitoshi Kohama

Komatsu Ltd.

Japanese Name	:	***Komatsu Seisakusho K.K.***
TSE Number	:	6301
Description	:	Manufacturer of construction machinery. Among the largest makers of construction machinery in the world, Komatsu is currently directing its energies to diversifying its product line to include plastics, electronics and robotics.
Address	:	2-3-6, Akasaka, Minato-ku, Tokyo 107
Tel & Fax	:	Tel: 03-5561-2616 Fax: 03-3505-9662
Statistics	:	Capital : 69662 million yen Employees : 14871
Officers	:	CEO : Mr. Tetsuya Katada/President Contact : Mr. Kenji Takahashi/General Manager, International
Sales	:	March, 1988 (3 months) - 123861 million yen March, 1989 - 566538 million yen March, 1990 - 628465 million yen March, 1991 - 674499 million yen
R & D Spending	:	29678 million yen Capital Investment: 49900 million yen
Sales Breakdown	:	SIC# Product Sales 3531 Construction machinery 67 % 356 Industrial machinery 9 %
Main Products	:	Construction Machinery: Bulldozers, Earth movers, Power shovels, Wheel loaders, Dump trucks, Motor graders Rough terrain cranes, Vibration rollers, Underground Maintenance machinery, Construction robots; Industrial Machinery: Industrial presses, Press brakes, Turret punch press, Laser processing tools, Plasma processing tools, Steel processing machinery, Industrial robots, Machine tools, Injection molding equipment; Other Areas: Engines, Compressors, Diesel generators, Hydraulic machinery, Cast/forged products, Armoured cars, Projectiles
Banks	:	Sumitomo Bank; Industrial Bank of Japan; Fuji Bank; Hokkoku Bank; Kyowa Saitama Bank
Branches	:	Hokkaido Regional Sales Office; Tohoku Regional Sales Office; Kanto Regional Sales Office; Tokyo Regional Sales Office; Chubu Regional Sales Office; Osaka Regional Sales Office; Chugoku Regional Sales Office; Kyushu Regional Sales Office
Factories	:	Awazu Plant; Himi Plant; Kawasaki Plant (Kanagawa-ken); Oyama Plant (Tochigi-ken); Indonesia Plant; Brazil Plant; U.S.A. Plant; U.K. Plant; Mexico Plant; Germany Plant
Laboratories	:	Research Division (Kanagawa-ken)
R&D Areas	:	Construction Machinery: Mini power shovel, 46 ton off-road truck, Shield equipment for tunnelling machinery; Industrial Machinery: CNC laser processing tool, High performance cam shaft mirror, Plasma process robot
Shareholders	:	Taiyo Mutual Life - 5.1% ; Daihayaku Mutual Life - 4.5% ; Sumitomo Bank [8318] - 4.2% ; Fuji Bank [8317] - 3.1% ; Industrial Bank of Japan [8302] - 3.1% ; Nippon Life - 2.9% ; Employees' Stockholders Assn. - 2.7%

Komatsu Seiren Co., Ltd.

Japanese Name : ***Komatsu Seiren K.K.***

TSE Number : 3580

Description : Engaged in finish processing of man-made textiles and synthetic leather products. A member of the Toray Industries group of companies, Komatsu Seiren is also active in merchandising and in developing new products, most notably synthetic leathers.

Address : Nu-167, Hamamachi, Neagarimachi, Nomi-gun, Ishikawa-ken 929-01

Tel & Fax : Tel: 0761-55-1111 Fax: 0761-55-4547

Statistics : Employees : 1140

Officers : CEO : Mr. Kenichi Nakayama/President

Sales : March, 1989 - 24376 million yen
March, 1990 - 27453 million yen
March, 1991 - 33166 million yen

Sales Breakdown :

SIC#	Product	Sales
2262	Synthetic fabrics processing	80 %
2295	Synthetic leather	15 %

Main Products : Finished synthetic textiles; Synthetic leather products

Banks : Hokkoku Bank; Hokuriku Bank; Fukui Bank; Sumitomo Trust and Banking

Overseas Subsidiaries : Siam Synthetic Textile Industry Ltd. (Thailand)

KSL Alcoa Aluminum

TSE Number	:	Not listed
Description	:	Manufacturer of aluminum can stock. A joint venture of Kobe Steel and Alcoa Company (U.S.), KSL Alcoa was established to serve the dual purpose of supplying Kobe Steel with a steady supply of aluminum ore and providing Alcoa with a Japanese partner and market entry. The company hopes to begin producing aluminum products for the automobile industry in the near future.
Address	:	1-8-2, Marunouchi, Chiyoda-ku, Tokyo 100
Tel & Fax	:	Tel: 03-3218-6055 Fax: 03-5252-7961
Sales Breakdown	:	SIC# Product 3334 Aluminum can stock
Main Products	:	Aluminum Products: Aluminum can stock
Factories	:	Moka Plant (Tochigi-ken)
Shareholders	:	Alcoa Co., Ltd. (U.S.A.) - 50% ; Kobe Steel, Ltd. [5406] - 50%

Kurabo Industries Ltd.

Japanese Name : ***Kurashiki Boseki K.K.***

TSE Number	:	3106
Description	:	The company has a long history in the spinning industry, and is continuing its efforts to develop original materials for textile production. Expanding into non-textiles, including information systems, chemicals and engineering.
Address	:	2-4-31, Kyutaro-machi, Chuo-ku, Osaka 541
Tel & Fax	:	Tel: 06-266-5111 Fax: 06-266-5555
Statistics	:	Capital : 22037 million yen Employees : 3747
Officers	:	CEO : Mr. Atsushi Fujita/President
Sales	:	March, 1988 (9 months) - 117059 million yen March, 1989 - 130700 million yen March, 1990 - 133000 million yen March, 1991 - 140800 million yen
Sales Breakdown	:	SIC# Product 2211 Yarn, cotton garments 2231 Woollen fabrics 3569 Precision filtration equipment
Main Products	:	Broadwoven Fabric Mills: Yarn, Fabrics, Knitted garments made of cotton, synthetic wool, polyurethane foam; Industrial Machinery: Precision filtration and pollution control equipment
Banks	:	Mitsui Taiyo Kobe Bank; Fuji Bank; Dai-Ichi Kangyo Bank; Kyowa Saitama Bank; Bank of Tokyo; Sumitomo Bank
Branches	:	Hong Kong Branch (Hong Kong); New York Representative Office (New York); Hamburg Representative Office (Hamburg); Tokyo Branch (Chuo-ku)
Laboratories	:	Technical Research Laboratory (Shimokida-cho, Osaka)
Shareholders	:	Japan Securities Clearing - 9.1% ; Nippon Mutual Life - 5.4% ; Mitsui Taiyo Kobe Bank [8314] - 4.9% ; Fuji Bank [8317] - 4.9% ; Sumitomo Life Insurance - 3.5% ; Chugoku Bank [8382] - 2.6%

Kuraray Co., Ltd.

Japanese Name	:	***Kurare K.K.***
TSE Number	:	3405
Description	:	A medium-scale manufacturer of synthetic fibers, and known as the first to commercialize vinylon in Japan. The company has expanded its business into plastics, fine chemicals, synthetic leather, and medical products. The company has become a comprehensive chemical manufacturer.
Address	:	1-12-39, Umeda, Kita-ku, Osaka 530
Tel & Fax	:	Tel: 06-348-2111 Fax: 06-348-2189
Statistics	:	Capital : 36697 million yen Employees : 5567
Officers	:	CEO : Mr. Hisao Nakamura/President
Sales	:	March, 1988 - 198795 million yen March, 1989 - 211065 million yen March, 1990 - 234205 million yen March, 1991 - 277000 million yen
R & D Spending	:	10600 million yen Capital Investment: 17100 million yen
Sales Breakdown	:	SIC# Product 2823 Various synthetic fibres 2819 Fine chemicals 2833 Medical and pharmaceuticals
Main Products	:	Fibers: Polyester, Vinylon, Rayon; Specialty Chemicals: Acetone cyanohydrin, Acetylene, Ethylene-vinyl acetate copolymer, Emulsion ethylene-vinyl acetate copolymer, Methacryl resin polymer, PVA sizing agent, Surface coating agent
Banks	:	Industrial Bank of Japan; Fuji Bank; Sumitomo Bank; Sumitomo Trust and Banking; Yasuda Trust and Banking
Branches	:	Tokyo Head Office (Chuo-ku); Nagoya Office (Nagoya-shi); Fukui Office (Fukui-shi); Fukuoka Office (Fukuoka-shi); Sapporo Office (Sapporo-shi); Sendai Office (Sendai-shi); Europe Office (Dusseldolf); Hong Kong Office (Hong Kong)
Factories	:	Kurashiki Plant (Kurashiki-shi); Saijo Plant (Saijo-shi, Ehime-ken); Okayama Plant (Okayama-shi); Nakajo Plant (Kitakanbara-gun, Niigata-ken); Tamashima Plant (Kurashiki-shi); Kashima Plant (Kashima-gun, Ibaraki-ken)
Laboratories	:	Central Research Laboratory (Kurashiki-shi); Technical Research Center (Kurashiki-shi)
Subsidiaries	:	Kuraray Chemical Co., Ltd. (Bizen-shi, Okayama-ken); Kuraray Plastics Co., Ltd. (Osaka-shi); and 17 other domestic subsidiaries
Overseas Subsidiaries	:	Kuraray International Corp. (New York, U.S.A); Clarino America Corp. (New York, U.S.A);

Kureha Chemical Industry Co., Ltd.

Japanese Name : ***Kureha Kagaku Kogyo K.K.***

TSE Number : 4023

Description : Major manufacturer of synthetic resin products and fine chemicals. Sales of mainline cancer immunotherapy drugs in severe slump accounting for steady decline in sales over recent years. Nonetheless, R&D levels remain high, concentrating on pharmaceuticals, engineering plastics and food wrapping films.

Address : 1-9-11, Nihonbashi-Horidome-cho, Chuo-ku, Tokyo 103

Tel & Fax : Tel: 03-3249-4681 Fax: 03-3639-4785

Statistics : Capital : 12460 million yen
Employees : 2363

Officers : CEO : Mr. Shunichiro Kodama/President
Contact : Mr. Kenji Tsunoda/Assistant General Manager

Sales : March, 1988 - 118403 million yen
March, 1989 - 115894 million yen
March, 1990 - 106431 million yen
March, 1991 - 101488 million yen

R & D Spending : 8800 million yen Capital Investment: 7836 million yen

Sales Breakdown :

SIC#	Product	Sales
308	Plastic products	28 %
2821	Synthetic resins	28 %
28	Specialty chemicals	27 %
281	Inorganic chemicals	11 %

Main Products : Plastic Products: PVDC films, PVC films, High barrier laminated films, Other films for food packaging, Plastic containers, Housing construction materials, PVDC fibers, Polyethylene fibers, Artificial turf, Synthetic fiber products; Synthetic Resins: PVC resins, HC polymers, BTA resins, HIA resins, Impact modifier for rigid PVD products, Other plastic modifiers, PVDC latex, Polyvinyl resins, KF polymers; Specialty Chemicals: Monochlorobenzene, Chlorobenzene, Other organic chemicals, Heat exchange petroleum compounds, Vinyl chloride monomer, Dye solvents, Carbon fibers, Polyphenylene sulfide engineering resins, Cancer immunotherapy agents, Veterinary drugs, Rice fungicides; Inorganic Chemicals: Caustic soda, Other soda products, Synthetic hydrochloric acid; Others: Paints, Food packaging machinery

Banks : Fuji Bank; Sumitomo Bank; Daiwa Bank

Branches : Osaka Branch (Chuo-ku, Osaka); Sapporo Sales Office (Sapporo-shi, Hokkaido); Iwaki Sales Office (Iwaki-shi; Fukushima-ken); Nagoya Sales Office (Nagoya-shi, Aichi-ken); Fukuoka Sales Office (Fukuoka-shi, Fukuoka-ken)

Factories : Nishiki Plant (Iwaki-shi, Fukushima-ken)

R&D Areas : Pharmaceuticals; Plastics: Engineering plastics (Fortron KPS), Plastic lenses, Laminated films

Kurosaki Refractories Co., Ltd.

Japanese Name	:	***Kurosaki Refractories K.K.***
TSE Number	:	5352
Address	:	1-1, Higashi-Hamamachi, Yahata-Nishi-ku, Kita Kyushu 806
Tel & Fax	:	Tel: 093-622-7224 Fax: 093-622-7200
Statistics	:	Capital : 4409 million yen Employees : 1705
Officers	:	CEO : Mr. Takuo Kono/President
Sales	:	March, 1988 - 44339 million yen March, 1989 - 49695 million yen March, 1990 - 52515 million yen March, 1991 - 52654 million yen
R & D Spending	:	1711 million yen Capital Investment: 3000 million yen
Sales Breakdown	:	

SIC#	Product	Sales
3255	Fire bricks and refractories	87 %
8711	Industrial design	13 %

Banks	:	Industrial Bank of Japan; Bank of Fukuoka; Sumitomo Trust and Banking; Fuji Bank; Nishi-Nippon Bank

Kyocera Corp.

Japanese Name : ***Kyosera K.K.***

TSE Number : 6971

Description : Manufacturer of semiconductors, electronic and optical products, bioceramics and fine ceramic products. The world's largest producer of IC ceramic packages, Kyocera has expanded its business operations to include information and optical equipment as well as office automation technology.

Address : 5-22, Kita-inoeucho, Higashino, Yamashina-ku, Kyoto 607

Tel & Fax : Tel: 075-592-3851 Fax: 075-501-2194

Statistics : Capital : 101407 million yen
Employees : 14031

Officers : CEO : Mr. Kensuke Ito/President

Sales : March, 1988 - 271165 million yen
March, 1989 - 299201 million yen
March, 1990 - 303760 million yen
March, 1991 - 330891 million yen

Capital Investment : 23820 million yen

Sales Breakdown :

SIC#	Product	Sales
3674	Semiconductors and related products	29 %
36	Electronic equipment	21 %
36	Electronic parts	21 %
3259	Bioceramics	10 %
3827	Precision optical equipment	9 %
3259	Fine ceramics	9 %

Main Products : Semiconductors and Related Products: Layer IC packages, Advanced IC packages, Ceramic modules, IC magazines, Other IC ceramic packages, Hybrid ICs; Electronic Equipment and Parts: Chip condensors, Chip capacitors, Thermal printheads, LED printheads, Chip resistors, CD players, Panel computers, Hard disc drives, Laser printers, Image scanners, Image sensors, Teleconference sytems, CD-I systems, Other communication equipment and peripheral equipment, Amorphous silicon drums; Optical Precision Equipment: Cameras, Video camcorders, Surveillance cameras; Bioceramics and Fine Ceramics: Artificial bones, Dental materials, Industrial solar power system materials, Ceramic knives and scissors, Various fine ceramic products

Banks : Sanwa Bank; Kyoto Bank; Daiwa Bank; Sumitomo Bank; Mitsui Taiyo Kobe Bank

Branches : Tokyo Branch (Tokyo)

Subsidiaries : Elco International K.K. (Yokohama-shi, Kanagawa-ku); Kyocera Denshi Kikai K.K. (Chuo-ku, Tokyo); Kyocera Kyosan K.K. (Shibuya-ku, Tokyo); Fuoka Optics K.K. (Tokyo); Kyocera Development K.K. (Kyoto)

Kyowa Hakko Kogyo Co., Ltd.

Japanese Name	:	***Kyowa Hakko Kogyo K.K.***
TSE Number	:	4151
Description	:	Among the world's largest manufacturers of amino acids. Active in biotechnology R&D and plant tissue culture.
Address	:	Otemachi Bldg., 1-6-1, Otemachi, Chiyoda-ku, Tokyo 100
Tel & Fax	:	Tel: 03-3201-7211 Fax: 03-3201-1968
Statistics	:	Capital : 26737 million yen Employees : 5400
Officers	:	CEO : Mr. Mikio Kato/President
Sales	:	December, 1988 - 255311 million yen December, 1989 - 274866 million yen December, 1990 - 288466 million yen
R & D Spending	:	16700 million yen Capital Investment: 13600 million yen
Sales Breakdown	:	SIC# Product 2834 Pharmaceuticals 2819 Inorganic chemicals 2085 Alcoholic spirits
Main Products	:	Pharmaceuticals: Acetyl spiramycin, Adenine hydrochloride (vitamin B4), Other pharmaceuticals; Specialty Chemicals: Acetaldehyde, Acetic acid, Acetone, Acetyl tributyl citrate, Acid protease, Adenosine and its salts, Amino acid, Citric acid, Organic phosphorous-based flame retardant, Flotation reagent, Monosodium L-glutamate, Phosphoric acid ester plasticizers, Aliphatic ester plasticizers, Adiphic acid plasticizers; Agricultural Chemicals: Fertilizers; Alcoholic Beverages: Wines, Spirits; Biotechnology products: Motilin
Banks	:	Industrial Bank of Japan; Norinchukin Bank
Branches	:	Tokyo Branch (Chuo-ku, Tokyo)
Factories	:	Hofu Factory (Hofu-shi, Yamaguchi-ken); Fuji Factory (Sunto-gun, Shizuoak-ken)
Laboratories	:	Tokyo Laboratories (Machida-shi); Pharmaceutical Laboratories (Sunto-gun)
Subsidiaries	:	Kyowa Petrochemical Co., Ltd. (Chiyoda-ku); Sainte Neige Wine Co., Ltd. (Yamanashi-shi); Kyowa Medex Co., Ltd. (Chuo-ku)

Kyushu Kogyo Gijitsu Shinkenjo

Japanese Name	:	***Kyushu Kogyo Gijitsu Shinkenjo***
TSE Number	:	Not listed
Description	:	Semi-public entity engaged in the promotion and development of industrial technologies. Kyushu Kogyo Gijitsu Shinkenjo is based in Saga Prefecture on the Southern island of Kyushu and works in conjunction with the Japanese Ministry of International Trade and Industry (MITI).
Address	:	Shyuku-machi, Tosu-shi, Saga 841
Tel & Fax	:	Tel: 0942-82-5161 Fax: 0952-83-0850
Sales Breakdown	:	SIC# Product 8731 Industrial technology development
Main Products	:	Development and promotion of industrial technologies

Marubeni Corp.

Japanese Name	:	***Marubeni K.K.***
TSE Number	:	8002
Description	:	One of Japan's leading general traders. Originally a textile specialist based in Osaka, the company plays a leading role in the Fuyo group. The company has over 140 overseas offices.
Address	:	1-4-2, Otemachi, Chiyoda-ku, Tokyo 100-88
Tel & Fax	:	Tel: 03-3282-2111 Fax: 03-3282-7456
Statistics	:	Capital : 180725 million yen Employees : 7185
Officers	:	CEO : Mr. Tomio Tatsuno/President
Sales	:	March, 1989 - 14189400 million yen March, 1990 - 18248200 million yen March, 1991 - 19015600 million yen
Sales Breakdown	:	SIC# Product 5051 Metals 5052 Minerals 5084 Machinery 5082 Construction equipment 517 Fuels 516 Chemicals 5131 Textiles
Main Products	:	Products Traded: Metals, Minerals, Machinery, Construction equipment, Energy, Chemicals, Textiles
Banks	:	Fuji Bank; Bank of Tokyo; Mitsui Taiyo Kobe Bank; Yasuda Trust and Banking; Industrial Bank of Japan
Overseas Subsidiaries	:	Marubeni Hong Kong Ltd. (Hong Kong); Marubeni America Corp. (New York); Marubeni Deutschland GmbH (Dusseldorf)

Matsumoto Yushi-Seiyaku Co., Ltd.

Japanese Name	:	***Matsumoto Yushi-Seiyaku K.K.***
TSE Number	:	Not listed
Description	:	Manufactures a range of surface treatment agents and surface processing products.
Address	:	2-1-3, Shibukawa-cho, Yao-shi, Osaka 581
Tel & Fax	:	Tel: 0729-91-1001 Fax: 0729-94-4596
Statistics	:	Capital : 3039 million yen Employees : 418
Officers	:	CEO : Mr. Goro Kimura/President Contact : Mr. Goro Kimura/President
Sales	:	November, 1988 - 17593 million yen March, 1989 - 6662 million yen March, 1990 - 18969 million yen
Sales Breakdown	:	SIC# Product 2843 Surfactants 2842 Synthetic materials 2842 Metal cleaning agents 2899 Fire retardants 2899 Synthetic resin treatments 3291 Synthetic diamonds
Main Products	:	Surface Active Agents: Surfactants; Polishes and Sanitation Goods: Synthetic paste, Cleaning agents for metals; Chemical Preparations: Alkyls, Antifoaming agent, Antistatic agents for fiber/plastics/rubber, Dispersion agents for cement/photography, Emulsifiers for foodstuffs, Gel agents, Stearic acid, Polyoxyethylene alkyl ether, Sea surface pollution prevention dispersant, Sequestering agent, Slip additive, Fire retardants, Rust proofing agents; Abrasive Products: Synthetic diamonds
Banks	:	Sanwa Bank
Branches	:	Osaka Office (Osaka-shi); Tokyo Office (Chuo-ku); Nagoya Office (Nagoya-shi); Hiroshima Office (Hiroshima-shi); Kanazawa Office (Kanazawa-shi)
Factories	:	Main Factory (Yao-shi); Shizuoka Factory (Fukuroi-shi)

Matsushita Electric Industrial Co., Ltd.

Japanese Name :	***Matsushita Denki Sangyo K.K.***
TSE Number :	6752
Description :	One of the world's largest consumer electronics manufacturers, the company is the nucleus of the Matsushita group. The company has about 75 overseas manufacturing subsidiaries and 26 sales companies worldwide. The company exports its products under the 'National', 'Panasonic', 'Technics', and 'Quasar' brands.
Address :	1006, Kadoma, Osaka 571
Tel & Fax :	Tel: 06-908-1121 Fax: 06-906-1762
Statistics :	Capital : 195258 million yen Employees : 42510
Officers :	CEO : Mr. Akio Tanii/President
Sales :	March, 1988 - 3277610 million yen March, 1989 - 4074670 million yen March, 1990 - 4248670 million yen

Sales Breakdown :

SIC#	Product	Sales
3651	Video equipment	26 %
366	Communication and industrial equipment	23 %
363	Home appliances	14 %
367	Electronic components	13 %
3651	Audio equipment	9 %
3692	Batteries and kitchen products	5 %

Main Products :	Consumer Electronics: Stereos, TVs, VCRs, Microwave ovens, Batteries, Home appliances; Other Areas: Communications equipment, Industrial equipment, Electronic components
Banks :	Sumitomo Bank; Kyowa Bank
Factories :	123 factories
Laboratories :	Central Research Laboratory (Moriguchi-shi)
Shareholders :	Sumitomo Bank [8318] - 4.66% ; Sumitomo Mutual Life - 4.19% ; Nippon Life - 3.92% ; Matsushita Kosan K.K. - 3.22%
Subsidiaries :	167 domestic subsidiaries
Overseas Subsidiaries :	Matsushita Electric Corp. of America (New Jersey, U.S.A.) Panasonic Europe Ltd. (Berks, U.K.); Asia Matsushita Electric (S) Pte., Ltd. (Singapore)

Matsushita Electronics Corp.

Japanese Name :	***Matsushita Denshi Kogyo K.K.***
TSE Number :	Not listed
Description :	Electronics manufacturer specializing in semiconductors and electron tubes. Matsushita Electronics Corp. is a joint venture between Matsushita Electric Industries and Philips (Netherlands).
Address :	1-1, Saiwai-cho, Takatsuki-shi, Osaka 569
Tel :	0726-82-5521
Statistics :	Capital : 41000 million yen Employees : 13000
Officers :	CEO : Mr. Kazuma Danjo/President
Sales :	March, 1988 - 395920 million yen March, 1989 - 431783 million yen March, 1990 - 444404 million yen March, 1991 - 471654 million yen

Sales Breakdown :

SIC#	Product	Sales
3674	Semiconductors and related equipment	51 %
3671	Electron tubes	34 %
364	Lighting equipment	15 %

Main Products :	Electron Tubes: Color television CRTs (cathode ray tube), Color display CRTs, Monochrome display CRTs, Projection CRTs, Oscilloscope CRTs, Plasma displays, Magnetrons; Semiconductors and Related Devices: VLSI (Microcomputer and ASIC), Memory devices, CCDs, Bipolar ICs, Logic ICs, GaAs devices, Laser diodes, Optoelectronic devices, Discrete transistors and diodes; Lighting Equipment: Incandescent lamps, Fluorescent lamps, Gas discharge lamps, Halogen lamps, Thermal protectors, Starters, Small lamps; Specialty Chemicals: Deodorizing catalyst
Banks :	Sumitomo Bank; Mitsubishi Bank; Kyowa Saitama Bank; Industrial Bank of Japan; Long Term Credit Bank of Japan
Factories :	Takatsuki Plant (Osaka); Nagaoka Plant (Niigata-ken); Hiraide Plant; Kiyohara Plant; Arai Plant (Niigata-ken); Uozu Plant (Toyama-ken); Ohayama Plant
Shareholders :	Matsushita Electric Industrial [6752] - 65%; Philips (Netherlands) - 35%
Subsidiaries :	Matsushita Electronic Instruments Corp.; West Electric Co., Ltd.; Toyo Denpa Co., Ltd.; Kagoshima MEC Co., Ltd.; Inai Electronics Co., Ltd.
Overseas Subsidiaries :	Matsushita Denshi Singapore Pte., Ltd. (Singapore); Beijing-Matsushita Color CRT Co., Ltd. (P.R.C); American Matsushita Electronics Corp. (Ohio, U.S.A.); Matsushita Semiconductor of America

Matsushita Electronics Corp.

Japanese Name	:	***Matsushita Denshi Kogyo K.K.***
TSE Number	:	Not listed
Description	:	Electronics manufacturer specializing in semiconductors and electron tubes. Matsushita Electronics Corp. is a joint venture between Matsushita Electric Industries and Philips (Netherlands).
Address	:	1-1, Saiwai-cho, Takatsuki-shi, Osaka 569
Tel	:	0726-82-5521
Statistics	:	Capital : 41000 million yen Employees : 13000
Officers	:	CEO : Mr. Kazuma Danjo/President
Sales	:	March, 1988 - 395920 million yen March, 1989 - 431783 million yen March, 1990 - 444404 million yen March, 1991 - 471654 million yen

Sales Breakdown :

SIC#	Product	Sales
3674	Semiconductors and related equipment	51 %
3671	Electron tubes	34 %
364	Lighting equipment	15 %

Main Products	:	Electron Tubes: Color television CRTs (cathode ray tube), Color display CRTs, Monochrome display CRTs, Projection CRTs, Oscilloscope CRTs, Plasma displays, Magnetrons; Semiconductors and Related Devices: VLSI (Microcomputer and ASIC), Memory devices, CCDs, Bipolar ICs, Logic ICs, GaAs devices, Laser diodes, Optoelectronic devices, Discrete transistors and diodes; Lighting Equipment: Incandescent lamps, Fluorescent lamps, Gas discharge lamps, Halogen lamps, Thermal protectors, Starters, Small lamps; Specialty Chemicals: Deodorizing catalyst
Banks	:	Sumitomo Bank; Mitsubishi Bank; Kyowa Saitama Bank; Industrial Bank of Japan; Long Term Credit Bank of Japan
Factories	:	Takatsuki Plant (Osaka); Nagaoka Plant (Niigata-ken); Hiraide Plant; Kiyohara Plant; Arai Plant (Niigata-ken); Uozu Plant (Toyama-ken); Ohayama Plant
Shareholders	:	Matsushita Electric Industrial [6752] - 65%; Philips (Netherlands) - 35%
Subsidiaries	:	Matsushita Electronic Instruments Corp.; West Electric Co., Ltd.; Toyo Denpa Co., Ltd.; Kagoshima MEC Co., Ltd.; Inai Electronics Co., Ltd.
Overseas Subsidiaries	:	Matsushita Denshi Singapore Pte., Ltd. (Singapore); Beijing-Matsushita Color CRT Co., Ltd. (P.R.C); American Matsushita Electronics Corp. (Ohio, U.S.A.); Matsushita Semiconductor of America

Matsushita Industrial Equipment Co., Ltd.

Japanese Name	:	***Matsushita Sangyo Kiki K.K.***
TSE Number	:	Not listed
Description	:	Manufacturer of welding and cutting machinery, electrical equipment and related products. A member of the Matsushita group of companies, Matsushita Industrial excels in the area of automated welding and cutting equipment.
Address	:	3-1-1, Inazu-cho, Toyonaka-shi, Osaka 561
Tel & Fax	:	Tel: 06-862-1121 Fax: 06-862-8047
Statistics	:	Capital : 3000 million yen Employees : 1350
Officers	:	CEO : Mr. Eizo Asakura/President
Sales	:	November, 1989 - 53800 million yen March, 1990 (4 months) - 16500 million yen March, 1991 - 55800 million yen
Sales Breakdown	:	SIC# Product 3548 Welding and cutting equipment 361 Electric distribution equipment 3675 Electronic condensors 3699 Misc. electrical equipment
Main Products	:	Welding and Cutting Equipment: 'Panastar RF200' series argon gas welders, TIG steel pipe welding robots, YAG laser equipped cutting robots, Carbon dioxide laser scalpels; Electric Distribution Equipment; Electronic Condensors; Misc. Electronic Equipment
Banks	:	Kyowa Saitama Bank; Sumitomo Bank
Overseas Subsidiaries	:	Panasonic Welding Industry (Thailand)

Matsushita Seiko Co., Ltd.

Japanese Name :	***Matsushita Seiko K.K.***
TSE Number :	6587
Description :	Manufacturer of household and industrial electrical products specializing in air-conditioning and ventilation equipment. Matsushita Seiko is a subsidiary of Matsushita Electric Industry Co., Ltd..
Address :	6-2-61, Nishi-Imafuku, Joto-ku, Osaka 536
Tel & Fax :	Tel: 06-939-1161 Fax: 06-939-6967
Statistics :	Capital : 12092 million yen Employees : 2027
Officers :	CEO : Mr. Tadao Suzuki/President
Sales :	March, 1989 - 84773 million yen March, 1990 - 86637 million yen March, 1991 - 93726 million yen
Capital Investment :	5635 million yen

Sales Breakdown :

SIC#	Product	Sales
3585	Household air-conditioners	51 %
363	Household electrical appliances	25 %
3535	Industrial air-conditioning equipment	22 %

Main Products :	Household Air-Conditioners and Electric Appliances: Unit air-conditioners, Residential air-conditioning systems, Heating units, Heat distribution units, Portable electric heaters, Electric fans, Food dryers; Industrial Air-Conditioning and Ventilation Equipment: Air conditioners for industrial and commercial use, Air handling units, Fan coil units, Chilling units, Building and plant ventilation units, Tunnel ventilation units, Clean room ventilation units, Fan motors, Office ventilation equipment
Banks :	Kyowa Saitama Bank; Sumitomo Bank; Tokai Bank; Bank of Tokyo; Industrial Bank of Japan
Branches :	Oveseas Offices: Hong Kong Office
Factories :	Osaka Plant (Same as home office); Fujisawa Plant (Fujisawa-shi, Kanagawa-ken); Kasugai Plant (Kasugai-shi, Aichi-ken)
R&D Areas :	Air-Conditioning and Purification Systems: System air handling unit; Sensor Technology
Subsidiaries :	Matsushita Seiko Engineering Co., Ltd. (Kasugai-shi, Aichi-ken)
Overseas Subsidiaries :	Matsushita Seiko Taiwan Co., Ltd. (Taiwan)

Mazda Motor Corp.

Japanese Name : ***Matsuda K.K.***

TSE Number : 7261

Description : Major automaker. A member of the Sumitomo group of companies, Mazda Motor Corp. maintains close ties with Ford Motor Company which is the company's largest single shareholder of Mazda stock. Over half of Mazda's production is sold overseas.

Address : 3-1, Shinchi, Fuchu-cho, Aki-gun, Hiroshima 730-91

Tel & Fax : Tel: 082-282-1111 Fax: 082-287-5237

Statistics : Capital : 92781 million yen
Employees : 31903

Officers : CEO : Mr. Norimasa Furuta/President

Sales : October, 1988 - 1844320 million yen
March, 1989 (5 months) - 784917 million yen
March, 1990 - 2045570 million yen
March, 1991 - 2225710 million yen

Sales Breakdown :

SIC#	Product	Sales
3711	Passenger automobiles	60 %
3714	Automotive parts	20 %
3713	Trucks	12 %

Main Products : Motor Vehicles: Automobiles, Trucks, Automotive parts

Banks : Sumitomo Bank; Sumitomo Trust and Banking; Mitsubishi Trust and Banking; Industrial Bank of Japan; Hiroshima Bank

Factories : Hofu Plant (Yamaguchi-ken); Hiroshima Plant (Hiroshima-ken); Miyoshi Factory (Hiroshima-ken)

Laboratories : Hiroshima Technical Research Center (Hiroshima-ken); Mazda R&D Center Yokohama (Kanagawa-ken)

Meidai Chemical Co., Ltd.

TSE Number	:	Not listed
Address	:	484-1, Sobara, Kurashiki-shi, Okayama 710-01
Tel	:	0864-85-1355

Minebea Co., Ltd.

Japanese Name : ***Minebea K.K.***

TSE Number : 6479

Description : Manufacturer of industrial equipment and parts, specializing in bearings and related products. The world's largest producer of miniature bearings, Minebea is also active in a number of other product areas including electronic equipment, a product category which the company has increasingly come to emphasize and which now accounts for the largest proportion of sales.

Address : Sports Shinkou Bldg. 5F, 3-7, Kanda-Ogawacho, Chiyoda-ku, Tokyo 101

Tel & Fax : Tel: 03-3293-9251 Fax: 03-3294-6750

Statistics : Employees : 3142

Officers : CEO : Mr. Goro Ogino/President

Sales : September, 1988 - 156521 million yen
September, 1989 - 170686 million yen
September, 1990 - 195482 million yen

Sales Breakdown :

SIC#	Product	Sales
36	Electronic equipment and parts	45 %
3562	Bearings	30 %
35	Machinery parts	15 %
37	Transportation equip. and parts	5 %

Main Products : Electronics Related Products; Bearings: Miniature bearings; Machinery parts; Transportation Related Products

Banks : Long Term Credit Bank of Japan; Mitsubishi Bank; Sumitomo Trust and Banking; Hokkaido Takushoku Bank; Mitsui Taiyo Kobe Bank

Factories : Overseas production facilities: Thailand, Singapore

Mitsu Kagaku

TSE Number	:	Not listed
Description	:	Manufacturer of chemical products, including a silicone dissolving agent which the company developed in 1991.
Address	:	Nitta-machi, Nitta-gun, Gunma 370-03
Tel & Fax	:	Tel: 0276-56-2807 Fax: 0276-56-2807
Sales Breakdown	:	SIC# Product 28 Chemical products 2899 Silicone dissolving agent
Main Products	:	Chemical Products: Silicone dissolving agent

Mitsubishi Electric Corp.

Japanese Name : **Mitsubishi Denki K.K.**

TSE Number	:	6503
Description	:	Japan's fifth ranking electronics maker.
Address	:	2-2-3, Marunouchi, Chiyoda-ku, Tokyo 100
Tel & Fax	:	Tel: 03-3218-2111 Fax: 03-5252-7119
Statistics	:	Capital : 174516 million yen Employees : 48616
Officers	:	CEO : Mr. Moriya Shiki/President Contact : Mr. Sozaburo Ihara/Senior Managing Director, Overseas
Sales	:	March, 1988 - 1954190 million yen March, 1989 - 2230100 million yen March, 1990 - 2387830 million yen March, 1991 - 2588840 million yen

Sales Breakdown :

SIC#	Product	Sales
36	Information, telecom., electronic devices	27 %
356	Heavy machinery	22 %
365	Consumer electronics	21 %
367	Industrial products and automation eqpmt.	20 %

Main Products	:	Industrial Machinery: Turbine generators; Electronic Equipment: Programmable controllers, Semiconductor ICs
Banks	:	Mitsubishi Bank; Mitsubishi Trust and Banking; Dai-Ichi Kangyo Bank
Branches	:	Kansai Branch Office (Osaka-shi) and 11 other branch offices, 30 sales offices and 116 overseas offices
Factories	:	Kobe Works (Kobe-shi); Itami Works (Amagasaki-shi); Nagasaki Works (Nagasaki-shi) and 33 other factories
Laboratories	:	Central Research Laboratory (Amagasaki-shi) and 12 other laboratories
Shareholders	:	Mitsubishi Trust and Banking [8402] - 4.53% ; Meiji Mutual Life - 4.14% ; Nippon Life - 3.59% ; Mitsubishi Bank [8315] - 3.37%
Subsidiaries	:	Mitsubishi Electric Building Techno Service Co., Ltd. (Chiyoda-ku); Mitsubishi Electric Service Engineering Co., Ltd. (Setagaya-ku); Mitsubishi Space Software Co., Ltd. (Minato-ku)
Overseas Subsidiaries	:	Mitsubishi Electric America, Inc. (U.S.A.); Mitsubishi Electric (UK) Ltd. (U.K).; Mitsubishi Electric (Singapore) Pte., Ltd. (Singapore)

Mitsubishi Gas Chemical Co., Inc.

Japanese Name	:	***Mitsubishi Gas Kagaku K.K.***
TSE Number	:	4182
Description	:	Major chemical company with strengths in basic chemicals such as methanol, ammonia and urea. In addition, the company is a top ranking producer of engineering plastics.
Address	:	Mitsubishi Bldg., 2-5-2, Marunouchi, Chiyoda-ku, Tokyo 100
Tel & Fax	:	Tel: 03-3283-5000 Fax: 03-3287-0833
Statistics	:	Employees : 3751
Officers	:	CEO : Mr. Reiji Nishikawa/President
Sales	:	March, 1989 - 199435 million yen March, 1990 - 205064 million yen March, 1991 - 225506 million yen
Capital Investment	:	21900 million yen
Sales Breakdown	:	SIC# Product 2819 Ammonia 2819 Urea 2869 Methanol 2821 Engineering plastics
Main Products	:	Inorganic Chemicals: Urea, Ammonia, etc.; Organic Chemicals: Xylene, Methanol, etc.; Plastics and Resins
Banks	:	Mitsubishi Bank; Mitsubishi Trust and Banking; Industrial Bank of Japan; Norinchukin Bank; Yokohoma Bank

Mitsubishi Heavy Industries, Ltd.

Japanese Name :	***Mitsubishi Jukogyo K.K.***
TSE Number :	7011
Description :	Japan's largest comprehensive heavy machinery maker and a key member within the Mitsubishi Industrial group. Leads in the production of ships, power plants, heavy machinery, and aircraft.
Address :	2-5-1, Marunouchi, Chiyoda-ku, Tokyo 100
Tel & Fax :	Tel: 03-3212-3111 Fax: 03-3284-1927
Statistics :	Capital : 253483 million yen Employees : 43914
Officers :	CEO : Mr. Kentaro Aikawa/President
Sales :	March, 1989 - 1711250 million yen March, 1990 - 2094260 million yen March, 1991 - 2327070 million yen
R & D Spending :	112200 million yen Capital Investment: 92811 million yen
Sales Breakdown :	SIC# Product 3599 Industrial mahinery 3731 Shipbuilding 1629 Power plants 3721 Aircraft
Main Products :	Shipbuilding; Steel Structures; Prime Movers; Machinery; Aircraft; Special Motor Vehicles
Banks :	Mitsubishi Bank; Mitsubishi Trust and Banking; Industrial Bank of Japan; Mistui Taiyo Kobe Bank; Tokai Bank
Branches :	Jakarta Liaison Office (Indonesia)
Laboratories :	Basic Research Laboratory (Yokohama-shi); Takasago Laboratory (Takasago-shi)
Overseas Subsidiaries :	MHI Forklift America, Inc. (Houston, U.S.A.),; Mitsubishi Heavy Industries Europe Ltd. (London, U.K.)

Mitsubishi Kasei Corp.

Japanese Name	:	***Mitsubishi Kasei K.K.***
TSE Number	:	4010
Description	:	The largest integrated chemical company in Japan, with main lines in cokes, petrochemicals, and other carbochemicals. Diversifying into pharmaceuticals, electronic materials, and biochemicals.
Address	:	Mitsubishi Bldg., 2-5-2, Marunouchi, Chiyoda-ku, Tokyo 100
Tel & Fax	:	Tel: 03-3283-6274 Fax: 03-3283-6287
Statistics	:	Capital : 108375 million yen Employees : 8964
Officers	:	CEO : Mr. Masahiko Furukawa/President Contact : Mr. Yasumasa Yamada/Manager, Overseas Office
Sales	:	September, 1988 - 439258 million yen March, 1988 (6 months) - 346300 million yen March, 1990 - 731800 million yen March, 1991 - 770525 million yen
R & D Spending	:	40000 million yen Capital Investment: 41000 million yen
Sales Breakdown	:	

SIC#	Product	Sales
299	Petrochemical products	39 %
0000	Functional products	32 %
281	Carbon and inorganic chemicals	29 %

Main Products	:	Mitsubishi Kasei Corp (MKC), a company which centers its business activities on coal and petrochemicals, has grown to be the largest integrated chemicals manufacturer in Japan. In 1981, MKC formulated a medium-to-long-term plan.
Banks	:	Mitsubishi Bank; Mitsubishi Trust and Banking; Industrial Bank of Japan; Norinchukin Bank; Dai-Ichi Kangyo Bank; Long Term Credit Bank of Japan
Factories	:	Kurosaki Factory (Kitakyushu-shi); Yokkaichi Factory (Yokkaichi-shi); Mizushima Factory (Kurashiki-shi); Sakaide Factory (Sakaide-shi); Kashima Factory (Ibaraki-ken)
Laboratories	:	Research Center (Yokohama-shi)
R&D Areas	:	Information Electronics: Recording materials, Rewritable optical disks; Pharmaceuticals; Functional Materials: Specialty polyolefin-based products, Processing technology for oxo alcohol manufacture
Subsidiaries	:	Mitsubishi-Kasei Institute of Life Science (Machida-shi); The Mitsubishi-Kasei Institute of Toxicological and Environmental Science (Minato-ku); Plantech Research Institute (Chiyoda-ku)
Overseas Subsidiaries	:	Mitsubishi Kasei America Inc. (New York, U.S.A.); Mitsubishi Kasei Europe GmbH (Dusseldorf, Germany); Mitsubishi Kasei do Brasil Ltda (Rio de Janeiro, Brazil)

Mitsubishi Materials Corp.

Japanese Name	:	***Mitsubishi Materiaru K.K.***
TSE Number	:	5711
Description	:	Major comprehensive manufacturer of metals and cement. A key member of the Mitsubishi group of companies, Mitsubishi Materials was formed through a recent merger between Mitsubishi Metal and Mitsubishi Mining & Cement. The company produces a broad range of products and is noted for its activities in electronics materials, nuclear engineering and sintered tools, all areas which are stressed in R&D expenditure.
Address	:	1-6-1, Ohtemachi, Chiyoda-ku, Tokyo 100
Tel & Fax	:	Tel: 03-5252-5200 Fax: 03-5252-5270
Statistics	:	Capital : 96845 million yen Employees : 9383
Officers	:	CEO : Mr. Masaya Fujimura/President
Sales	:	March, 1988 - 656441 million yen March, 1989 - 718466 million yen March, 1990 - 642753 million yen March, 1991 - 760996 million yen
R & D Spending	:	9778 million yen Capital Investment: 71545 million yen

Sales Breakdown :

SIC#	Product	Sales
3339	Gold and silver	16 %
3331	Copper	15 %
3241	Cement	14 %
3463	Sintered alloy products	11 %
3411	Aluminum cans	11 %
34	Special alloys and industrial machinery	9 %
3299	Non-metal and metal advanced materials	7 %
3339	Other nonferrous metals	5 %
29	Petroleum and coal products	4 %
3281	Construction materials	3 %

Main Products	:	Primary Non-Ferrous Metals: Copper, Copper wire rods, Copper cakes and billets, Zinc, Zinc die-casting alloys, Tin, Lead, Gold, Silver; Processed Metal Products: Aluminum cans, Cemented carbide products and tools, Diamond wheels and tools, Sintered machine parts, Sintered oil-impregnated sleeve bearings, Electrical contacts, Motor timers, Micro motors, Precision molding dies, Special corrosion and heat resistant alloys, Special copper alloys, Precision casting products, Copper tubes, Industrial machinery; Ceramics: Multilayer ceramic capacitors, Surge absorbers, NTC thermistors, Ring varistors, Ceramic artificial bones; Advanced Materials: Gold bonding wire, Special solders, Ribbons and pellets, Cladding materials, Sputtering targets, High purity materials for semiconductors, High purity selenium and tellurium, Selenium and tellurium alloys, Dental alloys, Quartz crucibles and other quartz products, Minor metal and inorganic salts, Pigments, Electroconductive powders, Flourine chemicals, Silicon wafers for semiconductors; Petroleum and Coal Products: Gasoline, Naptha, Kerosene, Gas oil, Heavy fuel oil, Asphalt, Lubricant oils, LPG, Steaming coal, Coking coal, Coke, Electrofused magnesia; Cement and Construction Materials: Portland cement, Soil stabilizing cement, Other cement products, Gravel, Sand, Rubble, Reclamation sand; Nuclear Energy: Nuclear R&D, Engineering, Construction and operation of nuclear fuel cycle facilities; Other: Real estate, Agriculture and forestry, Engineering, Consulting; Industrial Chemical Processing: Nuclear fuel processing
Banks	:	Mitsubishi Bank; Mitsubishi Trust and Banking; Industrial Bank of Japan
Laboratories	:	Central Research Institute (Omiya-shi, Saitama-ken); Cement Research Institute (Omiya-shi, Saitama-ken); New Products Development Center (Omiya-shi, Saitama-ken); Compound Semiconductor Center (Omiya-shi, Saitama-ken)

Mitsubishi Mining & Cement Co., Ltd.

Japanese Name	:	***Mitsubishi Kogyo Cement K.K.***
TSE Number	:	5238
Description	:	One of Japan's 'Big Four' cement producers. A member of the Mitsubishi group of companies, Mitsubishi Mining is diversifying into new materials including ceramics and a new artificial rhinoceros horn.
Address	:	Shin-Maru Bldg., 1-5-1, Marunouchi, Chiyoda-ku, Tokyo 100
Tel & Fax	:	Tel: 03-3211-7412 Fax: 03-3284-1480
Statistics	:	Employees : 1998
Officers	:	CEO : Mr. Masaya Fujimura/President
Sales Breakdown	:	SIC# Product 3241 Cement 2999 Petrochemicals
Main Products	:	Cement; Petroleum Products; Building Materials
Banks	:	Mitsubishi Bank; Mitsubishi Trust and Banking; Long Term Credit Bank; Industrial Bank of Japan, Nippon Credit Bank

Mitsubishi Motors Corp.

Japanese Name : ***Mitsubishi Jidosha Kogyo K.K.***

TSE Number	:	7211
Description	:	Major automobile manufacturer. A member of the Mitsubishi group of companies, Mitsubishi Motors maintains strong ties with Chrysler Corp. (U.S.) and is involved in overseas joint production ventures with Chrysler, Daimler-Benz (Germany) and Volvo (Sweden).
Address	:	Tamachi Bldg., 5-33-8, Shiba, Minato-ku, Tokyo 108
Tel & Fax	:	Tel: 03-3456-1111 Fax: 03-5232-7747
Statistics	:	Capital : 109477 million yen Employees : 25500
Officers	:	CEO : Mr. Hirokazu Nakamura/President
Sales	:	March, 1988 - 1752690 million yen March, 1989 - 1898830 million yen March, 1990 - 2025720 million yen March, 1991 - 2313640 million yen

Sales Breakdown :

SIC#	Product	Sales
3711	Passenger automobiles	38 %
3713	Trucks and buses	30 %
3714	CKD kits for overseas assembly	8 %
3711	Minicars	8 %

Main Products	:	Motor Vehicles: Passenger cars, Minicars, Trucks, Buses; Related Products: Automobile parts, CKD kits
Banks	:	Mitsubishi Bank; Mitsubishi Trust and Banking; Mitsui Taiyo Kobe Bank; Industrial Bank of Japan; Long Term Credit Bank of Japan; Tokai Bank; Dai-Ichi Kangyo Bank; Bank of Tokyo; Norinchukin Bank; Sumitomo Trust and Banking
Laboratories	:	Passenger Car Engineering Center (Okazaki-shi and Kyoto); Truck and Bus Engineering Center (Tokyo)
Subsidiaries	:	Tohoku Mitsubishi Automotive Components Co., Ltd.; Kureha Motors Corp. Mitsubishi Automotive Engineering Co., Ltd. and others

Mitsubishi Pencil Co., Ltd.

Japanese Name	:	***Mitsubishi Enpitsu K.K.***
TSE Number	:	7976
Description	:	Largest manufacturer of pens and pencils in Japan. Mitsubishi Pencil has begun to diversify into office automation machinery and sundry office goods. In addition to office related products, the company is also moving into the area of cosmetics and hair care products.
Address	:	5-23-37, Higashi-Ohi, Shinagawa-ku, Tokyo 140
Tel & Fax	:	Tel: 03-3458-6221 Fax: 03-3472-6214
Statistics	:	Employees : 990
Officers	:	CEO : Mr. Eiichiro Suhara/President Contact : Mr. Eiichiro Suhara/President
Sales	:	December, 1988 - 38359 million yen December, 1989 - 41645 million yen December, 1990 - 46348 million yen

Sales Breakdown :

SIC#	Product	Sales
3951	Ball-point pens	35 %
3952	Pencils	15 %
3951	Marking pens	15 %
357	Office automation equipment	10 %
3579	Misc. office goods	10 %
2844	Cosmetics	

Main Products	:	Pens and Pencils: Ball-point pens, Lead pencils, Mechanical pencils, Marking pens; Office Automation Equipment; Misc. Office Goods; Cosmetics
Banks	:	Industrial Bank of Japan; Bank of Yokohama; Mitsubishi Bank; Sumitomo Bank; Dai-Ichi Kangyo Bank
Overseas Subsidiaries	:	Key Wave Industrial Ltd. (Hong Kong); Mitsubishi Pencil Co., U.K. Ltd. (UK); Royal Sovereign (UK); Mitsubishi Pencil Corp. of America (California, U.S.A.); Unika (Delaware, U.S.A.)

Mitsubishi Petrochemical Co., Ltd.

Japanese Name : ***Mitsubishi Yuka K.K.***

TSE Number : 4184

Description : Mitsubishi Petrochemical is Japan's largest petrochemicals manufacturer and was jointly established in 1956 by the Mitsubishi and Royal Dutch/Shell groups. The company is boosting production of higher-value-added chemical products and fine chemicals, acrylic acid esters, and diversifying into biotechnology including diagnostic products. The push into new areas is strengthened by the activities of its consolidated subsidiaries and affiliates.

Address : Mitsubishi Bldg., 2-5-2, Marunouchi, Chiyoda-ku, Tokyo 100

Tel & Fax : Tel: 03-3283-5699 Fax: 03-3283-5472

Statistics : Capital : 53798 million yen
Employees : 3439

Officers : CEO : Mr. Masaki Yoshida/President

Sales : March, 1988 - 340000 million yen
March, 1989 - 344678 million yen
March, 1990 - 385358 million yen
March, 1991 - 434416 million yen

R & D Spending : 21000 million yen Capital Investment: 44400 million yen

Sales Breakdown :

SIC#	Product	Sales
28	General chemicals	46 %
2821	Synthetics	37 %
2869	Specialty chemicals	12 %
28	Fine chemicals	5 %

Main Products : Resin Products: High density polyethylene, Low density polyethylene, Polypropylene, Polyester, Functional resins; Chemical Products: Styrene monomer, Ethylene oxide, Ethylene glycol, Melamine, Phenol, Acetone, Bisphenol A, Ethylene, Benzene, Toluene, Xylene; Specialty Chemicals: Alkyl benzene (hard type), High purity alcohols, Acrylic acid, Ethyl acrylate, Aliphatic denatured polyamine, Alphatic denatured polyamine adduct, Aliphatic denatured polymercaptan, Preservatives, Bactericides for foods, Bactericides for fibers, Bactericides/preservatives for paints, Petroleum-based benzine, Epoxy resin curing agent, Ethylene, Ethylene-vinyl acetate copolymer, Higher alcohol for OXO process, Propylene, Sea surface pollution absorptive agent, Slip additive, Surface coating agents; Materials: Plastic Flooring; Toilet Preparations: Hair foam and mousses; Agricultural Chemicals: Soil conditioners; Diagnostic Substances: Diagnostic kits, Blood testing facilities

Banks : Mitsubishi Bank; Norinchukin Bank; Mitsubishi Trust and Banking; Long Term Credit Bank of Japan; Industrial Bank of Japan

Branches : Osaka Branch (Osaka-shi); Nagoya Branch (Nagoya-shi); Kyushu Branch (Fukuoka-shi); Hokkaido Branch (Sapporo-shi); Yokkaichi Division (Yokkaichi-shi); Kashima Division (Kashima-shi); London Office (London)

Factories : Yokkaichi Division (Yokkaichi-shi, Mie-ken)

Laboratories : Tsukuba Research Center (Inashiki-gun, Ibaraki-ken); Yokkaichi Research Center (Yokkaichi-shi)

Mitsubishi Rayon Co., Ltd.

Japanese Name	:	***Mitsubishi Reiyon K.K.***
TSE Number	:	3404
Description	:	Manufacturer of various types of synthetic fiber, related resin products and specialty chemicals.
Address	:	2-3-19, Kyobashi, Chuo-ku, Tokyo 104
Tel & Fax	:	Tel: 03-3272-4321 Fax: 03-3245-8781
Statistics	:	Capital : 51922 million yen Employees : 5145
Officers	:	CEO : Mr. Yataro Nagai/President
Sales	:	March, 1988 - 193973 million yen March, 1989 - 200325 million yen March, 1990 - 257918 million yen March, 1991 - 275413 million yen
ending	:	14400 million yen Capital Investment: 31400 million yen
Sales Breakdown	:	SIC# Product 2281 Synthetic fibers 2821 Resin products
Main Products	:	Fiber Products: Acrylic fibers, Acetate products, Polyester filaments, Polypropylene filaments, Suede-type fabrics; Resin Products: Methyl methacrylate resins, ABS resins, PBT resins, PET resins, Polycarbonate sheets, Polypropylene films, Coating resins, MBS resins, Plastic-fabricated products; Specialty Chemicals: Alkyl methacrylates, Fiber-use antistatic agent, Methacryl resin polymer, Sea surface pollution absorbent
Banks	:	Mitsubishi Bank; Mitsubishi Trust and Banking; Dai-Ichi Kangyo Bank; Industrial Bank of Japan; Long Term Credit Bank of Japan
Branches	:	Osaka Branch (Osaka-shi); Nagoya Branch (Nagoya-shi)
Factories	:	Toyohashi Production Center (Toyohashi-shi, Aichi-ken); Toyama Production Center (Toyama-ken)
Laboratories	:	Central Research Laboratories (Otake-shi); Products Development Laboratories (Nagoya-shi); Tokyo Research Laboratories (Kawasaki-shi)
Subsidiaries	:	Nitto Chemical Industry Co., Ltd. (Shinmaru Bldg. 4th Floor, 1-5-1, Marunouchi, Chiyoda-ku, Tokyo 100); Mitsubishi Rayon Engineering Co., Ltd. (3-1-1, Kyobashi, Chuo-ku, Tokyo 104)
Overseas Subsidiaries	:	Mitsubishi Rayon of America, Inc. (Los Angeles, U.S.A.)

Mitsui & Co., Ltd.

Japanese Name : ***Mitsui Bussan K.K.***

TSE Number : 8031

Description : Mitsui & Co. is Japan's second largest trading company vying with Mitsubishi Corp. Mitsui's mainstays are metals, machinery, foodstuffs, energy and chemicals, and ranks number one in chemicals and steel. The company is highly diversified with over 150 overseas branches and is active in overseas resource development. Exposure in the Middle East contributed to a decrease in sales for the current year but operating profits remain steady. Mitsui is advancing into financing and leasing, promoting the leasing of aircraft and engine parts. The company is pursuing trade in opening markets in Asia, with a strong foothold in the Chinese market. Also slated to participate in McDonnel Douglas Corp.'s MD12X 400-seat airliner project.

Address : 1-2-1, Otemachi, Chiyoda-ku, Tokyo 100

Tel & Fax : Tel: 03-3285-1111 Fax: 03-3285-9809

Statistics : Capital : 175803 million yen
Employees : 9145

Officers : CEO : Mr. Naohiko Kumagai/President

Sales : March, 1988 - 14131200 million yen
March, 1989 - 14827600 million yen
March, 1990 - 20330000 million yen
March, 1991 - 18234100 million yen

Sales Breakdown :

SIC#	Product	Sales
5084	Machinery and equipment	21 %
5051	Non-ferrous metals	20 %
5172	Petroleum	16 %
5051	Steel	12 %
5149	Foodstuffs	10 %
5162	Chemicals	9 %
5131	Textiles	5 %
4226	Warehousing	4 %

Main Products : Industrial Machines: Electrical machines, Plant and heavy machinery; Metals Service: Non-ferrous metals; Petroleum Products: Gas, Petrochemicals, LNG; Metals Service: Iron ore, Steel scrap; Groceries and Related Products: Various kinds of foods and food additives, Sugar, Livestock; Plastics Materials and Basic Chemicals: Plastic resins; Piece Goods: Textiles; Special Warehousing: Goods and commodities warehousing

Banks : Mitsui Taiyo Kobe Bank; Fuji Bank; Bank of Tokyo;Industrial Bank of Japan; Mitsui Trust and Banking; Long Term Credit Bank of Japan;

Subsidiaries : Mitsui Oil Co., Ltd.; Mikuni Coca-Cola Bottling Co., Ltd.; Mitsui Liquid Gas Co., Ltd.

Overseas Subsidiaries : Mitsui & Co. (U.S.A.); Mitsui & Co. Europe Ltd.

Mitsui Mining Co., Ltd.

Japanese Name :	***Mitsui Kozan K.K.***
TSE Number :	1501
Description :	Major coal mining company. Mitsui Mining Co. is also engaged in the production of various petroleum products as well as construction materials and industrial machinery.
Address :	2-1-1, Nihonbashi-Muromachi, Chuo-ku, Tokyo 103
Tel & Fax :	Tel: 03-3241-1334 Fax: 03-3241-8684
Statistics :	Employees : 1200
Officers :	CEO : Mr. Tadashi Harada/President
Sales :	March, 1988 - 251163 million yen March, 1989 - 249326 million yen March, 1990 - 253900 million yen March, 1991 - 289299 million yen

Sales Breakdown :

SIC#	Product	Sales
12	Coal	30 %
29	LPG and petroleum	25 %
3241	Cement and construction material	20 %
2999	Coke	10 %
35	Industrial machinery	

Main Products :	Coal and Petroleum Products: Coal, Petroleum, LPG, Coke, CNW (coal-water) fuels; Construction Materials: Cement, Ceramic materials; Industrial Equipment: Pollution control equipment
Banks :	Mitsui Trust and Banking; Mitsui Taiyo Kobe Bank; Industrial Bank of Japan; Long Term Credit Bank of Japan; Nippon Credit Bank
Factories :	Overseas coal mines: Canada; Australia
Overseas Subsidiaries :	Mitsui Mining Hong Kong Co., Ltd. (Hong Kong); Tokyo Cement Co., Ltd. (Sri Lanka); Quintette Coal Ltd. (Canada)

Mitsui Petrochemical Industries, Ltd.

Japanese Name	:	***Mitsui Sekiyu Kagaku K.K.***
TSE Number	:	4183
Description	:	A comprehensive chemicals manufacturer belonging to the Mitsui group.
Address	:	Kasumigaseki Bldg., 3-2-5, Kasumigaseki, Chiyoda-ku, Tokyo 100
Tel & Fax	:	Tel: 03-3580-3616 Fax: 03-3593-0028
Statistics	:	Capital : 32084 million yen Employees : 4461
Officers	:	CEO : Mr. Shogo Takebayashi/President
Sales	:	March, 1988 - 232465 million yen March, 1989 - 262194 million yen March, 1990 - 297527 million yen March, 1991 - 330826 million yen
R & D Spending	:	24500 million yen Capital Investment: 28200 million yen

Sales Breakdown :

SIC#	Product	Sales
282	Chemicals and raw materials	48 %
282	Synthetic resins	30 %
286	Specialty chemicals	22 %

Main Products : Basic Petrochemicals: Ethylene, Propylene, Budateine, Butene-1, C5 fraction, High-purity hydrogen, Acetaldehyde, Tert-Butyl Alcohol, Benzene, Toluene, Xylene, Cumene, Cymene, Diisopropyl benzene, Diisopropyl toluene, Acetone; Synthetic Resin Raw Materials: Purified terephthalic acid, Ethylene glycol, Hexylene glycol, Ethylene oxide; Phenol Products: Phenol, Bisphenol A, Acetophenone, alpha-Methylstyrene, meta.para-Cresol, Aniline, Isopropenul toluene; Other Industrial Chemicals: Hydroquinone, Resorcinol; Synthetic Resins: Polyethylene (all densities), Filler-reinforced polypropylene, Glass reinforced polypropylene, Adhesive polyolefin, Barrier resin, Low molecular weight polyethylene; Elastomers: Ethylene-propylene-dieneterpolymer, Olefinic thermoplastic elastomer, Polyolefin-based copolymer for resin modification, Olefin copolymer with heat sealability, High-performance olefin thermoplastic elastomer; Hydrocarbon Resins: High-quality hydrocarbon resin, Aliphatic hydrocarbon resin, Aromatic hydrocarbon resin; Fine Chemicals: Cumene hydroperoxide, Dicumyl peroxide, p-isopropyl phenol, p-isopropyl aniline, 1.3 bis and 1.4 bis [2-Hydroxy-2-propyl] benzene, m-Amino phenyl; Engineering Plastics: Glass-reinforced specialty polyamide resin, Polyamide polymer alloy, Glass-reinforced polymethylpentene, Transparent thermoset resin/ADC resin; Specialty Polyolefins: Polymethylpentene, Ultrahigh molecular weight polyethylene, High self-lubricating specialty polyolefin resin; High-performance Polymers: Hydrocarbon-based synthetic oil, Lubricant viscosity index improver, Liquid polyolefin adhesive; Epoxies and Compounded Resins: Epoxy resin and special curing agent, Epoxy resin formulations; Synthetic Pulp: Polyethylene/polypropylene pulp, Cotton-like polyethylene fiber (Asbestos alternative); Gas Piping: Pipe and fittings for gas distribution; Nonwoven Fabrics: Spun-bonded non-woven polypropylene matting and fabric, Filters and oil-absorbent materials made of nonwoven polypropylene fabric, Non-woven electrified filter; Agricultural Materials: Non-woven polypropylene fabric covering materials, Irrigation tubes, PVA film-based covering materials, Bird net, Hosing for powder spraying; Electronic Materials: Optical disks, WORM disks, Rewritable disks, Optical memory devices, Color image filing systems, CD-ROM verification systems; High Performance Materials: Synthetic crystals,

		Amorphous magnetic parts, Unmanned transporter system, Ultra high molecular weight polyethylene fiber, Ultra-high temperature carbon heat insulating material, High-purity chemical dispenser, Ultra-filtration modules; Biotech Products: Decorative houseplant set, Potted trumpet lillies, Mini-roses
Banks	:	Mitsui Taiyo Kobe Bank; Mitsui Trust and Banking; Long Term Credit Bank of Japan; Industrial Bank of Japan; Norinchukin Bank; Daiwa Bank
Branches	:	Osaka Branch (Osaka-shi); Nagoya Branch (Nagoya-shi); Fukuoka District Sales Office (Fukuoka-shi); Sapporo District Sales Office (Sapporo-shi); Hiroshima District Office (Hiroshima-shi); Beijing Office (Beijing, China)
Factories	:	Iwakuni-Ohtake Works; Chiba Works (Chiba-shi)
Overseas Subsidiaries	:	Mitsui Petrochemicals (America), Ltd. (U.S.A.); Mitsui Petrochemical Industries (Europa) GmbH (Germany)

Mitsui Toatsu Chemicals, Inc.

Japanese Name	:	**Mitsui Toatsu Kagaku K.K.**
TSE Number	:	4001
Description	:	A comprehensive chemical manufacturer belonging to the Mitsui group, the company is the third largest chemical maker in Japan. Originally specializing in fertilizers, Mitsui Toatsu now stresses agrochemicals, pharmaceuticals, chemicals and functional resins. Conducting biotechnical R&D into hybrid rice strains.
Address	:	3-2-5, Kasumigaseki, Chiyoda-ku, Tokyo 100
Tel & Fax	:	Tel: 03-3592-4111 Fax: 03-3592-4267
Statistics	:	Capital : 70247 million yen Employees : 5337
Officers	:	CEO : Mr. Haruo Sawamura/President
Sales	:	March, 1988 - 363338 million yen March, 1989 - 394766 million yen March, 1990 - 413322 million yen March, 1991 - 436603 million yen
R & D Spending	:	21200 million yen Capital Investment: 43000 million yen
Sales Breakdown	:	

SIC#	Product	Sales
28	Industrial chemicals	35 %
282	Synthetic resins and resin processing	29 %
0000	Functional products	26 %
28	Fine chemicals	10 %

Main Products	:	Industrial Chemicals: Phenol, Bis-phenol A, Methanol, Acrylonitryl, Acryl-amide, Acetone, Acid dyes, Phenol antioxidant, Bactericides/preservatives for foods/trees/bamboo/paints, Ammonia cooling agent, Methylene chloride cooling agent, Carbon dioxide cooling agent, Formalin, Iodine, Methacryl resin monomer, Phthalic anhydride, Vinyl chloride monomer, Caustic soda, Ethylene oxide, Ethylene glycol; Agrochemicals: Chlorpicrate, Agricultural intermediates, Chemical feeds; Fine |Chemicals: DPA, MR-6, Paranitro chlor-benzine, Raw materials for carcinostatic agent HCFU; Synthetic Resins and Plastics: PVC, Polypropylene, Polystyrene, Films for food wrapping, Agricultural plastic films, Urethane raw materials (PPG, TDI, MDI); Materials: Glass fiber building materials, Metal exterior walling, Functional resins, Electronic materials
Banks	:	Mitsui Taiyo Kobe Bank; Mitsui Trust and Banking; Industrial Bank of Japan; Long Term Credit Bank; Norinchukin Bank
Branches	:	Osaka Branch (Osaka); Nagoya Branch (Nagoya-shi, Aichi-ken); Fukuoka Branch (Fukuoka-ken); Sapporo Branch (Hokkaido); New York Office (New York); London Office (London); Hong Kong Office (Hong Kong)
Factories	:	Chiba Plant (Shigehara-shi, Chiba-ken); Nagoya Plant (Nagoya-shi, Aichi-ken); Osaka Plant (Takaishi-shi, Osaka); Hiroshima Plant (Shimonoseki-shi, Yamaguchi-ken); Omuta Plant (Omuta-shi, Shizuoka-ken)
Laboratories	:	General Laboratory (Yokohama-shi)
Overseas Subsidiaries	:	Mitsui Toatsu Chemicals (Asia) Pte., Ltd. (Singapore); MTC Industries, Inc. (U.S.A.); Mitsui Toatsu Chemicals (Europe) B.V. (Netherlands)

Mochida Pharmaceutical Co., Ltd.

Japanese Name	:	***Mochida Seiyaku K.K.***
TSE Number	:	4534
Description	:	Medium-standing pharmaceutical manufacturer noted for its diagnostics, hormone-related products and other metabolic preparations. In recent years the company has been sustaining relatively high R&D expenditures in its efforts to diversify into skin care products and medical equipment.
Address	:	1-7, Yotsuya, Shinjuku-ku, Tokyo 160
Tel & Fax	:	Tel: 03-3358-7211 Fax: 03-3358-7887
Statistics	:	Capital : 5021 million yen Employees : 2241
Officers	:	CEO : Mr. Susumu Watanabe/President Contact : Mr. Kunio Kokubu/Director, International Operations
Sales	:	March, 1988 - 45638 million yen March, 1989 - 49791 million yen March, 1990 - 46576 million yen March, 1991 - 49613 million yen
R & D Spending	:	6857 million yen Capital Investment: 6303 million yen
Sales Breakdown	:	SIC# Product Sales 2834 Pharmaceuticals 92 % 3845 Medical equipment 8 %
Main Products	:	Pharmaceuticals: Circulatory agents, Metabolic agents, Hormones, Antibiotics, Medicated cosmetics and other dermatological preparations, Diagnostic reagents; Medical Equipment
Banks	:	Mitsubishi Bank; Dai-Ichi Kangyo Bank; Mitsui Taiyo Kobe Bank; Tokai Bank; Yasuda Trust and Banking; Long Term Credit Bank of Japan
Factories	:	Shizuoka Plant (Shizuoka-ken); Oji Plant (Tokyo); Saitama Plant (Saitama-ken)
Laboratories	:	Fuji Central Research Laboratory (Shizuoka-ken); Toxicology Laboratory (Shizuoka-ken); Pharmaceutical Laboratory (Shizuoka-ken); Biosciences Research Laboratory (Tokyo)
R&D Areas	:	Pharmaceuticals
Shareholders	:	Ei Mochida - 8.3% ; Katsuro Mochida - 7.4% ; Mitsubishi Bank [8315] - 5.0% ; Kazue Mochida - 4.9% ; Mochida Memorial Foundation - 4.5% ; Dai-Ichi Kangyo Bank [8311] - 4.5% ; Susumu Watanabe - 4.4%
Subsidiaries	:	Mochida International Co., Ltd. (Tokyo); Tokai Chemical Co., Ltd. (Tokyo)
Overseas Subsidiaries	:	Mochida U.S.A., Inc. (Delaware, U.S.A.)

Murata Mfg. Co., Ltd.

Japanese Name	:	***Murata Seisakusho K.K.***
TSE Number	:	6981
Description	:	Leading producer of ceramic capacitors, filters and related components for use by the automotive, photographic and communications industries. R&D stresses mobile communications technologies such as cellular phones and a nation-wide radio paging system.
Address	:	2-26-10, Tenjin, Nagaokakyo City, Kyoto 617
Tel & Fax	:	Tel: 075-951-9111 Fax: 075-954-7720
Statistics	:	Capital : 51570 million yen Employees : 3800
Officers	:	CEO : Mr. Yasutaka Murata/President
Sales	:	September, 1988 (Adj. term) - 111805 million yen March, 1989 (6 months) - 103773 million yen March, 1990 - 213913 million yen March, 1991 - 236397 million yen
R & D Spending	:	11700 million yen Capital Investment: 15993 million yen
Sales Breakdown	:	

SIC#	Product	Sales
3675	Capacitors	36 %
3679	Piezoelectric components	18 %
3672	Coils	14 %
3672	Electronic modules	13 %
367	Other electronic products	11 %

Main Products	:	Capacitors: Ceramic capacitors, Trimmer capacitors, Electrical double-layer capacitors; Piezoelectric Components: Ceramic filters, Surface wave filters, Ceramic oscillators, Supersonic wave generators, Piezoelectric buzzers, Piezoelectric speakers; Electronic Coils: Flyback transformers, Assorted electronic coils; Electronic Modules: Television tuners, Cable television tuners, Hybrid ICs, Active filters, CR composite parts, High voltage power rectifiers, Antenna parts, LC filters, RF modulators, Microwave generators; Other Electronic Products: Microwave filters, EMI exclusion filters, Various sensors, Heat fuses, Ferrite cores, Connectors, Silver paste, Electrical resistant paste
Banks	:	Fuji Bank ; Sumitomo Bank; Yasuda Trust and Banking; Sumitomo Trust and Banking
Branches	:	Tokyo Branch (Kawasaki-shi, Kanagawa-ken)
Factories	:	Yokaichi Plant (Yokaichi-shi, Shiga-ken); Yasu Plant (Yasu-gun, Shiga-ken)
Laboratories	:	Yasu Plant (Yasu-gun, Shiga-ken); Yokohama Development Center
R&D Areas	:	Electronic Products and Components: Ceramic sensor gyroscopes, National paging service and supporting technologies, Ceramic antenna for GPS (global positioning system) application, Cellular phone systems and components.

Nachi-Fujikoshi Corp.

Japanese Name	:	***Fujikoshi K.K.***
TSE Number	:	6474
Description	:	Steel manufacturer specializing in bearings and cutting tools. Nachi-Fujikoshi is also active in industrial robotics where it is seen as a leader in the field.
Address	:	Sekai Boeki Senta Bldg. 25F, 2-4-1, Hamamatsu-cho, Minato-ku, Tokyo 105
Tel & Fax	:	Tel: 03-3435-5111 Fax: 03-3435-5144
Statistics	:	Employees : 4750
Officers	:	CEO : Mr. Kunio Owada/President
Sales	:	November, 1988 - 120317 million yen November, 1989 - 145127 million yen November, 1990 - 160272 million yen

Sales Breakdown :

SIC#	Product	Sales
3562	Bearings	30 %
354	Machine tools	23 %
34	Tools	20 %
35	Industrial equipment	15 %
34	Misc. steel products	12 %

Main Products	:	Bearings and Related Products; Machine Tools: Cutting tools; Tools; Industrial Equipment: Hydraulic equipment; Misc. Steel Products
Banks	:	Tokai Bank; Hokkuriku Bank; Mitsui Trust and Banking; Industrial Bank of Japan; Mitsubishi Trust and Banking; Kyowa Saitama Bank

Nakadai Kinzoku Co., Ltd.

Japanese Name	:	***Nakadai Kinzoku Co., Ltd.***
TSE Number	:	Not listed
Description	:	Manufacturer of metals and metal products.
Address	:	2-9-2, Kahei, Adachi-ku, Tokyo 121
Tel	:	03-3605-7730
Sales Breakdown	:	SIC# Product 33 Metal products
Main Products	:	Metals and Metal Products

NEC Corp.

Japanese Name : ***Nippon Denki K.K.***

TSE Number	:	6701
Address	:	5-7-1, Shiba, Minato-ku, Tokyo 108-01
Tel & Fax	:	Tel: 03-3454-1111 Fax: 03-3457-7249
Statistics	:	Capital : 18448 million yen Employees : 3800
Officers	:	CEO : Mr. Kenzo Nakamura/Chairman Contact : Mr. Tadahiro Sekimoto/President
Sales	:	March, 1989 - 2542050 million yen March, 1990 - 2760680 million yen March, 1991 - 2961100 million yen
R & D Spending	:	300000 million yen Capital Investment: 265000 million yen

Sales Breakdown :

SIC#	Product	Sales
3571	Computers and computer components	50 %
366	Communications equipment	30 %
36	Electronic instruments and equipment	18 %
36	Home electronics	
367	Semiconductors and ICs	

Banks	:	Sumitomo Bank; Sumitomo Trust; Industrial Bank of Japan; Yokohama Bank; Long Term Credit Bank of Japan; Mitsubishi Bank; Kyowa-Saitama Bank
Branches	:	Business Offices: Sanda; Tamagawa; Fuchu; Abiko; Otsuki; Branch Offices: Hokkaido; Tohoku; Shizuoka; Tokyo; Saitama; Chiba; Kanagawa; Tachikawa; Chubu; Hokuriku; Kansai; Kyoto; Kobe; Chugoku; Shikoku; Kyushu
Laboratories	:	Central Laboratory (Kanagawa-ken); Tsukuba Laboratory (Ibaraki-ken)

NGK Insulators, Ltd.

Japanese Name : ***Nippon Gaishi K.K.***

TSE Number : 5333

Description : Largest insulator manufacturer expanding into a wider range of ceramic parts production.

Address : 2-56, Suda-cho, Mizuho-ku, Nagoya-shi, Aichi-ken 467

Tel & Fax : Tel: 052-872-7171 Fax: 052-872-7103

Statistics : Capital : 32200 million yen
Employees : 4961

Officers : CEO : Mr. Toshihito Kohara/President
Contact : Mr. Shigekazu Iida/International Sales Division

Sales : March, 1988 - 171514 million yen
March, 1989 - 180925 million yen
March, 1990 - 195235 million yen
March, 1991 - 206625 million yen

Capital Investment : 22006 million yen

Sales Breakdown :

SIC#	Product	Sales
3264	Electric power products	39 %
2899	Ceramic products	29 %
8711	Engineering services	22 %
33	Metals	10 %

Main Products : Electric Power Division: Insulators, Power transmission equipment, Power transformer equipment, Power distribution equipment, Insulator cleaning equipment, Insulator fire-proofing equipment; Ceramics: Automotive exhaust catalytic converters, Oxygen sensors, Various engine parts, Enamel building materials, Frit; Engineering Services: Drainage and water supply equipment installation, Sound-proofing, Infra-red heating equipment, Radioactive waste processing equipment; Metal Products: Various mother alloys, Metal molds, Deep sea cable parts, Beryllium-based products

Banks : Mitsubishi Bank; Tokai Bank; Sanwa Bank; Mitsui Taiyo Kobe Bank; Bank of Tokyo

Branches : Tokyo Office (Chiyoda-ku, Tokyo); Osaka Office (Chuo-ku, Osaka); Kyushu Office (Chuo-ku, Fukuoka-shi)

Factories : Chita Plant (Handa-shi, Aichi-ken); Komaki Plant (Komaki-shi, Aichi-ken); Nagatsugawa Plant (Nakatsugawa-shi, Gifu-ken)

Laboratories : Technical Research Laboratory. (Handa-shi); Super High Pressure Research Laboratory (Komaki-shi); R&D Dept. of Electric Parts Division. (Komaki-shi); R&D Division (Nagoya-shi)

NGK Spark Plug Co., Ltd.

Japanese Name	:	***Nippon Tokusho Togyo K.K.***
TSE Number	:	5334
Address	:	14-18, Takatsuji-cho, Mizuho-ku, Nagoya, Aichi-ken 467
Tel & Fax	:	Tel: 052-872-5936 Fax: 052-872-5994
Statistics	:	Capital : 22841 million yen Employees : 4567
Officers	:	CEO : Mr. Teiichi Suzuki/President Contact : Mr. Torii Noboru/Director, International Sales Division
Sales	:	March, 1988 - 86436 million yen March, 1989 - 93077 million yen March, 1990 - 97799 million yen March, 1991 - 105975 million yen
R & D Spending	:	5825 million yen Capital Investment: 10883 million yen
Sales Breakdown	:	SIC# Product Sales 3694 Spark plugs and related products 53 % 3299 Ceramic electronics and other products 46 %
Banks	:	Mitsubishi Bank; Tokai Bank; Mitsubishi Trust Bank; Bank of Tokyo; Sanwa Bank
Factories	:	Main Plant (Nagoya-shi, Aichi-ken); Komaki Plant (Aichi-ken); Kagoshima-Miyanojo Plant (Kagoshima-ken)
Laboratories	:	R&D Center (Aichi-ken)
Subsidiaries	:	Kamioka Ceramic Co., Ltd. (Gifu-ken); Kani Ceramic Co., Ltd. (Kani-shi, Gifu-ken).

Nihon Cement Co., Ltd.

Japanese Name	:	***Nihon Semento K.K.***
TSE Number	:	5231
Description	:	Major manufacturer of cement and related products. Current R&D includes ceramics and electronic products.
Address	:	Otemachi Bldg. 6F, 1-6-1, Otemachi, Chiyoda-ku, Tokyo 100
Tel & Fax	:	Tel: 03-3201-4033 Fax: 03-3201-4034
Statistics	:	Capital : 43461 million yen Employees : 2115
Officers	:	CEO : Mr. Toru Kitaoka/President Contact : Mr. Yukimi Fujiwara/Manager, International Dept.
Sales	:	March, 1988 - 155138 million yen March, 1989 - 186961 million yen March, 1990 - 195626 million yen March, 1991 - 212900 million yen
R & D Spending	:	3500 million yen Capital Investment: 18514 million yen

Sales Breakdown :

SIC#	Product	Sales
3241	Cement and related products	95 %
35	Industrial machinery	3 %
0000	Other	

Main Products	:	Cement and Related Products: Portland cement, Alumina cement, White cement, Quick drying cement, Compound cements, Assorted other cements, Assorted concretes, Rockwall, 'Clean Set' ground and sludge solidifying agents; Industrial Machinery: Waste management equipment, Assorted other plant equipment; Other: Concrete colors, Real estate, Sports equipment, Engineering and consulting
Banks	:	Fuji Bank; Yasuda Trust and Banking; Long Term Credit Bank of Japan; Dai-Ichi Kangyo Bank
Branches	:	Hokkaido Branch (Sapporo-shi); Sendai Branch (Sendai-shi); Hokuriku Branch (Toyama-shi); Nagoya Branch (Nagoya-shi); Osaka Branch (Osaka-shi); Takamatsu Branch (Takamatsu-shi); Kyushu Branch (Fukuoka-shi)
Laboratories	:	Central Research Laboratory

Nihon Nohyaku Co., Ltd.

Japanese Name	:	***Nihon Noyaku K.K.***
TSE Number	:	4997
Description	:	Japan's second largest specialized agricultural chemical manufacturer. Furnishes hepatic drugs to Daiichi Seiyaku Co., Ltd..
Address	:	Eitaro Bldg., 1-2-5, Nihonbashi, Chuo-ku, Tokyo 103
Tel & Fax	:	Tel: 03-3274-3371 Fax: 03-3281-5462
Statistics	:	Capital : 8276 million yen Employees : 830
Officers	:	CEO : Mr. Tasuku Kodaira/President Contact : Mr. Yoshitake Kono/General Manager, Overseas Division
Sales	:	September, 1988 - 47834 million yen September, 1989 - 47038 million yen September, 1990 - 47228 million yen

Sales Breakdown :

SIC#	Product	Sales
2879	Herbicides	33 %
2879	Pesticides	28 %
2879	Disinfectants	23 %
2879	Disinfectant pesticides	9 %
2048	Chemical feedstuffs	

Main Products	:	Agricultural Chemicals: Herbicides, Pesticides, Fertilizers, Disinfectants, Feeds; Pharmaceuticals: Liver treatment drugs; Specialty Chemicals: Diazinon, Dimethoate, Nicotine sulphate, Preparation streptomycin; Medical accessories; Synthetic Resins; Landscaping/Construction/Civil Engineering Materials; Agricultural Materials
Banks	:	Dai-Ichi Kangyo Bank; Norinchukin Bank; Mitsubishi Bank; Kyowa-Saitama Bank; Sumitomo Bank; Mitsui Taiyo Kobe Bank
Branches	:	Tokyo Branch (Chuo-ku, Tokyo); Sapporo Branch (Sapporo-shi, Hokkaido); Sendai Branch (Sendai-shi, Miyagi-ken); Nagoya Branch (Nagoya-shi, Aichi-ken); Osaka Branch (Osaka); Fukuoka Branch (Fukuoka-shi, Fukuoka-ken)
Factories	:	Fukushima Factory (Nihonmatsu-shi, Fukushima-ken); Kashima Factory (Kashima-gun, Ibaraki-ken); Osaka Factory (Nishi-Yodogawa-ku, Osaka); Saga Factory (Sanyoki-gun, Saga-ken)
Laboratories	:	Daiichi Chemical Laboratory (Osaka); Daini Chemical Laboratory (Osaka); Biological Laboratory (Kawachinagano-shi, Osaka); Safety Laboratory (Kawachinagano-shi, Osaka)
Shareholders	:	Asahi Denka Kogyo K.K. [4401] - 27.93% ; Dai-ichi Kangyo Bank [8311] - 4.1% ; Norin Chukin Bank - 3.0% ; Asahi Mutual Life - 3.0% ; Taisei Fire and Marine [8765] - 1.6 %
Subsidiaries	:	Nichino Rokuka Co., Ltd.; Nichino Rec Co., Ltd.; Japan House Tech Co., Ltd.

Nikkiso Co., Ltd.

Japanese Name :	***Nikkiso K.K.***
TSE Number :	6376
Description :	Engaged in the manufacture of precision pumps and related equipment for use in medicine and the chemical industry. Nikkiso Co. also produces measuring and controlling systems for thermal power plants. The company's hemodialysis products enjoy an excellent reputation within the medical profession.
Address :	3-43-2, Ebisu, Shibuya-ku, Tokyo 150
Tel & Fax :	Tel: 03-3443-3711 Fax: 03-3473-4963
Statistics :	Capital : 6062 million yen Employees : 1300
Officers :	CEO : Mr. Masahiko Hatano/President
Sales :	September, 1988 - 31231 million yen March, 1989 - 16695 million yen March, 1990 - 33327 million yen March, 1990 - 35849 million yen

Sales Breakdown :

SIC#	Product	Sales
384	Medical equipment and materials	37 %
35	Pumps and related equipment	31 %
382	Measuring devices	22 %
3299	New materials	
3843	Dental equipment and materials	

Main Products :	Pumps and Related Equipment: Centrifugal pumps, Metering pumps, Self-priming non-clog pumps, Non-sealed cannes motor pumps, LNG pumps, Multiplex reciprocal plunger pumps, Flow meters, Valves, Blowers; Power Plant equipment; High purity water conditioning systems for carbon-based fuel and nuclear power plants, Thermal power plant boiler control systems; Medical Equipment and Supplies: Artificial kidney systems, Dialysate system patient monitors, Transfusion pumps, Disposable hemodialyzers, Hypodermic needles, Catheters, Vascular grafts; Dental Equipment and Supplies; New Materials: Carbon fibers, Carbon fiber reinforced plastics
Banks :	Dai-Ichi Kangyo Bank; Tokai Bank; Kyowa Saitama Bank; Fuji Bank; Bank of Tokyo; Mitsubishi Bank; Mitsui Taiyo Kobe Bank
Factories :	Higashimurayama Plant (Tokyo); Shizuoka Plant (Shizuoka-ken)
Laboratories :	Central Research Laboratory (Shizuoka-ken)
Subsidiaries :	Nikkiso Sundstrand K.K.; Nikkiso YSI K.K.; Nippon Kogyo Senjo K.K.; Nikkiso Tohoku Iko K.K.; Nikkiso M.E. Maintenance K.K.; Alpha Medical K.K.; Nikkiso Eiko K.K.; Nikkiso M.E.S. K.K.; Nikkiso Technica K.K.; Nikkiso Ryuki Techno K.K.
Overseas Subsidiaries :	Nikkiso Deutschland GmbH (Germany); Nikkiso Industria e Comercio Lda. (Brazil)

Nikkoshi Co., Ltd.

TSE Number	:	Not listed
Description	:	Manufacturer of analytical instruments.
Address	:	Nagai Bldg., 4-4-11, Nihonbashi, Honcho, Chuo-ku, Tokyo 103-91
Tel	:	03-3270-8851
Sales Breakdown	:	SIC# Product 3826 Analytical instruments
Main Products	:	Analytical Instruments: Metal composition analyzers

Nippon Canon Inc.

Japanese Name	:	***Nippon Canon K.K.***
TSE Number	:	Not listed
Description	:	Small-scale manufacturer of industrial machinery. A member of the Canon group of companies, Nippon Canon is known for its expertise in polyurethane plant technology.
Address	:	4610-1, Shimoniikura, Wako-shi, Saitama 351-01
Tel & Fax	:	Tel: 0484-66-9891 Fax: 0484-66-9894
Statistics	:	Capital : 49 million yen Employees : 21
Officers	:	CEO : Mr. Waichi Sato/President Contact : Mr. Waichi Sato/President
Sales Breakdown	:	SIC# Product 35 Industrial machinery
Main Products	:	Industrial Machinery: Industrial robots, Clumpers, Polyurethane plant and equipment
Banks	:	Fuji Bank; Dai-Ichi Kangyo Bank; Tokai Bank
Branches	:	Nagoya Office (Nagoya-shi, Aichi-ken)
Factories	:	Wako Factory (Wako-shi, Saitama-ken); Saitama Factory (Saitama-ken)
Laboratories	:	Wako Laboratory (Wako-shi, Saitama-ken); Saitama Laboratory (Saitama-ken)

Nippon Carbon Co., Ltd.

Japanese Name	:	***Nippon Kabon K.K.***
TSE Number	:	5302
Address	:	2-6-1, Hatchobori, Chuo-ku, Tokyo 104
Tel & Fax	:	Tel: 03-3552-6111 Fax: 03-3553-4694
Statistics	:	Capital : 7402 million yen Employees : 629
Officers	:	CEO : Mr. Masanori Fujii/President Contact : Mr. Masanori Fujii/President
Sales	:	December, 1988 - 16713 million yen December, 1989 - 17594 million yen December, 1990 - 18777 million yen
Sales Breakdown	:	SIC# Product 3624 Artificial graphite electrodes 2895 Carbon
Banks	:	Fuji Bank; Yasuda Trust and Banking; Yokohama Bank; Mitsui Taiyo Kobe Bank; Sanwa Bank; Long Term Credit Bank of Japan; Japan Development Bank; Industrial Bank of Japan
Branches	:	Osaka Branch (Osaka-shi); Nagoya Office (Nagoya-shi); Toyama Office (Toyama-shi)
Factories	:	Yokohama Factory (Yokohama-shi); Toyama Factory (Kaminiikawa-gun); Yamanashi Factory (Yamanashi-shi)
Laboratories	:	Laboratory (Yokohama-shi)
Subsidiaries	:	Shin Nippon Carbon Co., Ltd.; Nippon Carbon Seiko; Kyushu Tanso Kogyo; Keihan Tanso Kogyo; Nippon Carbon Engineering; Yamanashi Carbon; Nippon Carbon Shoji;
Overseas Subsidiaries	:	Nippon Carbon of America, Inc. (U.S.A.); Chuo Tanso (Taiwan)

Nippon Electric Glass Co., Ltd.

Japanese Name	:	***Nippon Denki Garasu K.K.***
TSE Number	:	5214
Address	:	2-7-1, Seiran, Otsu-shi, Shiga-ken 520
Tel & Fax	:	Tel: 0775-37-1700 Fax: 0775-34-4967
Statistics	:	Capital : 18374 million yen Employees : 4510
Officers	:	CEO : Mr. Kiyosaku Kishida/President
Sales	:	March, 1988 - 169005 million yen March, 1989 - 194564 million yen March, 1990 - 203133 million yen March, 1991 - 203415 million yen
R & D Spending	:	8079 million yen Capital Investment: 22434 million yen
Sales Breakdown	:	SIC# Product 3229 Glass tubes and bulbs 3231 Glass fibers
Banks	:	Sumitomo Bank; Sumitomo Trust and Banking; Nippon Credit Bank; Long Term Credit Bank of Japan; Shiga Bank
Branches	:	Osaka Office and Sales Headquarters (Osaka-shi)
Factories	:	Otsu Factory (Otsu-shi); Fujisawa Factory (Fujisawa-shi); Shiga Takatsuki Factory (Ika-gun); Notogawa Factory (Kanzaki-gun)
Subsidiaries	:	Nippon Electric Glass America, Inc. (Illinois, U.S.A.)

Nippon GE Plastics Co., Ltd.

TSE Number	:	Not listed
Description	:	Manufacturer of synthetic resins and related products. A subsidiary of General Electric Corp. (U.S.A.), Nippon GE Plastics is in the process of constructing a 4000 ton/year resin plant which will increase company production substantially and is scheduled for completion in 1992.
Address	:	3-7-2, Nihonbashi, Hon-cho, Chuo-ku, Tokyo 103-91
Tel	:	03-5695-4888
Sales Breakdown	:	SIC# Product 2821 Synthetic resins 3089 Misc. plastic products
Main Products	:	Synthetic Resins: PC (poly-carbonate) resins, Polypropylene resins; Misc. Plastic Products: Resin/glass fiber composite products
Shareholders	:	General Electric Corp. (U.S.A.)

Nippon Light Metal Co., Ltd.

Japanese Name	:	***Nippon Keikinzoku K.K.***
TSE Number	:	5701
Description	:	Nippon Light Metal Co., Ltd. is the only fully integrated aluminum company in Japan. The company has been an affiliate of Alcan Aluminum Ltd, Canada, since 1952. Produces and sells aluminum ingots, aluminum alloys, aluminum fabricated products and industrial machines. Diversifying into home products and appliances.
Address	:	3-13-12, Mita, Minato-ku, Tokyo 108
Tel & Fax	:	Tel: 03-3456-9211 Fax: 03-3456-9690
Statistics	:	Capital : 45163 million yen Employees : 4468
Officers	:	CEO : Mr. Shigeki Mukoyama/President Contact : Mr. Yuzuru Hayashi/Liaison Manager
Sales	:	March, 1988 - 248023 million yen March, 1989 - 267480 million yen March, 1990 - 269652 million yen
Sales Breakdown	:	SIC# Product 3334 Aluminum ingots and alloys 3354 Aluminum extruded products 3355 Aluminum rolling and drawing 3353 Aluminum sheets 3556 Ice-cream machines
Main Products	:	Aluminum Products: Aluminum ingots and alloys, Alumina, Aluminum hydrate, Aluminum rolled and extruded products, Aluminum finished products; Industrial Machinery: Ice-cream making machines, Cooling systems
Banks	:	Dai-Ichi Kangyo Bank; Industrial Bank of Japan; Sanwa Bank; Hokkaido Takushoku Bank; Mitsui Taiyo Kobe Bank
Branches	:	Osaka Regional Head Office (Osaka-shi); Sapporo Branch (Sapporo-shi); Fuji Branch (Fuji-shi); Nagoya Branch (Nagoya-shi); Hiroshima Branch (Hiroshima-shi); Fukuoka Branch (Fukuoka-shi)
Laboratories	:	Nikkei Teschno Research (Shizuoka-ken)
Subsidiaries	:	Shin Nikkei Co., Ltd. (Koto-ku); Nippon Fruehauf Co., Ltd. (Atsugi-shi); Nikkei Shoji Corp. (Minato-ku)

Nippon Metal Industry Co., Ltd.

Japanese Name	:	***Nippon Kinzoku Kogyo K.K.***
TSE Number	:	5479
Address	:	2-1-1, Nishi-Shinjuku, Shinjuku-ku, Tokyo 163
Tel & Fax	:	Tel: 03-3345-5451 Fax: 03-3345-5592
Statistics	:	Capital : 1197 million yen Employees : 1561
Officers	:	CEO : Mr. Tatsuo Saeki/President
Sales	:	March, 1988 - 79003 million yen March, 1989 - 100628 million yen March, 1990 - 100632 million yen March, 1991 - 101362 million yen
R & D Spending	:	1122 million yen Capital Investment: 13742 million yen
Sales Breakdown	:	SIC# Product Sales 3312 Primary steel products 80 % 34 Secondary steel products 20 %
Banks	:	Bank of Tokyo; Daiwa Bank; Fuji Bank; Long Term Credit Bank of Japan; Mitsui Taiyo Kobe Bank; Bank of Yokohama;
Branches	:	Osaka Branch; Nagoya Branch; Hokkaido Office; Tohoku Office; Niigata Office; Hokuriku Office; Hiroshima Office; Shikoku Office; Kyushu Office
Factories	:	Sagamihara Plant (Kanagawa-ken); Kinuura Plant
Overseas Subsidiaries	:	Nippon Metal Services Pte., Ltd. (Singapore)

Nippon Micalta Co., Ltd.

Japanese Name	:	***Nippon Maikaruta Kogyo K.K.***
TSE Number	:	Not listed
Description	:	Manufacturer of electrical insulation materials and equipment.
Address	:	2-2-26, Chuo, Joto-ku, Osaka-shi, Osaka 536
Tel	:	06-932-6281
Statistics	:	Capital : 12 million yen Employees : 45
Sales	:	August, 1989 - 386 million yen August, 1990 - 350 million yen
Sales Breakdown	:	SIC# Product 3699 Insulation materials and equipment
Main Products	:	Electrical Insulation Materials and Equipment
Banks	:	Mitsubishi Bank; Sanwa Bank; Bank of Osaka; Sumitomo Bank
Factories	:	Osaka Plant (Osaka); Shiga Plant (Shiga-ken)

Nippon Micalta Co., Ltd.

Japanese Name	:	***Nippon Maikaruta Kogyo K.K.***
TSE Number	:	Not listed
Description	:	Manufacturer of electrical insulation materials and equipment.
Address	:	2-2-26, Chuo, Joto-ku, Osaka-shi, Osaka 536
Tel	:	06-932-6281
Statistics	:	Capital : 12 million yen Employees : 45
Sales	:	August, 1989 - 386 million yen August, 1990 - 350 million yen
Sales Breakdown	:	SIC# Product 3699 Insulation materials and equipment
Main Products	:	Electrical Insulation Materials and Equipment
Banks	:	Mitsubishi Bank; Sanwa Bank; Bank of Osaka; Sumitomo Bank
Factories	:	Osaka Plant (Osaka); Shiga Plant (Shiga-ken)

Nippon Mining Co., Ltd.

Japanese Name :	***Nihon Kogyo K.K.***
TSE Number :	5712
Description :	Involved in mining and processing non-ferrous metals and petroleum products. Branching into pharmaceuticals and optical electronics.
Address :	2-10-1, Toranomon, Minato-ku, Tokyo 105
Tel & Fax :	Tel: 03-3505-8111 Fax: 03-3505-8157
Statistics :	Capital : 76990 million yen Employees : 5635
Officers :	CEO : Mr. Tatsuo Nakamura/President Contact : Mr. Nobuyasu Ishida/General Manager, Overseas Division
Sales :	March, 1988 - 709905 million yen March, 1989 - 718127 million yen March, 1990 - 822364 million yen March, 1991 - 989888 million yen
R& D Spending :	14700 million yen Capital Investment: 46700 million yen
Sales Breakdown :	SIC# Product 2911 Petroleum products 3331 Copper 3339 Zinc 2834 Pharmaceuticals 3851 Optical goods
Main Products :	Metals: Copper, Gold, Silver, Zinc, Lead, Sulfuric acid; Processed Metals: Drawn copper, Special steel products, Copper circuit materials; Other Materials Products: Advanced Materials, Optoelectronic products; Petroleum Products: Gasoline, Naphtha, Kerosene, Light oil, Fuel oil, Petrochemicals, Lubricant oils, Xylene, Sulfur dioxide cooling agent
Banks :	Industrial Bank of Japan; Mitsui Taiyo Kobe Bank; Dai-Ichi Kangyo Bank
Branches :	Osaka Office (Osaka-shi); Nagoya Office (Nagoya-shi); Kyushu Sales Office (Fukuoka-shi); New York Office; San Francisco Office; Australia Office (Melbourne); Beijing Office
Laboratories :	Electric Materials and Components Research Laboratory. (Toda-shi, Saitama-ken); Materials Development Research Laboratory (Toda-shi); Bioscience Research Laboratory (Toda-shi)

Nippon Nyukazai Co., Ltd.

Japanese Name	:	***Nippon Nyukazai K.K.***
TSE Number	:	Not listed
Address	:	Yoshizawa Bldg., 3-9-19, Ginza, Chuo-ku, Tokyo 104
Tel & Fax	:	Tel: 03-3543-8567 Fax: 03-3546-3174
Statistics	:	Capital : 100 million yen Employees : 290
Officers	:	CEO : Mr. Tetsuzo Koyama/President Contact : Mr. Masayuki Tsuchiya/Director, Overseas Dept.
Sales	:	March, 1988 - 11050 million yen March, 1989 - 11997 million yen March, 1990 - 12844 million yen March, 1991 - 14380 million yen
Sales Breakdown	:	SIC# Product 2869 Organic chemicals
Banks	:	Dai-Ichi Kangyo Bank; Shiga Bank; Tokai Bank; Mitsui Taiyo Kobe Bank; Shizuoka Bank
Branches	:	Osaka Office (Osaka-shi)
Factories	:	Kawasaki Plant (Kawasaki-shi, Kanagawa-ken)
Shareholders	:	Sankyo Co., Ltd. [4501]

Nippon Oil & Fats Co., Ltd.

Japanese Name : *Nihon Yushi K.K.*

TSE Number : 4403

Description : Nippon Oil & Fats is a major chemical manufacturer involved in the processing of oil and fats, gun powder and paint. An affiliate of Nissan Motors, it is strong in the sales of paints for the automotive and shipping industries. The company is beginning production of air bags and is diversifying into health foods, biotechnology, medical products and petrochemicals. Recent investments include condominium construction in Saitama Prefecture. The company operates nine domestic joint ventures, and seven others in Malaysia, Taiwan, Hong Kong and Thailand.

Address : 1-10-1, Yuraku-cho, Chiyoda-ku, Tokyo 100

Tel & Fax : Tel: 03-3283-7070 Fax: 03-3283-7120

Statistics : Capital : 15783 million yen
Employees : 2916

Officers : CEO : Mr. Kineo Okamoto/President

Sales : March, 1988 (4 months) - 36717 million yen
March, 1989 - 119839 million yen
March, 1990 - 127040 million yen
March, 1991 - 137398 million yen

R & D Spending : 5587 million yen Capital Investment: 9884 million yen

Sales Breakdown :

SIC#	Product	Sales
2899	Oleo and specialty chemicals	32 %
2851	Coatings	27 %
2899	Chemicals and explosives	25 %
2099	Food products	10 %
2999	Petrochemicals	4 %
2899	Welding materials	3 %

Main Products : Industrial Oils, Fats and Chemicals: Oleic acid, Lauric acid, Myristic acid, Palmitic acid, Stearic acid, Behenic acid, Other fatty acids, Glycerine, Metallic soaps, Anti-corrosive coatings, Antifoaming agents for paper and synthetic resin/rubber production, Sulfur antioxidant, Dispersion agents for rubber and plastics, Foaming agents, Dispersion agents for pesticides, Antistatic agents for plastics and rubbers, Antistatic agents for fibers, Calcium stearate, Dispersion agent for pulp and paper, Sludge dispersion agent, Emulsifiers for foodstuffs, Emulsifiers for asphalt, Emulsifiers for wax, Epoxy plasticizer, Epoxy resin curing agent, Epoxy resin enamels and varnishes, Halogen-based flame retardant, Organic phosphorous-based flame retardant, Flotation reagent, Gelling agent, Vulcanizing agent, Vulcanization accelerator, Organic antioxidant, Phosphoric acid ester plasticizers, Aliphatic ester plasticizers, Polyoxyethylene alkyl ether, Rust inhibitor, Rust-preventative oil, Marine pollution prevention dispersion agent, Slip additive, Surface coating agent, Alkyl compounds

Banks : Fuji Bank; Yasuda Trust and Banking; Industrial Bank of Japan; Norinchukin Bank; Dai-Ichi Kangyo Bank

R&D Areas : Oils: Natural oil-based feed for dairy cows; Industrial Explosives: Solid rocket propellent, Super precision electronic detonator

Nippon Oil Co., Ltd.

Japanese Name	:	***Nippon Sekiyu K.K.***
TSE Number	:	5001
Description	:	The nation's largest distributor of petroleum products. The company obtains its products from Koa Oil and Nippon Petroleum Refining and it purchases over 50% of its crude oil from Caltex. Nippon Oil has a business relationship with Mitsubishi Oil.
Address	:	Nisseki Honkan, 1-3-12, Nishi-Shimbashi, Minato-ku, Tokyo 105
Tel & Fax	:	Tel: 03-3502-1111 Fax: 03-3502-9352
Statistics	:	Capital : 20000 million yen Employees : 5000
Officers	:	CEO : Mr. Kentaro Iwamoto/President
Sales	:	March, 1989 - 1667030 million yen March, 1990 - 1853790 million yen March, 1991 - 2192940 million yen
R & D Spending	:	12000 million yen Capital Investment: 63000 million yen
Sales Breakdown	:	SIC# Product 2911 Gasoline 2999 Fuel oil 2999 Other petroleum-based products
Main Products	:	Petroleum-based Products: Gasoline, Naphtha, Fuel oil, Kerosene, Gas oil, Lubricants, Slip additive
Banks	:	Mitsui Taiyo Kobe Bank; Dai-Ichi Kangyo Bank; Fuji Bank
Factories	:	Shimomatsu Refinery (Shimomatsu-shi)
Laboratories	:	Central Technical Research Laboratory (Yokohama-shi)
Overseas Subsidiaries	:	Nippon Oil (Asia) Pte., Ltd. (Singapore); Nippon Oil (Delaware, U.S.A.) Ltd. (New York, U.S.A.); Nippon Oil (U.K.) Public Co., Ltd. (London, U.K.)

Nippon Paint Co., Ltd.

Japanese Name :	***Nippon Penki K.K.***
TSE Number :	4612
Description :	Major paint maker ranking second in sales but first in profitability. The company also holds the largest market share in cationic electro-deposition paints, and has an equal share of the domestic metal surface treatment agent market with Nihon Parkerizing.
Address :	2-1-2, Oyodokita, Kita-ku, Osaka 531
Tel & Fax :	Tel: 06-458-1111 Fax: 06-455-9260
Statistics :	Capital : 27630 million yen Employees : 2757
Officers :	CEO : Mr. Hiroshi Fujii/President
Sales :	March, 1988 (10 months) - 109874 million yen March, 1989 - 135311 million yen March, 1990 - 146857 million yen March, 1991 - 160187 million yen
Sales Breakdown :	SIC# Product 2851 Paints and thinners 2819 Surface treatment agents
Main Products :	Paints and Related Products: Synthetic resin paints, Thinners, Lacquers, Cationic electro-deposition paints, Anticorrosive coating, Epoxy resin enamels and varnishes; Industrial Chemicals: Metal surface treatment agents, Corrosion inhibitor
Banks :	Sumitomo Bank; Sanwa Bank; Fuji Bank; Mitsubishi Bank
Laboratories :	Technical Center (Neyagawa-shi)
Shareholders :	Sumitomo Mutual Life; Nippon Life; Sumitomo Trust and Banking [8403]
Subsidiaries :	Nippon Bee Chemical Co., Ltd.; Asahi Solvent Industry Co., Ltd. and 26 other domestic subsidiaries

Nippon Petrochemicals Co., Ltd.

Japanese Name	:	***Nihon Sekiyu Kagaku K.K.***
TSE Number	:	Not listed
Description	:	Supplier of petrochemical raw materials to petrochemical manufacturers of the Kawasaki Petrochemical Combinate and other companies. A wholly owned subsidiary of Nippon Oil Co., Ltd., Nippon Petroleum is expanding its business lines to include the biotechnology and functional resin fields.
Address	:	Saiwai Bldg., 1-3-1, Uchisaiwai-cho, Chiyoda-ku, Tokyo 100
Tel & Fax	:	Tel: 03-3501-7313 Fax: 03-3501-7630
Statistics	:	Capital : 6750 million yen Employees : 2193
Officers	:	CEO : Mr. Yutaka Katayama/President Contact : Mr. Kunihiko Haga/Asst. General Manager
Sales	:	March, 1988 - 161656 million yen March, 1989 - 163655 million yen March, 1990 - 190719 million yen March, 1991 - 220142 million yen

Sales Breakdown :

SIC#	Product	Sales
28	Chemicals	60 %
2821	Plastics	25 %
28	Specialty chemicals and materials	10 %

Main Products	:	Chemicals: Ethylene, Propylene, Butadiene, Benzene, Toluene, Xylenes, Isopropyl alcohol, Normal hexane, Normal paraffin; Plastics: Polyethylene, Polypropylene; Specialty Chemicals: Poly-butene, Polyisobutylene, Isoparaffin, Petroleum resins, Superaromatic solvents; Materials: High density carbon fiber, Low density carbon fiber, Linear low density carbon fiber, Non-woven fabrics
Banks	:	Mitsui Taiyo Kobe Bank; Dai-Ichi Kangyo Bank; Fuji Bank
Branches	:	Osaka Branch Office (Osaka); Nagoya Sales Office (Nagoya-shi, Aichi-ken); Fukuoka Sales Office (Fukuoka-shi, Fukuoka-ken); New York Office (New York, U.S.A.)
Factories	:	Kawasaki Operation Center (Kawasaki-shi, Kanagawa-ken); Kawasaki Plant (Kawasaki-shi, Kanagawa-ken); Kawasaki South Plant (Kawasaki-shi, Kanagawa-ken); Ukishima Plant (Kawasaki-shi, Kanagawa-ken)
Laboratories	:	Specialty Chemicals Laboratory (Kawasaki-shi, Kanagawa-ken); Plastics Laboratory (Kawasaki-shi, Kanagawa-ken); New Materials Laboratory (Kawasaki-shi, Kanagawa-ken); Tsukuba Laboratory (Tsukuba-shi, Ibaraki-ken)
Shareholders	:	Nippon Oil Co., Ltd. [5001] - 100%
Overseas Subsidiaries	:	Nisseki Chemical Texas Inc. (Texas, U.S.A.); Amoco Nisseki CLAF, Inc. (Georgia, U.S.A.); Kolon Chemical Co., Ltd.(Korea)

Nippon Petroleum Refining Co., Ltd.

Japanese Name	:	***Nippon Sekiyu Seisei K.K.***
TSE Number	:	Not listed
Address	:	1-3-12, Nishi-Shimbashi, Minato-ku, Tokyo 105
Tel & Fax	:	Tel: 03-3502-1111 Fax: 03-3502-9352
Statistics	:	Capital : 16000 million yen Employees : 1951
Officers	:	CEO : Mr. Jiro Yamamoto/President
Sales	:	December, 1988 - 523799 million yen December, 1989 - 575660 million yen
Sales Breakdown	:	SIC# Product 2911 Petrochemical products
Banks	:	Mitsui Taiyo Kobe Bank; Fuji Bank; Dai-Ichi Kangyo Bank
Factories	:	Yokohama Plant (Yokohama-shi, Kanagawa-ken); Kudamatsu Plant (Kudamatsu-shi, Yamaguchi-ken); Muroran Plant (Muroran-shi, Hokkaido); Negishi Plant (Yokohama-shi, Kanagawa-ken); Okinawa Plant (Nakato-gun, Okinawa)
Shareholders	:	Nippon Petrochemicals Co., Ltd.

Nippon Pillar Packing Co., Ltd.

Japanese Name	:	***Nippon Pira Kogyo***
TSE Number	:	6490
Description	:	Major manufacturer of mechanical seals and packings. Nippon Pillar Packing also produces fluorine resins and has been devoting an increasing amount of company resources to new product development, particularly in the area of new materials for use as packings and sealants.
Address	:	2-11-48, Nonaka-Minami, Yodogawa-ku, Osaka 532
Tel & Fax	:	Tel: 06-305-1781 Fax: 06-302-6404
Statistics	:	Employees : 521
Officers	:	CEO : Mr. Kiyohisa Iwanami/President Contact : Mr. Kiyohisa Iwanami/President
Sales	:	March, 1989 - 10374 million yen March, 1990 - 11600 million yen March, 1991 - 13028 million yen

Sales Breakdown :

SIC#	Product	Sales
3053	Mechanical seals	43 %
2821	'Pillaflon' fluorine resins	25 %
3053	Grand packing	17 %
3053	Vertical gaskets	

Main Products	:	Mechanical Seals: Petroleum seals, Chemical seals, Electrical seals, etc.; Grand packing: Electrical packings, Marine packings, Industrial machinery Packings; Vertical Gaskets: Automobile gaskets, Electrical gaskets, Industrial machinery gaskets; Synthetic Resin Products: 'Pillaflon' fluorine resins for use in semiconductor production
Banks	:	Mitsui Taiyo Kobe Bank; Sumitomo Bank; Industrial Bank of Japan; Dai-Ichi Kangyo Bank; Sanwa Bank; Sumitomo Bank; Fuji Bank
Branches	:	Tokyo Branch (Minato-ku, Tokyo); Nagoya Branch (Nagoya-shi, Aichi-ken); Kobe Sales Office (Kobe-shi, Hyogo-ken); Hiroshima Sales Office (Hiroshima-shi, Hiroshima-ken); Himeji Sales Office (Himeji-shi, Hyogo-ken)
Factories	:	Sanda Plant (Sanda-shi, Hyogo-ken); Fukuchiyama Plant (Fukuchiyama-shi, Kyoto)
R&D Areas	:	New Materials: Asbestos-free sealing and insulation materials; Industrial Equipment: Energy couplings, Non-contact gas seals
Subsidiaries	:	Tokyo Pillar K.K.; Pillar Service Trading K.K.: Chubu Pillar Trading Services K.K.; Hokuriku Pillar K.K.
Overseas Subsidiaries	:	Korea Pillar Packing Co., Ltd. (Korea)

Nippon Polyester Co., Ltd.

Japanese Name	:	***Nippon Poriesuteru K.K.***
TSE Number	:	Not listed
Description	:	Manufacturer of FRP (fiber reinforced polyester) products.
Address	:	Yatagai Bldg., 2-8-33, Shibata, Kita-ku, Osaka 530
Tel & Fax	:	Tel: 06-372-4551 Fax: 06-371-0164
Statistics	:	Capital : 100 million yen Employees : 250
Officers	:	CEO : Mr. Kenshiro Hakoi/President Contact : Mr. Kenshiro Hakoi/President
Sales	:	November, 1988 - 8698 million yen November, 1989 - 9491 million yen November, 1990 - 10532 million yen
Sales Breakdown	:	SIC# Product 308 FRP molded products
Main Products	:	FRP Products: Domes, Water storage tanks, Corrugated sheets, Flat sheets, Protective headgear, Materials for road paint, Highway signboards
Banks	:	Sanwa Bank; Daiwa Bank; Sumitomo Bank
Factories	:	Chiba Plant (Chiba-ken); Sanda Plant (Sanda-shi, Hyogo-ken)
Shareholders	:	Nippon Catalytic Chemical K.K. -177,000 Shares

Nippon Sanso K.K.

TSE Number	:	4091
Description	:	Japan's largest oxygen manufacturer and a leader in industrial gases for use in semiconductor production. Areas of specialization include super-high pressure, vacuum and cryogenic fields.
Address	:	1-16-7, Nishi-Shinbashi, Minato-ku, Tokyo 105
Tel & Fax	:	Tel: 03-3581-8200 Fax: 03-3581-9425
Statistics	:	Capital : 25852 million yen Employees : 1696
Officers	:	CEO : Mr. Hideo Mabuchi/President
Sales	:	March, 1989 - 147552 million yen March, 1990 - 164617 million yen March, 1991 - 179303 million yen
R & D Spending	:	3441 million yen Capital Investment: 23789 million yen
Sales Breakdown	:	SIC# Product 2813 Oxygen, nitrogen, argon 3599 Gas-related equipment
Main Products	:	Industrial Gases: Oxygen and nitrogen gases, Argon gas, Special-purpose gases, Acetylene, CFCs; Industrial Machinery: Gas-related equipment
Banks	:	Fuji Bank; Yasuda Trust and Banking; Industrial Bank of Japan; Long Term Credit Bank; Mitsui Taiyo Kobe Bank; Saitama Bank; Norinchunkin Bank
Branches	:	New York Rep. Office (New York, U.S.A.)

Nippon Seisen Co., Ltd.

Japanese Name	:	***Nippon Seisen K.K.***
TSE Number	:	5659
Description	:	The largest producer of wire material technology products in Japan. End use is mainly for household electrical goods and precision instruments in addition to the automotive industry. Strongly linked with Nippon Yakin Kogyo, one of Japan's three largest steel makers. Actively pursuing research in process technology and new product areas, alone and in co-operation with other firms.
Address	:	Annex Kogin Bldg., 4-2-7, Koraibashi, Chuo-ku, Osaka 541
Tel & Fax	:	Tel: 06-222-5432 Fax: 06-222-0718
Statistics	:	Capital : 4055 million yen Employees : 666
Officers	:	CEO : Mr. Takahashi Oda/President Contact : Mr. Tatsuo Tomiyama/General Manager, Export Department
Sales	:	May, 1988 - 21821 million yen May, 1989 - 25682 million yen May, 1990 - 26953 million yen May, 1991 - 27376 million yen
R & D Spending	:	47 million yen Capital Investment: 2348 million yen
Sales Breakdown	:	SIC# Product 3315 Stainless steel wires 3357 Non-ferrous wires 33 Metal fibers and filter elements
Main Products	:	Stainless Steel Products: Stainless steel round wire, Stainless steel shaped wire, Stainless steel cold rolled flat bars, Stainless steel bars, Stainless steel fiber 'Naslon', Stainless steel filter media; Other Wire Products: High nickel alloy wire, EDM wire, Titanium wire, Shape memory alloy wire, Wire netting, Kitchen utensils, Nails; Other Products: Injection molded precision parts, Powder metallurgy, Diamond dies
Banks	:	Industrial Bank of Japan; Nippon Credit Bank; Chuo Trust Bank; Tokai Bank; Ikeda Bank; Fuji Bank; 77 Bank; Dai-Ichi Life Insurance
Branches	:	Osaka Branch; Tokyo Branch; Nagoya Branch; Hiroshima Sales Office; Kyushu Sales Office (Fukuoka); Toyama Sales Office; Niigata Sales Office.
Factories	:	Hirakata Factory (Hirakata-shi, Osaka-ken)
Laboratories	:	R&D Laboratory (Hirakata Factory)
Overseas Subsidiaries	:	Thai Seisen Co., Ltd. (Thailand)

Nippon Sheet Glass Co., Ltd.

Japanese Name	:	***Nippon Itagarasu K.K.***
TSE Number	:	5202
Description	:	Japan's second largest manufacturer of sheet glass. Advancing into wool production through a joint venture with Nihon Cement. Also developing new materials for the electronics and precision equipment industries.
Address	:	3-5-11, Doshu-machi, Chuo-ku, Osaka 541
Tel & Fax	:	Tel: 06-222-7511 Fax: 06-222-7580
Statistics	:	Capital : 40700 million yen Employees : 3600
Officers	:	CEO : Mr. Tatsuji Nakajima/President Contact : Mr. Eiichi Futakado/Director, Overseas Dept.
Sales	:	March, 1988 - 203918 million yen March, 1989 - 217046 million yen March, 1990 - 234381 million yen March, 1991 - 249292 million yen
Sales Breakdown	:	SIC# Product 3211 Sheet glass 3229 Fine glass 3229 Glass building materials
Main Products	:	Glass and Glass Products: Flat glass, Glass fiber, Fine glass and fiber optics, Glass building materials
Banks	:	Sumitomo Bank; Sumitomo Trust and Banking; Industrial Bank of Japan
Branches	:	Tokyo Branch (Minato-ku); Osaka Branch (Osaka-shi); Kanto Branch (Minato-ku); Fukuoka Branch (Fukuoka-shi); Nagoya Branch (Nagoya-shi); Sapporo Branch (Sapporo-shi); Sendai Branch (Sendai-shi); Hiroshima Branch (Hiroshima-shi)
Factories	:	Yokkaichi Factory (Yokkaichi-shi); Maizuru Factory (Maizuru-shi); Chiba Factory (Ichihara-shi); Kyoto Factory (Kyoto-shi); Sagamihara Factory (Sagamihara-shi)
Laboratories	:	Central Research Laboratory (Itami-shi); Tsukuba Research Laboratory (Tsukuba-shi)

Nippon Shokubai Co., Ltd.

Japanese Name	:	***Nippon Shokubai K.K.***
TSE Number	:	4114
Description	:	Manufacturer of petrochemicals, fine and specialty chemicals, polymers, catalysts, resins and other chemical products.
Address	:	Kogin Bldg., 4-1-1, Koraibashi, Chuo-ku, Osaka 541
Tel & Fax	:	Tel: 06-223-9111 Fax: 06-201-3716
Statistics	:	Capital : 16514 million yen Employees : 2152
Officers	:	CEO : Mr. Shoso Tanaka/President Contact : Mr. Takashi Suemori/General Manager, International
Sales	:	November, 1988 - 117190 million yen November, 1989 - 132387 million yen November, 1990 - 137497 million yen November, 1991 - 150000 million yen

Sales Breakdown :

SIC#	Product	Sales
291	Petrochemicals	29 %
28	Chemical products	26 %
2869	Fine and specialty chemicals	23 %
2821	Polymers and resins	15 %
286	Catalysts	6 %

Main Products	:	Petrochemical Products: Ethylene oxide, Ethylene glycols, Ethanolamines; Chemical Products: Phthalic anhydride, Maleic anhydride, Fumalic acid, Arcylic acid and esters, Methacrylic acid and esters, Flocculants, Higher alcohols for Nisshoku process, Sequestering agent; Catalysts: Ozone deodorizing catalysts, Ammonia manufacture catalysts, Epoxy resin curing catalysts, Automotive exhaust catalysts, Chemical processing catalysts, De-nitrogen oxide (De-NOx) catalysts; Fine and Specialty Chemical Products: Polyethyleneimine, Athraquinone, Anisaldehyde, N-Phenylmaleimide, Super-absorbent polymers
Banks	:	Japan Development Bank; Industrial Bank of Japan; Japan Import and Export Bank
Branches	:	Tokyo Office (Chiyoda-ku); Nagoya Sales Office (Nagoya-shi, Aichi-ken); Hiroshima Sales Office (Hiroshima-ken); Fukuoka Sales Office (Fukuoka-ken)
Factories	:	Suita Plant (Suita-shi, Osaka); Kawasaki Plant (Kawasaki-shi); Himeji Plant (Himeji-shi, Hyogo-ken)
Laboratories	:	Central Research Laboratory (Suita-shi); Tsukuba Research Laboratory (Tsukuba-shi); Resins Research Laboratory (Suita-shi); Catalysts Research Laboratory (Himeji-shi); Himeji Research Laboratory (Himeji-shi); Kawasaki Research Laboratory (Kawasaki-shi)
Subsidiaries	:	Nippon Chemicals Co., Ltd. (Osaka-shi); Nisshoku Aro Chemicals Co., Ltd. (Osaka-shi); Nippon Polyester Co., Ltd. (Osaka-shi); Daiko Kaiun Co., Ltd. (Osaka-shi)
Overseas Subsidiaries	:	Nippon Shokubai America Inc. (New York, U.S.A.); Nippon Shokubai Kagaku Kogyo GmbH (Frankfurt, Germany); NA Industries, Inc. (Tennessee, U.S.A.)

Nippon Stainless Steel Co., Ltd.

Japanese Name	:	***Nippon Sutenresu K.K.***
TSE Number	:	5478
Address	:	8-2, Honshio-cho, Shinjuku-ku, Tokyo 160
Tel & Fax	:	Tel: 03-3358-9456 Fax: 03-3358-9497
Statistics	:	Capital : 9472 million yen Employees : 1593
Officers	:	CEO : Mr. Mitsunobu Kurita/President
Sales	:	March, 1988 - 69351 million yen March, 1989 - 97260 million yen March, 1990 - 102307 million yen March, 1991 - 104120 million yen
R & D Spending	:	1068 million yen Capital Investment: 6630 million yen

Sales Breakdown :

SIC#	Product	Sales
3312	Stainless steel	82 %
3325	Steel castings	15 %
3356	Titanium products and others	

Banks : Nippon Credit Bank; Sumitomo Bank; Sumitomo Trust and Banking; Long Term Credit Bank of Japan; Joyo Bank; Daishi Bank

Nippon Steel Chemical Co., Ltd.

Japanese Name : ***Shin Nittetsu Kagaku K.K.***

TSE Number	:	4363
Description	:	Nippon Steel Chemical was established as a chemical manufacturer in the Nippon Steel group, to utilize by-products such as coal tar and gases generated in coke furnaces. The company manufactures synthetic resins, carbon materials, electronics parts and cement.
Address	:	5-13-16, Ginza, Chuo-ku, Tokyo 104
Tel & Fax	:	Tel: 03-3542-1321 Fax: 03-3546-3575
Statistics	:	Capital : 20582 million yen Employees : 2689
Officers	:	CEO : Mr. Enji Moriguchi/President
Sales	:	March, 1988 - 202575 million yen March, 1989 - 232001 million yen March, 1990 - 254316 million yen March, 1991 - 306110 million yen
R & D Spending	:	8000 million yen Capital Investment: 34200 million yen

Sales Breakdown :

SIC#	Product	Sales
2999	Coke	38 %
2865	Toluene	20 %
2821	Synthetic resins	20 %
2816	Pigments	6 %
3241	Cement	6 %
2813	Argon	6 %

Main Products	:	Petroleum and Coal Products: Coke; Petrochemical and Specialty Chemical Products: Toluene, Xylene, Styrene monomer, Napthalene, Tar-based benzene, Ammonia cooling agent, Methanol, Phthalic acid anhydride; Other Chemical Products: Plastic materials, Synthetic resin: Inorganic Pigments: Paints, Adhesives, Sealants; Cement Products: Cement; Industrial Gases: Argon, Oxygen, Hydrogen
Branches	:	Osaka Branch (Kita-ku, Osaka); Nagoya Branch (Nakamura-ku, Nagoya-shi); Hiroshima Branch (Naka-ku, Hiroshima); Takamatsu Branch (Takamatsu-shi, Kagawa-ken); Hirohata Office (Himeji-shi, Hyogo-ken); Muroran Office (Muroran-shi, Hokkaido)
Factories	:	Tobata Plant (Kitakyushu-shi); Sakai Plant (Sakai-shi); Kimitsu Plant (Kimitsu-shi, Chiba-ken); Oita Plant (Oita-shi)
Laboratories	:	Technical Research Laboratory (Kitakyushu-shi, Fukuoka-ken); Products Development Center (Kisarazu-shi, Chiba-ken)
Subsidiaries	:	5 subsidiaries throughout Japan
Overseas Subsidiaries	:	Nippon Steel Chemical Corp. of America (U.S.A.); Thermofil Corp. (U.S.A.); Polymer General Inc. (U.S.A.); Envirotherm Inc. (U.S.A.)

Nippon Steel Corp.

Japanese Name : ***Shin Nihon Seitetsu K.K.***

TSE Number : 5401

Description : The world's largest steel manufacturer holding a 30% share in the domestic crude steel market. Due to the slump in the steel industry, the company has undergone extensive restructuring and has initiated several joint ventures in order to move into profitable new business areas and develop technology. The company has developed the TMCP control production process and is actively applying simulation models to improve steel production and develop new materials. Software models used include 'BRIGHT', 'NSMAC', and 'AMI' and 'NET', the latter two incorporating neural network technology. As part of its diversification strategies, the company is developing new materials for the electronics, food, and construction industries. In 1989, Nippon Steel, in a joint venture with Kitakyushu City, opened up 'Space World' as part of its efforts to advance into the services industry. It has moved into software marketing and restaurant chains.

Address : 2-6-3, Otemachi, Chiyoda-ku, Tokyo 100

Tel & Fax : Tel: 03-3242-4111 Fax: 03-3275-5607

Statistics : Capital : 412600 million yen
Employees : 54062

Officers : CEO : Mr. Hiroshi Saito/President

Sales : March, 1988 - 2147040 million yen
March, 1989 - 2385410 million yen
March, 1990 - 2573120 million yen
March, 1991 - 2603280 million yen

R & D Spending : 71100 million yen

Sales Breakdown :

SIC#	Product	Sales
3312	Steel products	70 %
3356	Titanium products	10 %
162	Heavy construction	5 %
8711	Engineering contracts	5 %
3264	Ferrite	5 %
7373	Software	3 %
2819	Aluminum and copper compounds	2 %

Main Products : Iron and Steel Products: Iron and steel sheets, Bars, Forgings, Rods, Wire products, Stainless steel plates, sheets and coils, Sections, Electrical steels, Pig iron, Ingots, Zn-Fe alloy-coated steel sheets; Titanium Products: Welded tubes for condensers, Coils, Wire rods, Plates, Fabricated products, Plates for aircraft production, Corrosion-resistant sheeting; Engineering and Heavy Construction: Industrial plants, Steelworks, Chemical plants, Oil/gas production facilities, Submarine pipelines; Building Construction: Intelligent buildings, Urban development, Hotel, Sports facilities, Steel frames for high-rise buildings, Residential buildings; Civil Engineering: Highways, Bridges, Airports, City gas works, Water supply, Marine construction, Sewerage systems, Offshore oil/gas development; New Materials: Gold-bonding wire, Resin coated wire, Soft ferrite cores, Metallic catalytic carriers, Fiber-reinforced thermoplastic stampable sheets, Three-dimensionally reticulated iron, High-temperature superconducting bulk magnet; Computer Integrated Systems: CAD-CAM software; Service Industry: Theme park, Restaurants, Retirement villages; High-Purity Industrial and Specialty Gases: Hydrogen, Argon, Nitrogen, Oxygen; Specialty Chemical

		Products: Gamma TiAl, Specialty monomers, Pharmaceuticals, Fertilizers, Information recording materials, Heat transfer fluids, Flavors and fragrances, Specialty solvents, Dyes and pigments, Specialty printing inks, Electronic chemicals, Tar-based benzene, Naphthalene, Toluene, Xylene; Catalysts: Nickel catalysts; Synthetic Lubricants: 'Shinlube' series; Phenolic Resins: Binders for refractories, Phenol foams, Molding materials; Carbon Specialties: Pantograph sliders, Graphite products for semiconductor production, EDM graphite, Carbon/Carbon composites, Thermal insulation products; Electronic Materials: Epoxy molding compounds for IC encapsulation, Low-alpha ray synthetic silica, Two-layer copper-clad laminates for FPC; Composite Materials: Thermoplastic compounds, Polymer alloys, Prepegs; Advanced building materials: Noise insulating floor materials, Free-access floor panels for OA rooms, Stone panels; Cultivation systems: Rock-wool based cultivation system
Banks	:	Industrial Bank of Japan; Fuji Bank; Mitsubishi Bank; Sumitomo Bank; Sanwa Bank
Laboratories	:	Steel Research Laboratories; Process Technology Research Laboratories; Advanced Materials & Technology Research Laboratories; Electronics R&D Laboratories (Sagamihara-shi); Plant Engineering & Technology Center; Steelworks R&D Laboratory

Nippon Telegraph and Telephone Corp.

Japanese Name	:	***Nippon Denshin Denwa K.K.***
TSE Number	:	9432
Description	:	Japan's largest telecommunications enterprise. Privatized in 1985, the company has begun to face competition for some of its services since deregulation of the telecommunications industry.
Address	:	1-1-6, Uchisaiwai-cho, Chiyoda-ku, Tokyo 100
Tel & Fax	:	Tel: 03-3509-3101 Fax: 03-3509-4290
Statistics	:	Capital : 780000 million yen Employees : 257600
Officers	:	CEO : Mr. Masashi Kojima/President Contact : Mr. Hiroyuku Kanamaru/Manager, Public Relations
Sales	:	March, 1988 - 5662000 million yen March, 1989 - 5841900 million yen March, 1990 - 6022400 million yen March, 1991 - 6251600 million yen
R & D Spending	:	280000 million yen Capital Investment: 1850000 million yen

Sales Breakdown :

SIC#	Product	Sales
4813	Telephone	77 %
4813	Leased circuits	6 %
4899	Data communications facilities	3 %
4812	Pagers	1 %
4822	Telegraph	1 %
4899	Digital data exchange	1 %

Main Products	:	Telecommunications: Fixed telephones, Mobile telephones, Telephone equipment, Facsimile machines, Pagers, Telephone circuit leasing, Digital data exchange, Telegraph, Data telecomms services
Banks	:	Dai-Ichi Kangyo Bank; Fuji Bank; Sanwa Bank; Sumitomo Bank; Mistui Taiyo Kobe Bank; Mitsubishi Bank; Tokai Bank
Overseas Subsidiaries	:	NTT Europe Ltd. (London, U.K.); NTT America Inc. (New York and Los Altos, U.S.A.); NTT Finance (UK) Ltd. (London, UK); NTT France S.A. (Paris, France); NTT Deutschland GmbH (Dusseldorf, Germany); NTT do Brasil Ltda. (Rio de Janeiro, Brazil)

Nippon Tungsten Co., Ltd.

Japanese Name : ***Nippon Tangusuten K.K.***

TSE Number : 6998

Address : 2-20-31, Shimizu, Minami-ku, Fukuoka-shi, Fukuoka-ken 815

Tel & Fax : Tel: 092-511-1111 Fax: 092-512-1222

Statistics : Capital : 2509 million yen
Employees : 722

Officers : CEO : Mr. Shozo Matsumoto/President
Contact : Mr. Shozo Matsumoto/President

Sales : March, 1989 - 10804 million yen
March, 1990 - 11079 million yen
March, 1991 - 12177 million yen

R & D Spending : 529 million yen Capital Investment: 1029 million yen

Sales Breakdown :

SIC#	Product	Sales
3313	Tungsten, electrical contacts	30 %
2819	Tungsten carbide	30 %
3643	Wiring rolls	25 %
3299	Ceramics	15 %

Banks : Fukuoka Bank; Kyowa Bank; Fuji Bank; Sanwa Bank; Long Term Credit Bank of Japan; Industrial Bank of Japan

Factories : Main Factory (Fukuoka-shi, Fukuoka-ken); Chigasaki Factory (Chigasaki-shi, Kanagawa-ken); Iizuka Factory (Iizuka-shi, Fukuoka-ken); Umi Factory (Kasuya-gun, Fukuoka-ken)

Shareholders : Toshiba Corp. [6502] - 18%; Kyushu Electric Power [9508] - 6.0%; Bank of Fukuoka [8326] - 5.0%; Toho Life - 4.0%; Kyoei Fire & Marine - 4.0%; Yasuda Trust and Banking [8404] - 2.5%; Hitachi Ltd [6501] - 2.0%

Nippon Yakin Kogyo Co., Ltd.

Japanese Name	:	***Nippon Yakin Kogyo K.K.***
TSE Number	:	5480
Address	:	1-5-8, Kyobashi, Chuo-ku, Tokyo 104
Tel & Fax	:	Tel: 03-3272-1511 Fax: 03-3273-4633
Statistics	:	Capital : 20733 million yen Employees : 1746
Officers	:	CEO : Mr. Yoshito Ishiguro/President
Sales	:	March, 1988 - 77885 million yen March, 1989 - 100133 million yen March, 1990 - 101672 million yen March, 1991 - 103748 million yen
R & D Spending	:	2669 million yen Capital Investment: 8926 million yen
Sales Breakdown	:	

SIC#	Product	Sales
3312	Stainless steel sheets	83 %
34	Stainless steel fabrications	4 %

Banks	:	Industrial Bank of Japan; Tokai Bank; Chuo Trust and Banking; Mitsubishi Trust and Banking; Japan Credit Bank; Fuji Bank
Factories	:	Kawasaki Plant (Kawasaki-ku, Kawasaki); Kanazawa Plant (Kanazawa-shi, Ishikawa-ken); Oeyama Branch (Miyatsu-shi, Kyoto-ken);
Overseas Subsidiaries	:	Thai Seisen Co., Ltd. (Thailand)

Nippon Zeon Co., Ltd.

Japanese Name	:	***Nippon Zeon K.K.***
TSE Number	:	4205
Description	:	Maintains half share of Japan's synthetic rubber market, with the other half taken by Japan Synthetic Rubber Co., Ltd.. Strength lies in specialty rubbers.
Address	:	2-6-1, Marunouchi, Chiyoda-ku, Tokyo 100
Tel & Fax	:	Tel: 03-3769-8623 Fax: 03-3769-8645
Statistics	:	Capital : 16943 million yen Employees : 2862
Officers	:	CEO : Mr. Takeshi Takizawa/President
Sales	:	March, 1988 - 102375 million yen March, 1989 - 110586 million yen March, 1990 - 118035 million yen March, 1991 - 129376 million yen
R & D Spending	:	5829 million yen Capital Investment: 13380 million yen
Sales Breakdown	:	SIC# Product Sales 2822 Synthetic rubber 62 % 2821 Synthetic resins 32 %
Main Products	:	Synthetic Rubber and Latex; Plastics; Ceramics; Specialty Chemicals: Methacryl resin polymer, Vinyl chloride monomer
Banks	:	Dai-Ichi Kangyo Bank; Industrial Bank of Japan; Fuji Bank; Kyowa Saitama Bank; Norinchukin Bank
Branches	:	Mizushima Branch (Okayama-ken); Osaka Branch (Kita-ku, Osaka); Nagoya Branch (Naka-ku, Nagoya); Okayama Sales Office (Okayama-shi)
Factories	:	Kawasaki Plant (Kawasaki-shi); Takaoka Plant (Takaoka-shi, Toyama-ken); Tokuyama Plant (Tokuyama-shi, Yamaguchi-ken); Mizushima Plant (Kurashiki-shi, Okayama-ken)
Laboratories	:	R&D Center (Kawasaki-shi)
Overseas Subsidiaries	:	Zeon Ceramics USA Inc. (Kentucky, U.S.A.); Nippon Zeon of America, Inc. (New York); Zeon Chemicals Europe Ltd. (U.K.); Zeon Europe GmbH (Dusseldorf, Germany)

Nippondenso Co., Ltd.

Japanese Name	:	***Nippondenso K.K.***
TSE Number	:	6902
Description	:	Major manufacturer of electronic parts for automobiles. A member of the Toyota group of companies, Nippondenso produces a variety of automobile components with airconditioners comprising the largest single product category. The company maintains a large network of foreign subsisidiaries through which it is rapidly expanding its overseas production capacity.
Address	:	1-1, Showa-cho, Kariya-shi, Aichi-ken 448
Tel & Fax	:	Tel: 0566-25-5511 Fax: 0566-25-4509
Statistics	:	Capital : 83091 million yen Employees : 38427
Officers	:	CEO : Mr. Tsuneo Ishimaru/President Contact : Mr. Satoru Takemoto/Director, International Sales
Sales	:	December, 1988 - 1115750 million yen December, 1989 - 1230680 million yen December, 1990 - 1379350 million yen
R & D Spending	:	83296 million yen Capital Investment: 116700 million yen

Sales Breakdown :

SIC#	Product	Sales
3585	Automobile heaters and air-conditioners	36 %
3714	Automotive electronic equipment	28 %
3714	Automotive fuel management systems	16 %
3714	Radiators	6 %
3714	Automotive meters	5 %
3714	Automotive filters	3 %

Main Products	:	Automotive Parts and Components: Air-conditioners, Electronic fuel injection systems, Other fuel management systems, Fuel injection pumps, Suspension control systems, Anti-lock brake systems, Heaters, Alternators, Starters, Spark plugs, Meters, Radiators, Air filters, Oil filters; Other Products: Bar code scanners, Magnetic card readers
Banks	:	Tokai Bank; Mitsui Taiyo Kobe Bank; Sanwa Bank
Branches	:	Tokyo Branch; Nagoya Branch; Osaka Branch; Hiroshima Branch; Sales offices in: Sapporo; Aomori; Sendai; Utsunomiya; Saitama; Chiba; Kanagawa; Niigata; Toyama; Shizuoka; Nagano; Kyoto; Kobe; Okayama; Takamatsu; Fukuoka; Miyazaki
Laboratories	:	Nippondenso Research Laboratories (Aichi-ken)
R&D Areas	:	Electronic Control Systems: Ceramics: Automotive Electronics: Fuel injection systems, Anti-lock brake control systems
Subsidiaries	:	Hamanakodenso Co., Ltd. (Shizuoka-ken); Anden Co., Ltd. (Aichi-ken); Kyosan Denki Co., Ltd. (Chuo-ku, Tokyo); Asmo Co., Ltd. (Shizuoka-ken); GAC Corp. (Nagano-ken)

Nishikawa Rubber Co., Ltd.

Japanese Name	:	***Nishikawa Gomu Kogyo K.K.***
TSE Number	:	Not listed
Description	:	Manufacturer of automotive trimming and stampings.
Address	:	2-2-8, Misasa-cho, Nishi-ku, Hiroshima-ken 733
Tel & Fax	:	Tel: 082-237-9371 Fax: 082-238-3051
Statistics	:	Capital : 851 million yen Employees : 1700
Officers	:	CEO : Mr. Masahiro Nishikawa/President Contact : Mr. Tamaki Eiso/Managing Director
Sales	:	March, 1988 - 36744 million yen March, 1989 - 40315 million yen March, 1990 - 43457 million yen
Sales Breakdown	:	SIC# Product 2396 Auto trimmings 3465 Auto moldings
Main Products	:	Automobile and Apparel Trimming: Weather stripping, Interior trim for automobiles; Automotive Stampings: Protective molding for automobile use
Banks	:	Hiroshima Bank; Sumitomo Bank; Fuji Bank
Factories	:	Hiroshima Factory (Hiroshima-shi, Hiroshima-ken); Shiriaki Factory (Hiroshima-shi, Hiroshima-ken); Asa Factory (Hiroshima-shi, Hiroshima-ken); Yoshida Factory (Takada-gun, Hiroshima-ken); Mihara Factory (Mihara-shi, Hiroshima-ken)
Subsidiaries	:	Nishikawa Bussan K.K. (Hiroshima-shi, Hiroshima-ken); Toyohira Nishikawa K.K.(Yamagata-gun, Hiroshima-ken); Seiwa Kogyo K.K. (Hiroshima-shi, Hiroshima-ken)

Nissan Motor Co., Ltd.

Japanese Name	:	***Nissan Jidosha K.K.***
TSE Number	:	7201
Description	:	Major manufacturer of automobiles and related products. Japan's second largest automaker, Nissan Motor Co. is actively expanding its already extensive overseas production network. The company is also active in the aerospace field where it enjoys a solid reputation for producing quality high-tech products.
Address	:	6-17-1, Ginza, Chuo-ku, Tokyo 104-23
Tel & Fax	:	Tel: 03-3543-5523 Fax: 03-3544-0109
Statistics	:	Capital : 203191 million yen Employees : 56873
Officers	:	CEO : Mr. Yutaka Kume/President
Sales	:	March, 1988 - 3418670 million yen March, 1989 - 3580110 million yen March, 1990 - 4005550 million yen March, 1991 - 4175010 million yen
Sales Breakdown	:	see table below

SIC#	Product	Sales
3711	Automobiles	81 %
3714	Automobile parts and accessories	16 %
35	Industrial machinery and marine	2 %
372	Aircraft and parts	1 %

Main Products	:	Transport Related: Automobiles, Automobile parts, Aircraft and parts, Marine craft; Machinery: Industrial machinery, Textile machinery
Banks	:	Fuji Bank; Industrial Bank of Japan; Sumitomo Bank; Yasuda Trust and Banking; Kyowa Saitama Bank
Branches	:	Osaka Branch (Osaka); Kyushu Branch (Fukuoka-shi, Fukuoka-ken)
Factories	:	Oppama Plant (Yokosuka-shi, Kanagawa-ken); Zama Plant (Zama-shi, Kanagawa-ken); Yokohama Plant (Yokohama-shi, Kanagawa-ken) and others
Laboratories	:	Central Engineering Laboratories (Yokosuka-shi, Kanagawa-ken); Nissan Technical Center (Atsugi-shi, Kanagawa-ken)

Nissei Electric Co., Ltd.

Japanese Name	:	***Nissei Denki K.K.***
TSE Number	:	Not listed
Description	:	Primarily engaged in the manufacture of facsimile equipment with an auxiliary product line in machine tools and related industrial machinery. Nissei Electric is probably best known for its small-sized and portable facsimile machines.
Address	:	Hiroo Office Bldg., 1-3-18, Hiroo, Shibuya-ku, Tokyo 150
Tel & Fax	:	Tel: 03-3442-8151 Fax: 03-3442-8692
Officers	:	CEO : Mr. Tamiji Imai/President Contact : Mr. Tamiji Imai/President
Sales Breakdown	:	SIC# Product 3661 Facsimile machines 354 Metal die DNC system
Main Products	:	Facsimile Machines; Metal Working Machinery: Metal die DNC system
Subsidiaries	:	Nissei OPT Co., Ltd.
Overseas Subsidiaries	:	Nissei America Corp. (New York, U.S.A.)

Nisshin Steel Co., Ltd.

Japanese Name	:	***Nisshin Seiko K.K.***
TSE Number	:	5407
Address	:	3-4-1, Marunouchi, Chiyoda-ku, Tokyo 100
Tel	:	03-3216-5511
Statistics	:	Capital : 79696 million yen Employees : 7336
Officers	:	CEO : Mr. Tsuyoshi Kai/President
Sales	:	March, 1988 - 363555 million yen March, 1989 - 414730 million yen March, 1990 - 428697 million yen March, 1991 - 429191 million yen
R & D Spending	:	11008 million yen Capital Investment: 60800 million yen
Sales Breakdown	:	

SIC#	Product	Sales
3398	Stainless and surface-treated products	75 %
3312	Hot and cold-rolled steel products	16 %
3317	High frequency pipes	8 %

Banks	:	Sanwa Bank; Long Term Credit Bank of Japan; Toyo Trust and Banking; Industrial Bank of Japan
Factories	:	Amagasaki Plant (Hyogo-ken); Kure Plant (Hiroshima-ken)
Overseas Subsidiaries	:	Nisshin Steel (Canada) Ltd. (Alberta, Canada); Nisshin Resources Canada Ltd. (British Columbia, Canada); Nisshin USA Inc. (New York, U.S.A.); Wheeling Nisshin Inc. (West Virginia, U.S.A.)

Nissho Iwai Corp.

Japanese Name :	***Nissho Iwai K.K.***
TSE Number :	8063
Description :	The sixth largest trading firm in Japan. Main items include machinery, steel products, ships and aircraft.
Address :	2-4-5, Akasaka, Minato-ku, Tokyo 107
Tel & Fax :	Tel: 03-3588-4016 Fax: 03-5661-7582
Statistics :	Capital : 79868 million yen Employees : 7482
Officers :	CEO : Mr. Akira Nishio/President
Sales :	March, 1988 - 10139000 million yen March, 1989 - 11016400 million yen March, 1990 - 15047500 million yen March, 1991 - 13343200 million yen

Sales Breakdown :

SIC#	Product	Sales
505	Metal products	39 %
5172	Energy	22 %
5084	Machinery	18 %
5199	General commodities	12 %
514	Foodstuffs and food products	7 %
5131	Textiles	3 %

Main Products :	Metal Products; Industrial Machinery; Transportation Equipment and Parts: Aircraft, Ships; Textile products; Construction Materials: Lumber; Heavy Construction: Oil tanks, Natural gas plant; Chemical Products
Banks :	Sanwa Bank; Dai-Ichi Kangyo Bank
Branches :	48 domestic branches; 157 overseas branches
Shareholders :	Sanwa Bank [8320] - 4.96% ; Dai-Ichi Kangyo Bank [8311] - 4.90% ; Daiwa Bank [8319] - 3.62%
Subsidiaries :	314 domestic subsidiaries
Overseas Subsidiaries :	157 overseas subsidiaries

Nissin Chemical Industry Co., Ltd.

Japanese Name	:	***Nissin Kagaku Kogyo K.K.***
TSE Number	:	Not listed
Address	:	2-17-33, Kita-ku, Takefu-shi, Fukui-ken 915
Tel & Fax	:	Tel: 0778-22-5100 Fax: 0778-24-0657
Statistics	:	Capital : 500 million yen Employees : 163
Officers	:	CEO : Mr. Yoshiro Yoshikawa/President Contact : Mr. Yoshiro Yoshikawa/President
Sales	:	February, 1988 - 6086 million yen February, 1989 - 6189 million yen February, 1990 - 6037 million yen
Sales Breakdown	:	SIC# Product 2821 Copolymer resins 2822 Acrylic rubber 3069 Prophylactics 3086 Insulation materials
Banks	:	Mitsubishi Bank; Long Term Credit Bank of Japan; Fukui Bank; Hokuriku Bank; Norinchukin Bank; Dai-Ichi Kangyo Bank
Branches	:	Osaka Sales Office (Osaka-shi); Hokuriku Office (Takefu-shi); Sales Dept. (Chiyoda-ku)
Factories	:	Main Factory (Takefu-shi)
Shareholders	:	Shin-Etsu Chemical Co., Ltd. [4063]

Nissin Electric Co., Ltd.

Japanese Name	:	***Nissin Denki K.K.***
TSE Number	:	6641
Description	:	Manufacturer of power substation and semiconductor production related equipment. Member of the Sumitomo group and has a 90% share of the market for power capacitors used by electric power companies in Japan.
Address	:	47, Umezu-takase-cho, Ukyo-ku, Kyoto 615
Tel & Fax	:	Tel: 075-861-3151 Fax: 075-872-0742
Statistics	:	Capital : 9959 million yen Employees : 3500
Officers	:	CEO : Mr. Yasuo Tanaka/President Contact : Mr. Eisuke Kobayashi/Manager, International Business
Sales	:	March, 1988 - 60054 million yen March, 1989 - 71136 million yen March, 1990 - 80187 million yen March, 1991 - 88683 million yen
R & D Spending	:	4200 million yen Capital Investment: 6171 million yen

Sales Breakdown :

SIC#	Product	Sales
361	Substation equipment	49 %
3613	Reactive power compensation equipment	23 %
382	Control systems, charged ptcl beam eqpt.	18 %
3559	Other electrical equipment	10 %

Main Products	:	Electronic and Electrical Equipment: Power capacitors, Substation equipment, Reactive power compensation equipment, Computerized control systems, Thin film formation systems, Electron processing systems, Ion implantation systems
Banks	:	Sumitomo Bank; Sumitomo Trust and Banking; Kyoto Bank
Branches	:	Tokyo Office (Chiyoda-ku); Kansai Office (Osaka-shi); Chubu Office (Nagoya-shi)
Factories	:	Maebashi Works (Maebashi-shi); Kuze Factory (Kyoto-shi)
Laboratories	:	R&D Division (Kyoto-shi)
R&D Areas	:	Electrical Equipment: High energy ion doping equipment, Excimer laser processing equipment, 400Hz power supply for parked aircraft
Subsidiaries	:	Nissin High Voltage Co., Ltd. (Kyoto); Nissin Systems Co., Ltd. (Osaka); Nissin Tec Co., Ltd. (Kyoto); Nissin High-tech Co., Ltd. (Kyoto); Nissin Engineering Co., Ltd. (Kyoto); K.K. System Design (Osaka); Nissin Denki Shoji K.K. (Osaka)
Overseas Subsidiaries	:	Nissin Electric (Thailand) Co., Ltd. (Bangkok, Thailand); Nissin Electric U.S.A. Inc (U.S.A.); NHV America Inc. (U.S.A.)

Nitivy Co., Ltd.

Japanese Name	:	***Nichibi K.K.***
TSE Number	:	Not listed
Description	:	Manufacturer of synthetic fiber products, selling both domestically and overseas.
Address	:	Katakura Bldg., 3-1-2, Kyobashi, Chuo-ku, Tokyo 104
Tel & Fax	:	Tel: 03-3272-6901 Fax: 03-3281-2729
Statistics	:	Capital : 650 million yen Employees : 226
Officers	:	CEO : Mr. Hiroshi Suzuki/President Contact : Mr. F. Yamaguchi/Export Division
Sales	:	September, 1988 - 2949 million yen September, 1989 - 3571 million yen September, 1990 - 4503 million yen

Sales Breakdown :

SIC#	Product	Sales
2824	Water soluble eliminable fiber 'Solvron'	90 %
2824	Ion exchange fiber	
2824	Continuous alumina fiber	

Main Products : Vinylon — a synthetic fiber first developed in Japan. Vinylon is usually produced using a wet spinning method. Nitivy Co., however, has the distinction of being the only company in the world which uses a dry spinning method to produce Vinylon. Nitivy Co. developed this revolutionary process, and it has become the cornerstone of the company's production and marketing activities. The dry spinning technology allows for a number of innovative production possibilities. For example, in one of the company's products it is possible, through the selection of polyvinyl alcohol for use as a raw material, to produce a remarkable fiber which is soluble throughout a broad spectrum of water temperatures, ranging from cold water to temperatures near the boiling point. Nitivy Co. is continuously conducting research to develop new types of synthetic fiber, and, in addition to extended fiber created through the use of its dry spinning technology, it also produces various types of ion exchange and alumina continuous fibers, both of which have enhanced the company's competitive standing in the synthetic fiber industry. At present, Nitivy Co. is continuing to increase its efforts in R&D in the area of special fiber technology.

Banks	:	Fuji Bank; Mitsui Taiyo Kobe Bank
Branches	:	Osaka Branch (Osaka-shi)
Factories	:	Shizuoka Factory (Fujieda-shi)
Laboratories	:	Laboratory (Fujieda-shi)
Shareholders	:	Katakura Kogyo Ltd.

Nitta Zerachin K.K.

Japanese Name	:	***Nitta Zerachin K.K.***
TSE Number	:	Not listed
Description	:	The company is a leader in the domestic gelatin market.
Address	:	1-8-12, Hon-machi, Chuo-ku, Osaka 541
Tel & Fax	:	Tel: 06-266-1911 Fax: 06-266-1900
Statistics	:	Capital : 375 million yen Employees : 447
Officers	:	CEO : Mr. Yasushi Kurata/President Contact : Mr. Norimichi Soga/Manager, Overseas Office
Sales	:	August, 1989 - 17623 million yen August, 1990 - 18512 million yen
Sales Breakdown	:	SIC# Product 2899 Gelatine 2899 Collagen 3843 Dental cement 2099 Food additives 2891 Adhesives 3535 Pneumatic conveying systems
Main Products	:	Gelatine Products: Collagen gel, Gelatine; Chemical Products: Food additives, Hot melt adhesives; Dental Supplies: Dental cement; Machinery: Pneumatic conveying systems
Banks	:	Mitsui Taiyo Kobe Bank
Branches	:	Tokyo Branch (Taito-ku)
Factories	:	Osaka Factory (Yao-shi, Osaka); Nara Factory (Shiki-gun, Nara-ken)
Laboratories	:	Central Research Laboratory (Yao-shi)
Overseas Subsidiaries	:	Kerala Chemicals & Proteins Ltd. (India); Nitta Corp. of America (U.S.A.); Cangel Inc. (Canada)

Nitta Veneer

Japanese Name :	***Nitta Veniya***
TSE Number :	Not listed
Description :	Manufacturer of plywood, laminated panels and related products. Nitta Veneer is a subsidiary of Nitta Gelatine Inc.
Address :	8-21-1, Ginza, Chuo-ku, Tokyo 104
Tel & Fax :	Tel: 03-571-9218 Fax: 03-571-0332
Statistics :	Capital : 90 million yen Employees : 161
Sales :	September, 1989 - 4125 million yen September, 1990 - 3992 million yen
Sales Breakdown :	SIC# Product 2435 Plywood 2439 Compressed laminated panels 2499 Misc. wooden building materials 3089 Synthetic ivory
Main Products :	Plywood and Compressed Laminated Panels; Misc. Building Materials; Misc. Plastics Products: 'Translarm' synthetic ivory
Banks :	Hokkaido Takushoku Bank; Obihiro-Shinyo Bank; Tokyo-Sogo Bank; Tokai Bank; Dai-Ichi Kangyo Bank
Branches :	Osaka Sales Office (Osaka); Nagoya Sales Office (Nagoya-shi, Aichi-ken); Sapporo Sales Office (Sapporo-shi, Hokkaido)
Factories :	Makubetsu Plant (Hokkaido); Chiba Plant (Chiba-ken)
Shareholders :	Chiba Nitta Bokujo - 167,390 Shares; Akira Nitta - 130,660 Shares; Tadashi Nitta - 121,000 Shares

Nittetsu Micrometal Corp.

TSE Number	:	Not listed
Description	:	Manufacturer of micro wires. Nittetsu Micrometal, in conjunction with Oki Electric Co., Ltd., is in the process of completing a 10 micron wire production plant with a projected capacity of 15 million meters per month. The company is a subsidiary of Nippon Steel Corp..
Address	:	
Tel	:	0429-34-6101
Sales Breakdown	:	SIC# Product 33 Micro wire products
Main Products	:	Micro Wire Products: 10 micron diameter wire
Shareholders	:	Nippon Steel Corp. [6703]

Nittetsu Mining Co., Ltd.

Japanese Name : ***Nittetsu Kogyo K.K.***

TSE Number : 1515

Description : Mining company specializing in mineral resources, including non-metallic minerals, metallic minerals and fuels. A member of the Nippon Steel group of companies, Nittetsu Mining Co. is also engaged in chemicals and machinery manufacture and is making initial steps in the field of real estate development.

Address : Yusen Bldg. 6F, 2-3-2, Marunouchi, Chiyoda-ku, Tokyo 100

Tel & Fax : Tel: 03-3284-0516 Fax: 03-3215-8480

Statistics : Capital : 4176 million yen
Employees : 1124

Officers : CEO : Mr. Masanobu Nakagami/President
Contact : Mr. Akiharu Shiokawa/Manager, Overseas Department

Sales : March, 1988 - 57038 million yen
March, 1989 - 62735 million yen
March, 1990 - 66084 million yen
March, 1991 - 72842 million yen

Sales Breakdown :

SIC#	Product	Sales
1422	Mineral mining - limestone, etc.	55 %
106	Metallic minerals	17 %
355	Industrial machinery	14 %
0000	Fuels and other products	10 %
28	Chemical products	3 %
651	Real estate	1 %

Main Products : Non-metallic Minerals: Limestone, Dolomite, Crushed stone, Limestone powder, Silica stone, Diatom earth, Serpentine, Iron ore concentrates, Fluorite, Quicklime, Slaked lime, Lightweight aggregates, Cement, Gypsum, Granulated blast furnace slag; Metallic Minerals: Electrolytic copper, Electrolytic gold, Electrolytic silver, Tungsten concentrates, Sulfuric acid; Fuels & Other Products: Petroleum products, Coal, LPG, Wood chips for pulp use, Timber; Machinery: Steel mill machinery, Spalling and crushing machinery, Dusters, Heat exchangers for comfort control application, Monorail passenger cars; Chemical Products: Polyferric sulfate 'Polytetsu', Inorganic paper, Fire retardant paper, Limestone powdered paper, Specialty papers, Freshness protection wrapping paper, Freshness protection wrapping film; Real Estate: Commercial building leasing, Residential building leasing

Banks : Mitsui Taiyo Kobe Bank; Industrial Bank of Japan; Mitsubishi Bank; Dai-Ichi Kangyo Bank; Fuji Bank

Laboratories : R&D Center (Mitaka-shi, Tokyo)

Shareholders : Nippon Steel Corp. [5401] - 27.70% ; Nittetsu Mining Shogakukai - 6.45% ; Industrial Bank of Japan [8302] - 4.98% ; Mitsui Taiyo Kobe Bank [8314] - 3.48%; Dai-Ichi Kangyo Bank [8311] - 2.90% ; Seiko Gakuen - 2.60% ; Mantoyo Gakuen - 1.32%

Overseas Subsidiaries : Sherkate Sahami Khas Minakan (Iran); Nittetsu Panama Inc. (Panama); El Roble Exploracion y Explotacion S.A. (Columbia)

NKK Corp

Japanese Name	:	***NKK K.K.***
TSE Number	:	5404
Description	:	Among the world's top crude steel producers, specializing in steel piping. The company has recently diversified into electronics and chemicals.
Address	:	1-1-2, Marunouchi, Chiyoda-ku, Tokyo 100
Tel & Fax	:	Tel: 03-3212-7111 Fax: 03-3214-8404
Statistics	:	Capital : 233732 million yen Employees : 22346
Officers	:	CEO : Mr. Yoshinari Yamashiro/President
Sales	:	March, 1988 - 1050330 million yen March, 1989 - 1261230 million yen March, 1990 - 1277600 million yen March, 1991 - 1326800 million yen
R & D Spending	:	31237 million yen Capital Investment: 118000 million yen

Sales Breakdown :

SIC#	Product	Sales
3312	Steel plating	40 %
8711	Engineering	21 %
0000	Ferroalloys, new materials, related prod.	18 %
3316	Steel tubing	15 %
3312	Steel bars	7 %

Main Products	:	Steel Division: Steel piping, Steel tubing, Steel plating, Steel structures, Ferroalloys and related products, Semi-finished metal goods, Slag-based products, Surface treated steel plating, Rails, Electronics products, Advanced materials; Energy Engineering: Gas and oil pipelines, LNG and LPG low temperature tanks, Storage tanks, Chemical plant, Maritime structures, Starter motors for deisel gas and turbine engines, Generators; Environmental Engineering: Municipal refuse incinerators, Water treatment facilities, Water supply pipes, Water purification facilities; Steel-making Engineering: Mini-mill installation, Cold rolling process lines, Iron and steel-making facilities; Steel Structure and Machinery Engineering: Floodgates, Bridge frames, Multi-level parking building machinery, Steel silos, Fermentation tanks, Brewing vats, Food manufacturing equipment; Maritime Engineering: Oil tankers, LPG tankers, LNG tankers, Various boats, Sea-floor oil development machinery, Ship repairs and refitting; Other Engineering: Civil engineering, Factory automation, Leisure facilities, Nuclear power equipment
Banks	:	Fuji Bank; Industrial Bank of Japan; Yasuda Trust and Banking; Dai-Ichi Kangyo Bank; Long Term Credit Bank of Japan
Branches	:	Osaka Branch (Osaka-shi); Nagoya Branch (Nagoya-shi); Hokkaido Branch (Sapporo-shi); Chiba Branch (Chiba-shi); Bangkok Office (Bangkok, Thailand); Beijing Office (Beijing, China); Dusseldorf Office (Dusseldorf, Germany)
Factories	:	Keihin Works (Kawasaki-shi); Fukuyama Works (Fukuyama-shi, Hiroshima-ken); Tsurumi Works (Yokohama-shi); Tsu Works (Tsu-shi, Mie-ken); Shimizu Works (Shimizu-shi, Shizuoka-ken); Toyama Works (Toyama-ken); Niigata Works (Niigata-ken)
Laboratories	:	Central Research Laboratories; Iron and Steel Laboratories; Applied Technology Laboratories and Electronics Laboratories (Kawasaki-shi)

NOK Corp.

Japanese Name :	***NOK K.K.***
TSE Number :	7240
Description :	The company manufactures, imports and sells seal products, industrial mechanical parts, fundamental hydraulic and pneumatic equipment, synthetic chemical products, electronic products and various other goods. Currently stepping up its R&D activities.
Address :	1-12-15, Shibadaimon, Minato-ku, Tokyo 106
Tel & Fax :	Tel: 03-3432-4211 Fax: 03-3432-3919
Statistics :	Capital : 15911 million yen Employees : 3645
Officers :	CEO : Mr. Masato Tsuru/President
Sales :	March, 1988 - 144636 million yen March, 1989 - 159751 million yen March, 1990 - 178659 million yen March, 1991 - 196656 million yen
Capital Investment :	15261 million yen

Sales Breakdown :

SIC#	Product	Sales
306	Industrial rubber products	38 %
3053	Oil seals	22 %
36	Electronics	19 %
3592	Mechanical seals	5 %

Main Products :	Gaskets and Seals: Radial shaft seals, O-rings, Packings, Mechanical face seals, Noise and vibration dampers for automotive and industrial machinery; Other Equipment: Hydraulic and pneumatic equipment, Electronic products
Banks :	Sumitomo Bank; Sanwa Bank; Mitsui Taiyo Kobe Bank; Yokohama Bank; Bank of Tokyo; Kyowa-Saitama Bank
Laboratories :	Tsukuba Research Center (Tsukuba-shi, Ibaraki-ken); Technical Center (Fujisawa-shi, Kanagawa-ken)
Overseas Subsidiaries :	Singapore Oilseal Co., (Pte) Ltd. (Singapore); Freudenberg Componentes Ltd. (Sao Paulo, Brazil); Thai NOK Co., Ltd. (Thailand); NOK Inc. (U.S.A.); NOK International Co., Ltd. (U.K.); NOK UK Ltd. (U.K.)

Noritake Co., Ltd.

Japanese Name	:	***Noritake Kanpani Rimitedo***
TSE Number	:	5331
Description	:	Manufacturer of tableware, industrial materials, and electronic parts and materials. A core member of the Morimura group of companies, Noritake Co. produces a variety of products derived from ceramics and other non-metal minerals. The company's high quality tableware products enjoy an international reputation and provide a steady source of revenue. New ceramic materials are the central focus of ongoing R&D activities.
Address	:	3-1-36, Noritake-Shinmachi, Nishi-ku, Nagoya, Aichi 451
Tel & Fax	:	Tel: 052-561-7111 Fax: 052-582-5376
Statistics	:	Capital : 10302 million yen Employees : 2171
Officers	:	CEO : Mr. Susumu Saeki/President
Sales	:	November, 1988 - 83542 million yen March, 1989 - (4 months) 29621 million yen March, 1990 - 96500 million yen March, 1991 - 107871 million yen
R & D Spending	:	739 million yen Capital Investment: 4576 million yen
Sales Breakdown	:	

SIC#	Product	Sales
32	Industrial materials	40 %
32	Tableware	30 %
367	Electronic parts and components	15 %
325	Ceramic products	15 %

Main Products	:	Industrial Abrasives and Related Products: Coated abrasives, Grinding and polishing wheels, Grinding and polishing machines, Diamond tools, CBN Tools, Industrial ceramics; Thermal Products: Thermal heads, Thermal pens; Electronic Products: Fluorescent display tubes, Network rheostats, Thick membrane IC substrates, Static mixers; Ceramic Products: Ceramic pipes, Preformed ceramic cores; Tableware: Glassware, Crystal, Ceramic stoneware
Banks	:	Mitsubishi Bank; Bank of Tokyo; Tokai Bank; Kyowa Saitama Bank; Mitsubishi Trust and Banking
Factories	:	Tsushima Resinoid Products Plant (Aichi-ken); Atsugi Crystalware Plant (Kanagawa-ken); Nishi-Kamo Plant (Aichi-ken); Nitto Gypsum Plant (Nagoya-shi, Aichi-ken)
Laboratories	:	Nishi-Kamo Research Center (Aichi-ken)
Subsidiaries	:	20 domestic subsidiaries

NSK, Ltd.

Japanese Name	:	***Nippon Seiko K.K.***
TSE Number	:	6471
Description	:	Major bearings manufacturer. Nippon Seiko maintains subsidiaries and overseas production facilities worldwide and has been placing increased emphasis upon exploring new product lines outside of the company's mainline bearing products.
Address	:	1-6-3, Osaki, Shinagawa-ku, Tokyo 141
Tel	:	03-3779-7111
Statistics	:	Employees : 9800
Officers	:	CEO : Mr. Toshio Arata/President
Sales	:	April, 1988 - 265225 million yen March, 1989 (11 months) - 275828 million yen March, 1990 - 344219 million yen March, 1991 - 368633 million yen
Sales Breakdown	:	see below

SIC#	Product	Sales
3562	Bearings	60 %
35	Machinery parts and other	40 %

Main Products	:	Bearings; Industrial Machinery and Related Products
Banks	:	Industrial Bank of Japan; Fuji Bank; Bank of Yokohama; Mitsubishi Bank; Yasuda Trust and Banking
Factories	:	Overseas production facilities: Brazil; U.S.A.; U.K.

NTT, Electron Application Institute

Japanese Name	:	***NTT Denshi Oyo Kenkyujo (Kiokusochi Kenkyubu)***
TSE Number	:	Not listed
Description	:	Research subsidiary of NTT Corp. engaged in electronics R&D.
Address	:	162, Oaza-Shirakata, Tokai-mura, Naka-gun, Ibaraki-ken 319-11
Tel	:	0292-87-7046
Sales Breakdown	:	SIC# Product 8731 Electronics R&D
Main Products	:	Electronics R&D
R&D Areas	:	Communications Equipment
Shareholders	:	Nippon Telegraph and Telephone Corp. [9432]

Obayashi Corp.

Japanese Name : ***Obayashi Gumi***

TSE Number : 1802

Description : Major building and engineering contractor based in Tokyo. A member of the Sanwa group of companies and one of Japan's five largest contractors, Obayashi Corp. is involved in projects throughout Asia and the rest of the world.

Address : 2-3, Kanda Tsukasa-cho, Chiyoda-ku, Tokyo 101

Tel : 03-3292-1111

Statistics : Capital : 57566 million yen
Employees : 11085

Officers : CEO : Mr. Yoshiro Obayashi/President

Sales : March, 1988 - 820288 million yen
March, 1989 - 944655 million yen
March, 1990 - 1151980 million yen
March, 1991 - 1131810 million yen

Sales Breakdown :

SIC#	Product	Sales
15	Building construction	76 %
8711	Civil engineering	21 %
65	Real estate	3 %

Main Products : Building Construction; Civil Engineering; Real Estate

Banks : Sanwa Bank; Mitsui Taiyo Kobe Bank; Mitsubishi Bank; Sumitomo Bank; Industrial Bank of Japan; Toyo Trust and Banking

Laboratories : Technical Research Institute (Kiyose-shi, Tokyo)

Okamoto Industries, Inc.

Japanese Name	:	***Okamoto K.K.***
TSE Number	:	5122
Description	:	Manufacturer of plastic and rubber goods. Okamoto Industries produces a variety of rubber and plastic products with tires and condoms figuring prominently.
Address	:	3-27-12, Hongo, Bunkyo-ku, Tokyo 113
Tel & Fax	:	Tel: 03-3817-4111 Fax: 03-3817-0465
Statistics	:	Capital : 6920 million yen Employees : 1010
Officers	:	CEO : Mr. Takehiko Okamoto/President
Sales	:	March, 1989 - 74193 million yen March, 1990 - 79499 million yen March, 1991 - 89863 million yen
R & D Spending	:	341 million yen Capital Investment: 5020 million yen

Sales Breakdown :

SIC#	Product	Sales
30	Building and industrial materials	25 %
3021	Footwear	25 %
3011	Automobile tires	25 %
3069	Condoms and sundry goods	10 %
2869	Vinyl and vinyl products	10 %

Main Products	:	Plastic and Rubber Materials for Construction and Industry; Automobile Tires: 'Riken' brand tires; Medical and Sundry Goods: Condoms, Adhesive tapes; Footwear; Vinyl and Vinyl Products: PVC films, PVC imitation leathers, Olefinic film
Banks	:	Dai-Ichi Kangyo Bank; Fuji Bank; Yasuda Trust and Banking; Sanwa Bank; Mitsui Taiyo Kobe Bank; Tokai Bank; Bank of Tokyo
Subsidiaries	:	15 domestic subsidiaries
Overseas Subsidiaries	:	Hong Kong Riken Yoko Co. (Hong Kong); Riken America Inc. (California, U.S.A.); Okamoto U.S.A. (Conneticut, U.S.A.)

Oki Electric Industry Co., Ltd.

Japanese Name :	***Oki Denki Kogyo K.K.***
TSE Number :	6703
Description :	Major manufacturer of communications equipment and electronic components. With a dramatic expansion of IC production capacity in the works, Oki Electric Industry hopes to expand its presence in the international semiconductor market.
Address :	1-7-12, Toranomon, Minato-ku, Tokyo 105
Tel & Fax :	Tel: 03-3501-3111 Fax: 03-3581-5522
Statistics :	Employees : 14100
Officers :	CEO : Mr. Nobumitsu Kosugi/President
Sales :	March, 1989 - 503786 million yen March, 1990 - 552162 million yen March, 1991 - 582183 million yen

Sales Breakdown :

SIC#	Product	Sales
357	Information processing systems	45 %
366	Telecommunications equipment	30 %
3679	Electronic parts and components	20 %

Main Products :	Information Processing Systems and Equipment: Printers, Mainframe terminals, Notebook-size PCs; Telecommunications Equipment: Mobile phones; Electronic Parts and Components: CMOS semiconductors, Memory ICs
Banks :	Long Term Credit Bank of Japan; Fuji Bank; Yasuda Trust and Banking; Tokai Bank; Mitsui Taiyo Kobe Bank
Factories :	Miyazaki Plant (Miyazaki-ken); Overseas production facilities: U.S.A.; Scotland (U.K.)

Omi Goukin Co., Ltd.

Japanese Name	:	***Omi Goukin K.K.***
TSE Number	:	Not listed
Description	:	Manufacturer of aluminum alloy molding materials. As of 1991, the company's total production capacity for aluminum alloy molding material was approximately 200 tons/month.
Address	:	185, Shimotago, Himi, Toyama 935
Tel & Fax	:	Tel: 0766-91-1111 Fax: 0766-91-1110
Statistics	:	Capital : 80 million yen Employees : 70
Sales	:	May, 1989 - 811 million yen May, 1990 - 856 million yen
Sales Breakdown	:	SIC# Product 3334 Aluminum alloys and related products
Main Products	:	Aluminum Alloys and Related Products: Aluminum alloy molding materials
Banks	:	Toyama-Daiichi Bank; Hokuriku Bank; Toyama-ken Shinyo Kumiai
Branches	:	Tokyo Branch (Tokyo); Osaka Branch (Osaka)
Factories	:	Himi Plant (Toyama)
Shareholders	:	Mr. Hisashi Omi

Onoda Cement Co., Ltd.

Japanese Name	:	***Onoda Semento K.K.***
TSE Number	:	5233
Description	:	Major cement manufacturer. A member of the Mitsui group of companies, Onoda Cement is active in the export of cement technology and enjoys strong domestic sales in building materials. Acting through its subsidiaries, the company is expanding business lines to include engineering and electronics.
Address	:	6276, Oaza-Onoda, Onoda-shi, Yamaguchi-ken 756
Tel & Fax	:	Tel: 0836-83-3331 Fax: 0836-83-2011
Statistics	:	Capital : 2901800 million yen Employees : 1700
Officers	:	CEO : Mr. Kazusuke Imamura/President
Sales	:	March, 1988 - 186887 million yen March, 1989 - 193116 million yen March, 1990 - 205165 million yen March, 1991 - 218971 million yen
R & D Spending	:	4100 million yen Capital Investment: 41565 million yen
Sales Breakdown	:	SIC# Product Sales 3241 Cement 73 % 1422 Limestone and other 27 %
Main Products	:	Cement Products; Limestone
Banks	:	Industrial Bank of Japan; Mitsui Bank; Mitsui Trust and Banking; Kyowa Bank; Yamaguchi Bank
Branches	:	Overseas offices: Australia Rep. Office (Sydney, Australia); Hong Kong Rep. Office (Hong Kong); Beijing Rep. Office (Beijing, China)
Laboratories	:	Tsukuba Research Laboratory (Tsukuba-shi, Ibaraki)
Shareholders	:	Nippon Life - 6.4%; Mitsui Trust and Banking [8401] - 6.1%; Mitsui Life - 4.6%; Industrial Bank of Japan [8302] - 4.4%; Mitsui Taiyo Kobe Bank [8314] - 4.4%
Subsidiaries	:	Onoda A.L.C. Co., Ltd. (Nagoya-shi, Aichi-ken); Oriental Construction Co., Ltd. (Chiyoda-ku, Tokyo); Tokai Ren K.K. (Chiyoda-ku, Tokyo)
Overseas Subsidiaries	:	Onoda U.S.A., Inc. (California, U.S.A.)

Optoelectronics Technology Research Corp.

Japanese Name :	***Hikari Gijutsu Kenkyu Kaihatsu K.K.***
TSE Number :	Not listed
Description :	Private research firm engaged in optoelectronics R&D with emphasis upon technologies used in the production of semiconductors and other microelectronic products.
Address :	5-5, Tokodai, Tsukuba, Ibaraki 300-06
Tel :	0298-47-4331
Sales Breakdown :	SIC# Product 8731 Optoelectonics R&D
Main Products :	Optoelectronics R&D: Semiconductor production lasers, DB laser diodes, 'Areal' semiconductor light modulation systems, Optical NAND logic systems, Photolithography techniques, Optical measurement instruments
R&D Areas :	Optoelectronics Technologies; Semiconductor and Microelectronics Production Technology

Osaka Diamond Industrial

Japanese Name :	***Osaka Diamondo Kogyo K.K.***
TSE Number :	6126
Description :	Osaka Diamond is one of the leading diamond-tipped tool makers in Japan, mainly producing diamond wheels, dies and tools for the automotive, electronics and ceramics industries. The company is showing firm profits and is expanding abroad with a joint venture in Taiwan and new offices in Asia.
Address :	2-80, Ohtori-Kitamachi, Sakai-shi, Osaka 593
Tel & Fax :	Tel: 0722-62-1061 Fax: 0722-64-4881
Statistics :	Capital : 2590 million yen Employees : 713
Officers :	CEO : Mr. Keiichi Ohura/President Contact : Mr. Keiichi Ohura/President
Sales :	March, 1988 - 10947 million yen March, 1989 - 12559 million yen March, 1990 - 14183 million yen March, 1991 - 15718 million yen
R & D Spending :	390 million yen Capital Investment: 746 million yen

Sales Breakdown :

SIC#	Product	Sales
3545	Diamond wheels	60 %
3546	Drills, handtools	25 %
3544	Diamond-tipped tools	15 %

Main Products :	Machine Tools and dies: Diamond wheels, Drills, Dies, Diamond-tipped tools
Banks :	Sumitomo Trust and Banking; Sumitomo Bank; Industrial Bank of Japan
Factories :	Harima Plant
Shareholders :	Sumitomo Electric Industries [5802] - 53.6% ; Mitsui & Co. [8031] - 3.5% ; Sumitomo Trust and Banking [8403] - 2.3% ; Yasuda Trust and Banking [8404] - 1.2% ; Kas Depositary Trust - 1.1% ; OID - 0.9% ; Company's Kyoeikai Assn. - 0.7%

Osaka Gas Co., Ltd.

Japanese Name :	***Osaka Gasu K.K.***
TSE Number :	9532
Description :	Japan's second largest urban gas utility. Serving the Kansai area (Osaka, Kyoto and Kobe), Osaka Gas has converted a large portion of its operations to natural gas. The company is also active in the manufacture and sale of household gas appliances and has been utilizing ultra-low temperature LNG to develop new product lines such as liquefied industrial gases and frozen foods.
Address :	4-1-2, Hirano-cho, Chuo-ku, Osaka 541
Tel & Fax :	Tel: 06-202-2221 Fax: 06-222-6195
Statistics :	Employees : 3549
Officers :	CEO : Mr. Masafumi Onishi/President
Sales :	March, 1988 - 552401 million yen March, 1989 - 563293 million yen March, 1990 - 559354 million yen

Sales Breakdown :

SIC#	Product	Sales
492	Gas	75 %
363	Gas appliances	11 %
492	Gas supply work	6 %
2813	Industrial gases	

Main Products :	Gas: Petroleum gas, LNG; Industrial Gases: Carbonic acid gas; Specialty Chemicals: Tar-based benzene, Naphthalene, Styrene monomer, Toluene, Xylene; Other Products: Gas appliances, Gas supply work
Banks :	Mitsui Taiyo Kobe Bank; Tokai Bank; Industrial Bank of Japan
Branches :	Overseas offices: New York Rep. Office (New York. U.S.A.); London Office (London. U.K.)
Laboratories :	Research Laboratory (Osaka)
Shareholders :	Daiwa Bank [8319] - 4.9%; Nippon Life - 4.9%; Sanwa Bank [8320] - 3.5%; Sumitomo Life - 3.1%

Osaka Sanso Kogyo, Ltd.

Japanese Name	:	***Osaka Sanso Kogyo K.K.***
TSE Number	:	4089
Description	:	Osaka Sanso is a major producer of industrial gases. The company is expanding into supplies of absorptive gas generators and special gases for high-tech industries.
Address	:	Sumitomo Seimei Shinosaka Kita Bldg., 7th Floor, 1-14, Miyahara 4-chome, Yodogawa-ku, Osaka 532
Tel & Fax	:	Tel: 06-396-3181 Fax: 06-396-3199
Statistics	:	Capital : 4899 million yen Employees : 673
Officers	:	CEO : Mr. Masataka Tamura/President Contact : Mr. Masataka Tamura/President
Sales	:	September, 1988 - 41902 million yen September, 1989 - 46365 million yen September, 1990 - 51398 million yen
Sales Breakdown	:	SIC# Product 2813 Oxygen, nitrogen, argon
Main Products	:	Industrial Gases: Oxygen, nitrogen, argon
Branches	:	Tokyo Branch (Minato-ku); Nagoya Branch (Nagoya-shi); Kinki Branch (Amagasaki-shi); Chugoku Branch (Kurashiki-shi); Kyushu Branch (Kitakyushu-shi)
Factories	:	Amagasaki Factory (Amagasaki-shi); Himeji-Factory (Himeji-shi)
Shareholders	:	The B.O.C Group plc.
Subsidiaries	:	Daisan Gas Shoji (Amagasaki-shi); Bocho Daisan (Kudamatsu-shi, Yamaguchi-ken); Himeji Daisan (Himeji-shi); Fukuoka Daisan (Fukuoka-shi); A-tech (Akashi-shi); Kurosaki Sanso

Osaka Titanium Co., Ltd.

Japanese Name	:	***Osaka Chitaniumu K.K.***
TSE Number	:	5725
Description	:	Produces titanium and semiconductor materials.
Address	:	1, Higashihama-cho, Amagasaki-shi, Hyogo-ken 660
Tel & Fax	:	Tel: 06-411-1121 Fax: 06-413-3435
Statistics	:	Capital : 2535 million yen Employees : 9152
Officers	:	CEO : Mr. Shigeru Tamamoto/President
Sales	:	March, 1988 - 42277 million yen March, 1989 - 49520 million yen March, 1990 - 53279 million yen March, 1991 - 58829 million yen
R & D Spending	:	827 million yen Capital Investment: 5028 million yen
Sales Breakdown	:	SIC# Product 3339 Titanium 3674 Semiconductor silicon
Main Products	:	Primary Non-ferrous Metals: Metallic titanium; Semiconductor Materials: Semiconductor silicon
Banks	:	Sumitomo Bank; Sanwa Bank; Sumitomo Trust and Banking; Industrial Bank of Japan
Branches	:	Tokyo Office (Minato-ku); Kyushu Office (Fukuoka-shi); Tohoku Office (Sendai-shi)
Factories	:	Factory (Amagasaki-shi)

Otsuka Chemical Co., Ltd.

Japanese Name : ***Otsuka Kagaku K.K.***

TSE Number : Not listed

Description : Manufactures, researches and develops chemical products, processed food and beverages, and furniture.

Address : 3-2-27, Otedori, Chuo-ku, Osaka 540

Tel & Fax : Tel: 06-943-7711 Fax: 06-946-0860

Statistics : Capital : 2800 million yen
Employees : 1650

Officers : CEO : Mr. Yujiro Otsuka/President
Contact : Mr. Makio Okamoto/Director, Overseas Office

Sales : September, 1988 - 94700 million yen
September, 1989 - 99600 million yen
September, 1990 - 106700 million yen

Sales Breakdown :

SIC#	Product	Sales
209	Foods and beverages	64 %
289	Chemical products	15 %
25	Furniture	11 %
287	Agricultural chemicals	9 %

Main Products : Chemical Products: Industrial chemicals, Blowing agents, Agrichemicals; Foods and beverages; Furniture

Factories : Tokushima Factory; Naruto Factory; Osaka Factory; Kishiwada Factory

Laboratories : Tokushima Research Laboratories (Tokushima-shi); Naruto Research Center (Naruto-shi, Tokushima-ken); Food Research Laboratories (Tokushima-shi)

Subsidiaries : Riken EMC Inc.; Japan Whisker Co., Ltd.; Otsuka Omi Ceramics Co., Ltd.; Shinkoyo Glass Co., Ltd.; Otsuka Furniture Sales Co., Ltd.

Overseas Subsidiaries : Hebron S.A. (Barcelona, Spain); P.T. Lautan Otsuka Chemical (Indonesia); Korea Otsuka (Pusan, Korea)

Polyplastics Co., Ltd.

Japanese Name	:	***Poripurasuchikkusu K.K.***
TSE Number	:	Not listed
Description	:	Manufacturer of synthetic resin products. Polyplastics Co., Ltd. is a joint venture of Daicel Chemical Industries, Ltd and Hoechst Celanese Corp.
Address	:	Kasumigaseki Bldg., 3-2-5, Kasumigaseki, Chiyoda-ku, Tokyo 100
Tel & Fax	:	Tel: 03-3593-2444 Fax: 03-3580-0629
Statistics	:	Capital : 3000 million yen Employees : 770
Officers	:	CEO : Mr. Takuzo Kasuga/President Contact : Mr. Ikuo Izumi/Manager, Export Division
Sales	:	September, 1989 - 43000 million yen September, 1990 - 45000 million yen September, 1991 - 49000 million yen
Sales Breakdown	:	SIC# Product 2821 Synthetic resins
Main Products	:	Synthetic Resin Materials: Polyacetal 'Duracon', Polybutylene terephthalate 'Duranex', Polyphenylene sulfide resin 'Fortron', Liquid crystal polymer 'Vectra', Long fiber reinforced thermoplastic molding materials 'Celstran'
Banks	:	Mitsui Bank; Sanwa Bank; Mitsui Trust and Banking; Tokai Bank
Branches	:	Osaka Office (Osaka); Nagoya Office (Nagoya-shi); Sendai Office (Sendai-shi); Shinshu Office (Matsumoto-shi)
Factories	:	Fuji Plant (Fuji-shi)
Laboratories	:	Research Center (Fuji-shi); Application Development Center (Fuji-shi); Technical Service Center (Fuji-shi)
Shareholders	:	Daicel Chemical Industries, Ltd [4202] - 55% ; Hoechst Celanese Corp. - 45%
Overseas Subsidiaries	:	Taiwan Engineering Plastics Co., Ltd. (Taipei, Taiwan)

Sakai Chemical Industry Co., Ltd. (Osaka)

Japanese Name : ***Sakai Kagaku Kogyo K.K.***

TSE Number : 4078

Description : Sakai Chemical Industry is a mid-ranking chemical manufacturer with the mainlines being inorganic chemicals and pharmaceuticals. Sakai was a pioneer in the field of titanium dioxide production and is applying this technology in the painting, printing, textile and paper industries. The company is increasing investment into pharmaceuticals and marketing an X-ray medium, 'Baritop' and cold remedies under the trade name 'Kaigen'. The company will soon offer for public subscription shares of Sakai Shoji.

Address : 1-1-23, Ebisunocho-Nishi, Sakai-shi, Osaka 590

Tel & Fax : Tel: 0722-23-4111 Fax: 0722-23-8355

Statistics : Capital : 18900 million yen
Employees : 1012

Officers : CEO : Mr. Hiroshi Matsumoto/President

Sales : March, 1988 - 38426 million yen
March, 1989 - 40960 million yen
March, 1990 - 44008 million yen
March, 1991 - 46332 million yen

R & D Spending : 1977 million yen Capital Investment: 6305 million yen

Sales Breakdown :

SIC#	Product	Sales
3341	Titanium dioxide	28 %
2819	Stabilizers	18 %
2819	Barium products	15 %
2834	Pharmaceuticals	10 %
2819	Zinc products	10 %
2819	Catalysts	8 %

Main Products : Secondary Non-ferrous Metals: Titanium dioxide; Industrial Chemicals: Organo-tin-based stabilizer, Lead compound-based polyvinyl chloride stabilizer, Metallic soap-based polyvinyl stabilizers, Sequestering agent, Barium and strontium based products; Metals Products: Zinc oxide and zinc dust, Magnetic recording materials, Catalysts; Pharmaceuticals: Digestives, Tonics, Cold remedies

Banks : Mitsubishi Trust and Banking; Industrial Bank of Japan; Sanwa Bank; Norinchukin Bank; Joyo Bank

Branches : Tokyo Branch (Chuo-ku, Tokyo); Osaka Branch (Nishi-ku, Osaka)

Laboratories : Research Center (Sakai-shi, Osaka)

Overseas Subsidiaries : Sakai Trading New York Inc. (U.S.A.)

Sakai Textile Manufacturing

Japanese Name :	***Sakai Sen-i Kogyo K.K.***
TSE Number :	3408
Description :	Major textile manufacturer ranked among the top dyeing and finishing textile firms in Japan. With synthetic fibers experiencing a slowdown, the company is advancing into new lines of business, including apparel and industrial materials. The company is undergoing reconstruction and will change its name to Sakai Ovex.
Address :	2-15-1, Hanandonaka, Fukui-shi, Fukui-ken 910
Tel & Fax :	Tel: 0776-36-5800 Fax: 0776-36-5827
Statistics :	Capital : 2800 million yen Employees : 805
Officers :	CEO : Mr. Norio Igarashi/President Contact : Mr. Norio Igarashi/President
Sales :	March, 1988 - 17333 million yen March, 1989 - 16944 million yen March, 1990 - 20833 million yen March, 1991 - 25800 million yen
Capital Investment :	890 million yen

Sales Breakdown :

SIC#	Product	Sales
2262	Textile dyeing	80 %
2221	Apparel	10 %
3699	Factory automation machines	10 %

Main Products :	Man-made Fibers Finishing Plants: Textile dyeing, Apparel; Electrical Equipment and Supplies: Factory automation machines
Banks :	Fukui Bank; Industrial Bank of Japan; Japan Development Bank; Nippon Credit Bank; Mitsui Trust and Banking

Sanyo Chemical Industries, Ltd.

Japanese Name : ***Sanyo Kasei Kogyo K.K.***

TSE Number : 4471

Description : Leading fine chemicals manufacturer specializing in synthetic resins and related chemical products. An affiliate of Tomen Corp., Sanyo Chemical Industries produces an extremely broad range of synthetic chemical products. R&D continues to stress mainline surfactants, polymers and synthetic resins.

Address : 11-1, Ikkyo-Nomoto-cho, Higashiyama-ku, Kyoto 605

Tel & Fax : Tel: 075-541-4311 Fax: 075-551-2557

Statistics : Capital : 13026 million yen
Employees : 1100

Officers : CEO : Mr. Takehiko Fujimoto/President
Contact : Mr. Tokio Yamada/General Manager, Overseas Division

Sales : March, 1988 - 55391 million yen
March, 1989 - 59349 million yen
March, 1990 - 63498 million yen
March, 1991 - 68514 million yen

R & D Spending : 3459 million yen Capital Investment: 4300 million yen

Sales Breakdown :

SIC#	Product	Sales
2821	Plastic specialty products	27 %
2843	Surfactants	26 %
2821	Polyurethane products	25 %
2821	High polymers	21 %

Main Products : Plastic Specialty Products: Soil stabilizers, Waterproofing agents, Floor covering materials, Epoxy resin hardeners, Synthetic inks, Toner binders; Surface Active Agents: Super-absorbent polymers, Other surfactants; Polyurethane Products: Polyurethane foam materials, Polyurethane elastomers, Polyurethane solutions for synthetic leathers; High Polymers and Related Products: High-polymer flocculants, Lubricating oil additives, Paper sizing chemicals, Polymerpolyols for polyurethane foams; Other Specialty Chemicals: Alkenyl succinic acid, Alkyl allyl sulfonic salt, Alkenyl succinic anhydride, Alkyl ethers, Antistatic agents for fiber, Dispersion agents for rubber/plastic/pulp and paper, Epoxy resin curing agent, Flocculant, Flotation reagent, Phosphoric acid ester plasticizers, Polyoxyethylene alkyl ether, Rust inhibitor, Cationic sizing agent, Anionic sizing agent

Banks : Mitsui Trust and Banking; Mitsui Taiyo Kobe Bank; Nippon Credit Bank; Kyoto Bank

Factories : Nagoya Factory (Tokai-shi, Aichi-ken); Kyoto Factory (Kyoto-shi); Kashima Factory (Kashima-gun, Ibaraki-ken)

Laboratories : Research Division (Kyoto-shi)

R&D Areas : Synthetic Resins: Surfactants, Polyester resins, Silicon resins, Urethane resins, Urethane resin plates and coatings, Laminated films, MIM (metal injection mold) products

Overseas Subsidiaries : Produtos Quimicos Sanyo Kasei do Brasil LTDA (San Paulo, Brazil); Dah Shyang Chemical Co., Ltd. (Taipei); SANAM Corp. (Delaware, U.S.A.); Hercules-Sanyo Inc. (Delaware, U.S.A.)

Sanyo Electric Co., Ltd.

Japanese Name	:	***Sanyo Denki K.K.***
TSE Number	:	6764
Description	:	Major manufacturer of electronic consumer products, most notably refrigerators and audio-video equipment. A world leader in sales of cadmium-nickel rechargeable batteries, Sanyo also excels in solar cell technology and is stressing its environmentally friendly products such as CFC-free refrigeration systems. The company also produces a number of industrial and commercial products including heating and cooling systems, factory automation equipment and biotechnical equipment.
Address	:	2-18, Keihan-Hondori, Moriguchi-shi, Osaka 570
Tel & Fax	:	Tel: 06-991-1181 Fax: 06-991-5411
Statistics	:	Capital : 168337 million yen Employees : 28554
Officers	:	CEO : Mr. Satoshi Iue/President
Sales	:	November, 1988 - 987539 million yen November, 1989 - 1040150 million yen November, 1990 - 1104520 million yen
R & D Spending	:	74800 million yen Capital Investment: 87300 million yen

Sales Breakdown :

SIC#	Product	Sales
363	Home appliances	24 %
357	Information systems and electronics	22 %
3651	Video equipment	20 %
3651	Audio equipment	10 %
362	Industrial and commercial equipment	9 %
3691	Batteries and other products	8 %

Main Products : Home Appliances: Microwave ovens, Clothes washers and dryers, Refrigerators, Air-conditioners, Heating equipment, Cooking appliances, System kitchens, TV intercoms, Lighting equipment, Various home appliances incorporating 'fuzzy logic' systems; Information Systems and Electronic Devices: Personal computers, 'Medicom' medical computer systems, Portable fax and telephone components and systems, Desktop telephone/computer integrated communication terminals, Copiers, Capacitors, Transistors, Bipolar ICs, Thick-film ICs, LSIs, Optical pick-ups; Audio/Video Equipment: BS (satellite broadcast) tuner integrated televisions, Other televisions, High definition LCD AV systems, 'Fuzzy logic' 8 mm video camcorders, Other camcorders, VCRs, Compact component stereo systems, Portable CD/tape players, Car stereos incorporating voice recognition technology, Other consumer audio equipment; Industrial and Commercial Equipment: High speed VQFP-type IC mounter, Other FA equipment, Clean rooms, Incubators, Clean benchs, Ultralow-temperature freezers, Other culturing equipment for biotechnical applications, Absorption heat pumps, Heating and cooling systems, Air compressors, CFC alternative 'clean' refrigeration systems, Industrial refrigeration equipment, Supermarket and related commercial refrigerated display cases, Commercial freezers, Card system vending machines, Other vending machines; Batteries and Related Products: Nickel-cadmium rechargeable batteries, Manganese dioxide-lithium batteries, High-performance alkaline bateries, Nickel-metal hydride rechargeable batteries, 'Amorton' solar cells

Banks	:	Sumitomo Bank; Kyowa Saitama Bank
Subsidiaries	:	Tottori Sanyo Electric Co., Ltd.; Niigata Sanyo Electronic Co., Ltd.; Sanyo Life Electronics Co., Ltd.; Sanyo Electric Tokki Co., Ltd.; Sanyo Electric Trading Co., Ltd.; Sanyo Electric Credit Co., Ltd.; Sanyo Manufacturing Corp.
Overseas Subsidiaries	:	Sanyo Electronics (Singapore) Pte. Ltd.; Sanyo Semiconductor (Hong Kong) Co., Ltd.; Sanyo Fisher (U.S.A.) Corp.; 100 other subsidiaries throughout the world

Seiko Epson Corp.

Japanese Name	:	***Seiko Epson K.K.***
TSE Number	:	Not listed
Description	:	Manufacturer of computer components, electronic devices and optical products. Known globally as the introducer of the quartz crystal watch, Seiko Epson now produces a variety of electronic and optical goods with computers and computer peripherals comprising the largest single product category in terms of sales. R&D continues to stress mainline computer and semiconductor products including disk drives and related magnetic memory storage devices.
Address	:	3-3-5, Owa, Suwa-shi, Nagano-ken 392
Tel & Fax	:	Tel: 0266-58-1705 Fax: 0266-52-8775
Statistics	:	Capital : 12530 million yen Employees : 13300
Officers	:	CEO : Mr. Hideaki Yasukawa/President
Sales	:	March, 1989 - 400000 million yen March, 1990 - 470000 million yen March, 1991 - 456900 million yen
Sales Breakdown	:	see table below

SIC#	Product	Sales
3577	Computers and printers	45 %
3679	Electronic components and equipment	28 %
3873	Watches	16 %

Main Products	:	Computers and Peripheral Equipment: Computers, Printers; Electronic Components and Equipment: LCDs, CMOS LSIs, Factory automation systems, Compact precision assembly robots, Floppy disk drives, Hard disk drives; Others: Watches, Plastic corrective lenses, Plastic contact lenses, Optoelectronic products, Other optical products, Engineering plastics, Ceramic electronic materials
Banks	:	Fuji Bank; Dai-Ichi Kangyo Bank
Factories	:	Hirooka Plant (Nagano-ken); Fujimi Plant (Nagano-ken); Suwa South Plant (Nagano-ken); Shiojiri Plant (Nagano-ken); Ina Plant (Nagano-ken); Okaya Plant (Nagano-ken); 10 other domestic plants
R&D Areas	:	Computers and Peripherals: Disk drives; Electronic Components: CMOS LSIs;
Shareholders	:	Seiko Corp. [8050]
Subsidiaries	:	Tohoku Epson Corp.; Okaya Precision Co., Ltd.; Hino C Mos Corp.; Epson Hanbai Co., Ltd.

Seiko Instruments Inc.

Japanese Name	:	***Seiko Denshi Kogyo K.K.***
TSE Number	:	Not listed
Description	:	Comprehensive precision electronics maker and leader in watch production. The company manufactures data processing equipment, and analysis and measurement equipment. In addition, the company is developing an automatic DNA basic sequence analyzer and an alcohol sensor.
Address	:	6-31-1, Kameido, Koto-ku, Tokyo 136
Tel & Fax	:	Tel: 03-3682-1111 Fax: 03-3682-1102
Officers	:	CEO : Mr. Reinosuke Hara/President Contact : Mr. Reinosuke Hara/President
Sales Breakdown	:	SIC# Product 3873 Timepieces 3826 Processing equipment 3845 Alcohol sensors
Main Products	:	Watches: Various types of timepieces; Analytical Instruments: Processing apparatus; Electromedical Equipment: Alcohol sensors, Sequencing analyzers

Sekaicho Rubber Co., Ltd.

TSE Number	:	5143
Description	:	A major manufacturer of canvas and rubberized shoes. The company also produces sealing materials.
Address	:	8-39, Tsuto-Ogocho, Nishinomiya-shi, Hyogo-ken 663
Tel & Fax	:	Tel: 0798-26-5711 Fax: 0798-35-1756
Statistics	:	Capital : 1732 million yen Employees : 519
Officers	:	CEO : Mr. Taichi Zaizen/President Contact : Mr. Taichi Zaizen/President
Sales	:	March, 1988 - 14798 million yen March, 1989 - 14985 million yen March, 1990 - 15466 million yen March, 1991 - 16655 million yen
R & D Spending	:	56 million yen Capital Investment: 544 million yen
Sales Breakdown	:	SIC# Product 3021 Rubber footwear 2891 Sealants
Main Products	:	Footwear: Rubber footwear; Chemical Products; Sealants
Banks	:	Sanwa Bank; Daiwa Bank; Mitsui Taiyo Kobe Bank; Mitsubishi Bank; Dai-Ichi Kangyo Bank; Toyo Trust and Banking; Sumitomo Trust and Banking
Branches	:	Tokyo Branch (Shibuya-ku); Nagoya Office (Nagoya-shi); Maebashi Office (Maebashi-shi); Sendai Office (Sendai-shi); Fukuoka Office (Fukuoka-shi); Hiroshima Office (Hiroshima-shi)
Factories	:	Nishinomiya Factory (Nishinomiya-shi); Tokuyama Factory (Shinnanyo-shi)

Sekisui Chemical Co., Ltd.

Japanese Name : ***Sekisui Kagaku Kogyo K.K.***

TSE Number : 4204

Description : Sekisui Chemical is a leading chemical manufacturer and pioneer in resin processing. Sekisui is ranked number one in PVC resins and has swiftly moved into prefabrication of houses in which it is now ranked fifth. The company produces its own line of high-impact pipes under the brand name 'Eslon', and is achieving commercial success with its line of industrial tapes, foams and films sold under the 'Techno Goods' label. The company has diversified into new products including pharmaceuticals and medical equipment, where it has launched new products including 'Insepak', polyethylene terephthalate vacuum tubes for blood transfusions, latex reagents, and 'Esmedica V' a compound used in a spectrum of medical products.

Address : 2-4-4, Nishitenma, Kita-ku, Osaka 530

Tel & Fax : Tel: 06-365-4122 Fax: 06-365-4370

Statistics : Capital : 92624 million yen
Employees : 5524

Officers : CEO : Ms. Kaoru Hirota/President

Sales : March, 1988 - 441488 million yen
March, 1989 - 504220 million yen
March, 1990 - 554000 million yen
March, 1991 - 617100 million yen

R & D Spending : 21600 million yen Capital Investment: 32000 million yen

Sales Breakdown :

SIC#	Product	Sales
3448	Housing	49 %
3498	Pipe fittings	16 %
3444	Building materials	10 %
2821	Molding compounds	7 %
2869	Plasticizers	5 %
3081	Films	5 %
2891	Adhesive tapes	5 %

Main Products : Prefabricated Buildings: Prefabricated housing; Fabricated Pipe and Fittings: PVC pipes; Plastics Materials and Resins: Molding Compounds; Sheet Metalwork: Building materials; Industrial Organic Chemicals: Phthalic acid ester plasticizers, Alophatic acid plasticizers, Adiphic acid plasticizers; Plastics Films: PVC film; Adhesives and Sealants: Adhesive tape

Banks : Sanwa Bank; Daiwa Bank; Industrial Bank of Japan; Toyo Trust and Banking; Norinchukin Bank

Branches : Hokkaido Branch (Sapporo-shi); Tohoku Branch (Sendai-shi); Tokyo Branch (Minato-ku); Chubu Branch (Nagoya-shi); Kinki Branch (Osaka-shi); Chu-Shikoku Branch (Hiroshima-shi); Kyushu Branch (Fukuoka-shi)

Laboratories : Kyoto Technical Center (Kyoto-shi); Basic Reseach Center (Mishima-gun, Osaka); Housing Research Center (Tsukuba-shi); Applied Electric Research Center (Tsukuba-shi)

Subsidiaries : Sekisui Seikei Co., Ltd.; Sekisui Machinery Co., Ltd.; Toyo Sekisui Shoji Co., Ltd.; Sekisui Leasing Co. Ltd.; Sekisui Finance Co. Ltd.

Sekisui Plastics Co., Ltd.

Japanese Name	:	***Sekisui Kaseihin Kogyo K.K.***
TSE Number	:	4228
Description	:	Top manufacturer of plastic foam products. Styrol foam and polyethylene foam account for 99% of total sales.
Address	:	2-4-4, Nishi-Tenman, Kita-ku, Osaka 530
Tel & Fax	:	Tel: 06-365-3014 Fax: 06-365-3114
Statistics	:	Capital : 15967 million yen Employees : 1260
Officers	:	CEO : Mr. Shohei Masuda/President
Sales	:	March, 1988 - 67897 million yen March, 1989 - 71604 million yen March, 1990 - 75620 million yen
Sales Breakdown	:	

SIC#	Product	Sales
2821	Synthetic resins	45 %
3086	Foam sheets	40 %
3089	Plastic materials	7 %
3089	Building materials	7 %

Main Products	:	Plastic Materials: Synthetic resins; Plastics Foam Products: Foam sheets, Foam boards; Plastics Products: Industrial materials, Building materials
Banks	:	Sanwa Bank; Industrial Bank of Japan; Daiwa Bank; Sumitomo Bank; Norinchukin Bank; Toyo Trust and Banking
Branches	:	Main Office (Tokyo); Branch Offices and Sales Offices: Nagoya; Fukuoka; Sapporo; Sendai; Hamamatsu; Okinawa
Factories	:	Kanto Factory (Sarushima-gun, Ibaraki-ken); Sakai Factory (Sarushima-gun, Ibaraki-ken); Nara Factory (Nara-ken); Tenri Factory (Nara-ken); Shiga Factory (Koga-gun); Hokkaido Factory (Chitose-shi); Oita Factory (Nakatsu-shi)
Laboratories	:	Central Laboratory (Tenri-shi, Nara-ken)

Sharp Corp.

Japanese Name	:	***Shaapu K.K.***
TSE Number	:	6753
Description	:	Engaged in R&D, manufacturing and sales/marketing of consumer electronics products, information systems and electronic components.
Address	:	22-22, Nagaike-cho, Abeno-ku, Osaka 545
Tel & Fax	:	Tel: 06-621-1221 Fax: 06-628-1667
Statistics	:	Capital : 171455 million yen Employees : 65000
Officers	:	CEO : Mr. Haruo Tsuji/President Contact : Mr. Katsuhiko Machida/General Manager, Int'l Business
Sales	:	March, 1988 - 1225190 million yen March, 1989 - 1238400 million yen March, 1990 - 1344800 million yen March, 1991 - 1496110 million yen
R & D Spending	:	89351 million yen Capital Investment: 108900 million yen

Sales Breakdown :

SIC#	Product	Sales
3674	Electronic and computer components	47 %
3651	Audio visual products	37 %
363	Household appliances	16 %

Main Products	:	Electronics Division: Color TVs, Combo TV/videos, Video cassette recorders, Liquid crystal color TV, LCD displays, Personal workstations, Combination TV/game computer, Video printer, HDTV systems; Audio Division: Radio cassette tape recorders, Headphone stereos, Stereo systems, Audio components, CD players, Laser disc players, DAT decks, Car stereos, Multi-function telephones, Cordless telephones, DAT memory machines, Optical disk equipment; Electrochemical Equipment Division: Air-conditioners, Coolers, Fans, Humidifiers, Dehumidifiers, Oil stoves, Oil fan heaters, Ceramic fan heaters, Electric carpets, Electric blankets, Kotatsu (Japanese heated tables), Central heating systems, Central water heating systems, Solar power systems, Ice making machines, Water purifiers, Wine conditioners, Microwaves, Commercial microwaves, Refrigerators, Freezers, Toaster ovens, Dishwashers, Dishdryers, Electric cookware, Ovens, Roasters, Blenders, Mixers, Hand mixers, Coffee makers, Hot plates, Water coolers, Washing machines, Clothes dryers, Vacuum cleaners, Hand-held vacuum cleaners, Futon dryers, Irons, Sewing machines, Trouser presses, Hair dryers, Hair curlers, Shavers, Lamps, Flake ice dispenser, Vending machines, Home automation systems; Information Electronics and Electronic Parts Division: Pocket computers, Electronic diaries, Electronic dictionaries, Cameras, Office computers, English-Japanese machine translation systems, Personal computers, SS-NET information interface systems, Word processors, Electronic typewriters, Facsimile machines, Color facsimile machines, Laser printers, Ink jet printers, Scanners, Optical fiber systems, Electronic cash registers, Bank machines, POS systems, Computer software, Electronic medical equipment, Traffic information systems, Ultrasonic cleaning and other ultrasonic equipment, Automatic water quality inspection systems, Factory automation systems, CAD systems, LSIs, ICs, Optoelectronic components, Power components, Micro color LC panels and units, EL and LED components, Semiconductor laser diodes, Satellite broadcasting system parts, Electronic tuners, Printed circuit boards, Solar cells, Magneto-optical disks
Banks	:	Fuji Bank; Daiwa Bank; Sanwa Bank
Branches	:	Tokyo Branch (Shinjuku-ku)

Shibata Scientific Technology Ltd.

Japanese Name : ***Shibata Kagaku Kikai Kogyo K.K.***

TSE Number	:	Not listed
Description	:	Engaged in the development and manufacture of a variety of glass products and scientific instruments for research, medical and industrial applications.
Address	:	3-1-25, Ikenohata, Taito-ku, Tokyo 110
Tel	:	03-3822-2111
Statistics	:	Capital : 100 million yen Employees : 290
Sales	:	December, 1989 - 5178 million yen December, 1990 - 4900 million yen
Sales Breakdown	:	SIC# Product 3821 Laboratory glassware and glass apparatus 384 Medical and sanitary equipment 382 Scientific instruments 3829 Pollution measuring equipment 3229 Industrial glass equipment 3821 Biological testing equipment 3559 Plastics production equipment
Main Products	:	Laboratory Supplies and Equipment: Laboratory glassware and apparatus; Medical and Sanitary Equipment; Scientific Instruments; Environmental Pollution Measuring Equipment; Industrial Glass Equipment; Biological Testing Equipment; Plastics Production Machinery: L-B film production system
Banks	:	Fuji Bank; Mitsui Taiyo Kobe Bank; Sanwa Bank; Yasuda Trust and Banking
Branches	:	Osaka Sales Office (Osaka); Nagoya Sales Office (Nagoya-shi, Aichi-ken); Tsukuba Sales Office (Ibaraki-ken); Sapporo Sales Office (Hokkaido)
Factories	:	Soka Plant (Soka-shi, Saitama-ken)

Shikoku Chemicals Corp.

Japanese Name	:	***Shikoku Kasei Kogyo K.K.***
TSE Number	:	4099
Description	:	Major producer of mirabilite anhydride, synthetic detergents, and intermediates used in the manufacture of carbon disulfide. Currently concentrating on the development of fine chemicals used in the production of resin hardening agents for ICs.
Address	:	147-1, Minato-machi, Marugame-shi, Kagawa-ken 763
Tel & Fax	:	Tel: 0877-22-4111 Fax: 0877-23-9372
Statistics	:	Capital : 6852 million yen Employees : 666
Officers	:	CEO : Mr. Atsushi Akazawa/President Contact : Mr. Kunihiko Yamashita/Executive Managing Director
Sales	:	March, 1989 - 28662 million yen March, 1990 - 32300 million yen March, 1991 - 33667 million yen
R & D Spending	:	937 million yen Capital Investment: 3622 million yen
Sales Breakdown	:	

SIC#	Product	Sales
32	Building materials	43 %
281	Inorganic chemical products	24 %
286	Organic chemical products	18 %
28	Fine chemicals	14 %

Main Products	:	Chemical Products: Carbon disulfide, Sodium sulfate, Chlorinated isocyanurates, Imizadole derivatives, Rust inhibitor; Building Materials: Fiberboard walling, Exterior decorative walls, Concertina gates, General purpose doors, Aluminum shutters, Fencing
Banks	:	Hyakujuyon Bank; Industrial Bank of Japan, Fuji Bank; Mitsubishi Bank
Branches	:	Tokyo Office (Chuo-ku, Tokyo); Osaka Office (Osaka)
Factories	:	Marugame Factory (Marugame-shi, Kagawa-ken); Tokushima Daiichi Factory (Tokushima-shi, Tokushima-ken); Tokushima Daini Factory (Tokushima-shi, Tokushima-ken)
Shareholders	:	Nisshinbo Industries [3105] - 7.5% ; Nippon Life Insurance - 6.5% ; Industrial Bank of Japan [8302] - 3.9%
Subsidiaries	:	Nihon Kenso Co., Ltd.; Eartechs Co., Ltd.; Shikoku Kosan Co., Ltd.

		Shimadzu Corp.
Japanese Name	:	**_Shimadzu Seisakusho K.K._**
TSE Number	:	7701
Description	:	Major manufacturer of precision equipment. Associated with Mitsubishi Corp.
Address	:	1, Nishinokyo-Kuwabaracho, Nakagyo-ku, Kyoto 604
Tel & Fax	:	Tel: 075-823-1111 Fax: 075-811-3188
Statistics	:	Capital : 16700 million yen Employees : 4200
Officers	:	CEO : Mr. Minoru Nishihachijo/President Contact : Mr. Sadao Arai/Director, International Division
Sales	:	March, 1988 - 133453 million yen March, 1989 - 149561 million yen March, 1990 - 158959 million yen March, 1991 - 163384 million yen
Sales Breakdown	:	SIC# Product 3826 Precision medical apparatus 366 Optical communications equipment. 3823 Industrial control equipment 384 Biotechnology equipment 38 Aerospace-related equipment 382 Measuring equipment
Main Products	:	Precision Medical Apparatus: High performance liquid chromatograph 'LC-94', Superconductive NMR imaging systems 'SMT-50X, -100X, -150X', Intelligent X-ray TV system 'Shimavision Series'; Measuring Equipment; Process Automation Equipment; Industrial Precision Equipment; Biotechnological Apparatus; Optical Communications Equipment; Aerospace-Related Equipment
Banks	:	Mitsubishi Bank; Mitsubishi Trust and Banking; Kyoto Bank
Branches	:	Tokyo Office (Chiyoda-ku, Tokyo); Osaka Office (Osaka); Fukuoka Office (Fukuoka-shi, Fukuoka-ken); Nagoya Office (Nagoya-shi, Aichi-ken); Sapporo Office (Sapporo-shi, Hokkaido); 12 other offices throughout Japan
Factories	:	Sanjo Works (Kyoto); Atsugi Works (Atsugi-shi, Kanagawa)
Laboratories	:	Central Research Laboratory (Kyoto)
Shareholders	:	Meiji Life Insurance - 7.8% ; Mitsubishi Bank [8315] - 4.9% ; Nippon Life Insurance - 4.8% ; Mitsubishi Trust and Banking [8402] - 3.9% ; Sumitomo Trust and Banking [8403] - 3.6% ; Tokio Marine & Fire [8751] - 3.2%

Shimizu Corp.

Japanese Name	:	***Shimizu Kensetsu K.K.***
TSE Number	:	1803
Description	:	A prominent general contractor, especially in private construction, which is making efforts to broaden its civil engineering operations, and to move into cleanroom construction and other high-technology areas.
Address	:	1-2-3, Shibaura, Minato-ku, Tokyo 105-07
Tel & Fax	:	Tel: 03-5441-4111 Fax: 03-5441-0070
Statistics	:	Capital : 74226 million yen Employees : 11073
Officers	:	CEO : Mr. Harusuke Imamura/President
Sales	:	March, 1988 - 1101710 million yen March, 1989 - 1254640 million yen March, 1990 - 1476620 million yen March, 1991 - 1883480 million yen
Sales Breakdown	:	SIC# Product 161 Road construction 152 Housing construction
Main Products	:	Construction and Civil Engineering: Road construction, Housing; High-Technology Installations: Clean rooms
Laboratories	:	Institute of Technology (Koto-ku, Tokyo)

Shin-Etsu Chemical Co., Ltd.

Japanese Name	:	***Shin-Etsu Kagaku Kogyo K.K.***
TSE Number	:	4063
Description	:	A top manufacturer of PVC and semiconductor silicon. Stepping up its business activities in superconductivity, biotechnology, and other high-technology fields.
Address	:	Asahi Tokai Bldg., 2-6-1, Otemachi, Chiyoda-ku, Tokyo 100
Tel & Fax	:	Tel: 03-3246-5111 Fax: 03-3246-5350
Statistics	:	Capital : 35370 million yen Employees : 3547
Officers	:	CEO : Mr. Chihiro Kanagawa/President
Sales	:	March, 1988 (10 months) - 185090 million yen March, 1989 - 247279 million yen March, 1990 - 260202 million yen March, 1991 - 283593 million yen
R & D Spending	:	10700 million yen Capital Investment: 24700 million yen

Sales Breakdown :

SIC#	Product	Sales
2821	Synthetic resins	44 %
3674	Electronic materials	38 %
286	Organic chemicals	8 %
334	Alloys	2 %
287	Chemical fertilizers	2 %
281	Inorganic chemicals	1 %

Main Products	:	Synthetic Resins: PVC, Silicone, Silicone-based products, Vinyl acetate monomer, PVA; Organic Chemicals: Methanol, Chloromethane, Cellulose derivatives, Acetic ester, Methylene chloride cooling agent, Medical coating agents; Inorganic Chemicals: Caustic soda, Agricultural chemicals; Medical Products: Pharmaceuticals, Medical supplies and equipment; Metal Products and Electronic Materials: Semiconductor silicon, New metals, Metals, Metal alloys
Branches	:	Osaka Branch (Osaka-shi); Nagoya-Branch (Nagoya-shi); Fukuoka Branch (Fukuoka-shi); Niigata Sales Office (Niigata-shi); Hokuriku Sales Office (Fukui-shi); Hiroshima Sales Office (Hiroshima-shi)
Factories	:	Naoetsu Plant (Niigata-ken); Takefu Plant (Takefu-shi, Fukui-ken); Isobe Plant (Annaka-shi, Gunma-ken); Nanyo Plant (Shin Nanyo-shi, Yamaguchi-ken); Kashima Plant (Kashima-gun, Ibaraki-ken)
Laboratories	:	Corporate Research Center (Kawasaki-shi); PVC Research Center (Kashima-gun); Polymers & High Performance Materials Research Center (Kashima-gun); Silicone-Electronic Materials Research Center (Annaka-shi) and 3 other research centers
Shareholders	:	Nippon Life - 7.71% ; Mitsubishi Bank [8315] - 4.56% ; Long Term Credit Bank of Japan [8303] - 4.08% ; Dai-Ichi Kangyo Bank [8311] - 4.08% ; Mitsui Taiyo Kobe Bank [8314] - 3.53% ; Hachijuni Bank [8359] - 3.48%

Shin-Etsu Handotai Co., Ltd.

Japanese Name	:	***Shin-Etsu Handotai K.K.***
TSE Number	:	Not listed
Description	:	Produces semiconductors and materials used in their manufacture.
Address	:	Togin Bldg., 1-4-2, Marunouchi, Chiyoda-ku, Tokyo 100
Tel & Fax	:	Tel: 03-3214-1831 Fax: 03-3215-9040
Statistics	:	Capital : 2400 million yen Employees : 2000
Officers	:	CEO : Dr. Suguru Igarashi/President Contact : Dr. Suguru Igarashi/President
Sales	:	March, 1988 - 68457 million yen March, 1989 - 73905 million yen March, 1990 - 83128 million yen
Sales Breakdown	:	SIC# Product 2869 Silicon for semiconductors 3674 Semiconductors
Main Products	:	Industrial Chemicals: High-purity silicon for semiconductors; Electronic Components: Compound semiconductors
Banks	:	Mitsubishi Bank; Dai-Ichi Kangyo Bank; Citibank; Bank of Tokyo; Mitsubishi Trust & Banking
Factories	:	Isobe Factory (Annaka-shi); Shirakawa Factory (Nishi-Shirakawa-gun); Takefu Factory (Takefu-shi, Fukui-ken)
Laboratories	:	Isobe Laboratory (Annaka-shi); Shirakawa Laboratory (Nishi Shirakawa-gun)
Shareholders	:	Shin-Etsu Chemical Co., Ltd. [4063] - 100%
Overseas Subsidiaries	:	SEH America Inc. (U.S.A.); SEH Malaysia Co., Ltd. (Malaysia); SEH Europe Co., Ltd. (Scotland)

Shin-Etsu Polymer Co., Ltd.

Japanese Name	:	***Shin-Etsu Porimaa K.K.***
TSE Number	:	7970
Description	:	Major plastics manufacturer specializing in PVC resins and silicon resin molded rubber products. A subsidiary of Shin-Etsu Chemical Co., Ltd., Shin-Etsu Polymer is a market leader in the area of silicon resins, especially silicon-rubber electro-conductive products, and controls a substantial share of the PVC processing market. R&D continues to stress mainline silicon and PVC materials and products.
Address	:	Shin-Etsu Bldg., 4-3-5, Nihonbashi-Honcho, Chuo-ku, Tokyo 103
Tel & Fax	:	Tel: 03-3279-1712 Fax: 03-3246-2529
Statistics	:	Capital : 11279 million yen Employees : 825
Officers	:	CEO : Mr. Masami Hashimoto/President Contact : Mr. Masami Hashimoto/President
Sales	:	March, 1988 - 47173 million yen March, 1989 - 50112 million yen March, 1990 - 51843 million yen March, 1991 - 55263 million yen
R & D Spending	:	639 million yen Capital Investment: 2149 million yen
Sales Breakdown	:	

SIC#	Product	Sales
308	Plastic industrial materials	48 %
3081	Plastic films and sheets	29 %
2821	Construction and engineering materials	23 %
35	Industrial machinery	

Main Products	:	Plastic Industrial Materials: Injection molded precision plastic products, Silicon rubber products, PVC compounds, Silicon resin molded rubber contacts and interconnectors; Plastic Films and Sheets: PVC stretch films, PVC sheets, Floppy disks; Plastic Construction and Engineering Materials: PVC pipe, PVC plates, Plastic floor tiles; Industrial Machinery
Banks	:	Yasuda Trust and Banking; Long Term Credit Bank of Japan; Mitsubishi Bank; Musashino Bank; Hachijuni Bank; Norinchukin Bank; Dai-Ichi Kangyo Bank
Branches	:	Osaka Branch (Osaka-shi); Nagoya Branch (Nagoya-shi); Fukuoka Branch (Fukuoka-shi); Tohoku Branch (Sendai-shi)
Factories	:	Tokyo Factory (Omiya-shi); Namyo Factory (Shimnamyo-shi); Kodama Factory (Kodama-gun)
Laboratories	:	Products Laboratory (Omiya-shi)
R&D Areas	:	Synthetic Resins: PVC resins, Silicon resins, Vinyl chloride functional compounds, Vinyl chloride plates
Subsidiaries	:	Shinano Polymer K.K.; Second Shinano Polymer K.K.; Niigata Polymer K.K.; Urawa Polymer K.K.; Polymer Higashi-Nihon K.K.; Polymer Kaseihin Hambai K.K.; Polymer Nagoya Hambai K.K.; Shin-Etsu Sapporo K.K.
Overseas Subsidiaries	:	Shin-Etsu Polymer America, Inc. (California, U.S.A.)

Shinko Wire Co., Ltd.

Japanese Name : **Shinko Kosen Kogyo K.K.**

TSE Number : 5660

Description : Major manufacturer of steel wires, steel cables and related products. A subsidiary of Kobe Steel, Ltd., Shinko Wire is also engaged in engineering where the construction of suspension bridges has been providing increasing earnings in recent years.

Address : 10-1, Nakahama-cho, Amagasaki-shi, Hyogo 660

Tel & Fax : Tel: 06-411-1051 Fax: 06-411-1056

Statistics : Capital : 1838 million yen
Employees : 845

Officers : CEO : Mr. Seiichi Kojima/President
Contact : Mr. Seiichi Kojima/President

Sales : March, 1989 - 24960 million yen
March, 1990 - 26466 million yen
March, 1991 - 28088 million yen

R & D Spending : 569 million yen

Sales Breakdown :

SIC#	Product	Sales
3315	Steel wires	48 %
3315	Steel cables	23 %
3315	Stainless steel wire products	19 %
8711	Engineering services	11 %

Main Products : Steel Wire Products: PC steel wires, Concrete reinforcing wires, Aerial wires, Hard steel wires, Spoke wires, Valve spring wires, Wires for use in communications equipment, Bead wires for use in tires; Steel Cables: Bridge cables, Misc. engineering cables, Marine cables; Stainless Steel Wires and Related Products: Stainless steel wires, Stainless steel cables, Stainless steel abrasive pads and sponges; Engineering Services: Suspension bridge construction, Wire drawing plant engineering

Banks : Dai-Ichi Kangyo Bank; Sanwa Bank; Mitsui Taiyo Kobe Bank; Yasuda Trust and Banking; Toyo Trust and Banking

Branches : Tokyo Branch (Chuo-ku, Tokyo); Osaka Sales Office (Osaka); Fukuoka Sales Office (Fukuoka-ken); Sapporo Sales Office (Hokkaido); Hiroshima Sales Office (Hiroshima-ken); Nagoya Sales Office (Nagoya-shi, Aichi-ken)

Factories : Amagasaki Plant (Amagasaki-shi, Hyogo-ken); Kakogawa Plant (Kakogawa-shi, Hyogo-ken); Izumi Sano Plant (Izumi Sano-shi, Osaka)

R&D Areas : Wires and Conductive Materials; Aluminum Alloys and Applications

Overseas Subsidiaries : Shinko Wire America Inc. (Texas, U.S.A.)

Shizuoka NEC Ltd.

Japanese Name	:	***Shizuoka Nippon Denki K.K.***
TSE Number	:	Not listed
Description	:	Engaged in the manufacture and development of communications and information processing equipment. Shizuoka NEC is an affiliate of Japanese electronics giant NEC Corp.
Address	:	4-2, Shimomata, Kakegawa-shi, Shizuoka 436
Tel & Fax	:	Tel: 0537-23-1111 Fax: 0537-24-7820
Statistics	:	Capital : 200 million yen Employees : 1299
Sales	:	March, 1989 - 56903 million yen March, 1990 - 52286 million yen
Sales Breakdown	:	SIC# Product 366 Communications equipment 357 Image processing equipment 357 Information processing equipment 3559 Electronics production machinery 8731 Communications equipment R&D
Main Products	:	Communications Equipment: Data communications equipment, Image communications equipment, Misc. communications equipment; Image Processing Equipment; Information Processing Equipment; Electronics Production Machinery: 'Micro-Coat HY-10' CFC-free flux application system for PCB soldering
Banks	:	Sumitomo Bank; Shizuoka Bank; Sumitomo Trust and Banking; Industrial Bank of Japan
Factories	:	Kanaya-cho Plant (Kanaya-cho, Shizuoka-ken)
Shareholders	:	NEC Corp. [6701] - 100%

Showa Denko K.K.

Japanese Name	:	***Showa Denko K.K.***
TSE Number	:	4004
Description	:	Comprehensive chemical manufacturer producing a broad range of chemical products with concentrations in olefins, synthetic resins and other petroleum based chemicals. A member of the Fuyo group of companies, Showa Denko is Japan's fourth largest chemical producer. The company has been diversifying into synthetic resins, ceramics, metals, carbon, gases, and electronic materials. Beginning in 1990, a number of civil suits have been brought against Showa Denko in the U.S. asserting an association between the fatal disease EMS and an L-tryptophan product produced by the company. Litigation is still pending. Showa Denko is currently studying the possibility of a joint venture with the U.S. company Himont for the production of polymer alloys.
Address	:	1-13-9, Shiba-Daimon, Minato-ku, Tokyo 105
Tel & Fax	:	Tel: 03-5470-3533 Fax: 03-3436-2625
Statistics	:	Capital : 105448 million yen Employees : 5497
Officers	:	CEO : Mr. Makoto Murata/President
Sales	:	December, 1988 - 471288 million yen December, 1989 - 525471 million yen December, 1990 - 569651 million yen
R & D Spending	:	21500 million yen Capital investment: 48600 million yen

Sales Breakdown :

SIC#	Product	Sales
28	Olefins and organic chemicals	48 %
3299	Ceramics and other materials	35 %
28	Chemicals and carbons	17 %
8711	Engineering services	

Main Products	:	Olefins and Synthetic Resins: Ethylene, Propylene, High and low density polyethylene, Polypropylene, Polystyrene, Chlorinated polyethylene, ACS resins; Plastics: Conductive plastic, Static-free film, Polyamide 66I, Engineering plastics, OA equipment housings, Medical products packaging; Organic Chemicals: Acetal, Acetaldehyde derivatives, Propylene glycol, Propylene oxide, Acrylic esters, Acrylonitrile, Amino acid, Ammonia cooling agent, Sulfur dioxide cooling agent, CFC cooling agent, Epoxy resin curing agent, Ethylene, Ethylene-vinyl acetate copolymer, CFC gas, Lactic acid, Styrene monomer, Ascetic acid, Ascetic acid vinyl monomer, Ethyl ascetic acid, Epichlorohydrin, Allyl alcohol; Ceramics: Abrasive ceramic materials, Fire retardant ceramic materials, Deoxidized ceramics, Fine ceramics, Other ceramics products; Alumina Products: Aluminum hydroxide, Alumina, High purity alumina; Metals and Electronic Materials: Ferrochrome, Electrolytic iron, Aluminum and aluminum alloys, Gallium, Indium, Silicon wafers; Construction Materials: Premix mortars and cements; High polymers; Chemicals and Carbons: Caustic soda, Sulfuric acid, Sodium chlorate, Potassium chlorate, Ascorbic acid PM, Amino acids, Industrial use enzymes, High-compound fertilizers, Herbicides, Agricultural chemical intermediates and raw materials, Pharmaceutical intermediates and raw materials; Gases: NO_2, CFC, Argon, Acetylene, Gases for use in the electronics industry; Carbon Porducts: Synthetic graphite electrodes, Fine carbon products; Engineering Services
Banks	:	Fuji Bank; Industrial Bank of Japan; Yasuda Trust and Banking; Norinchukin Bank; Japan Data Bank; Long Term Credit Bank of Japan

Branches	:	Osaka Branch; Nagoya Branch; Fukuoka Branch; Sapporo Sales Office; Sendai Sales Office; Toyama Sales Office; Hiroshima Sales Office
Factories	:	Ohita Works (Ohita); Kawasaki Works (Kanagawa); Higashinagahara Works (Fukushima); Yokohama Works (Kanagawa); Omachi Works (Nagano-ken); Nagano Works (Nagano-ken); Shiojiri Works (Nagano); Chichibu Works (Saitama)
R&D Areas	:	Chemicals and Synthetic Resin Products: Biochemicals, Fine chemicals, Liquid chromatography, High performance plastics; Electronics Materials: IC packages, Alumina fillers

Showa Electric Wire & Cable Co., Ltd.

Japanese Name	:	**Showa Densen Denran K.K.**
TSE Number	:	5805
Address	:	1-1-18, Toranomon, Minato-ku, Tokyo 105
Tel & Fax	:	Tel: 03-3597-7011 Fax: 03-3503-4506
Statistics	:	Capital : 18416 million yen Employees : 2663
Officers	:	CEO : Mr. Zennosuke Matsui/President
Sales	:	March, 1988 - 104464 million yen March, 1989 - 121396 million yen March, 1990 - 128706 million yen March, 1991 - 141117 million yen
R & D Spending	:	5000 million yen Capital Investment: 7526 million yen

Sales Breakdown :

SIC#	Product	Sales
3357	Optical and glass fibers	26 %
3350	Bare wires	19 %
3315	Covered wires and windings	19 %
3679	Oscillators and OA equipment	17 %
3355	Aluminum wires	19 %

Banks : Industrial Bank of Japan; Kyowa Saitama Bank; Bank of Yokohama; Mitsubishi Trust and Banking; Mitsui Taiyo Kobe Bank

Showa Highpolymer Co., Ltd.

Japanese Name	:	***Showa Kobunshi Kogyo K.K.***
TSE Number	:	4214
Description	:	A manufacturer of resins and paints, the company absorbed Showa Union Gosei and is currently registering solid earnings. The company markets UV vinyl ester and is advancing into high quality paints and lacquers.
Address	:	3-20, Kanda-Nishikicho, Chiyoda-ku, Tokyo 101
Tel & Fax	:	Tel: 03-3293-8844 Fax: 03-3233-0137
Statistics	:	Capital : 7741 million yen Employees : 577
Officers	:	CEO : Mr. Teruo Hiromoto/President Contact : Mr. Teruo Hiromoto/President
Sales	:	March, 1988 - 25600 million yen March, 1989 - 28400 million yen March, 1990 - 29500 million yen March, 1991 - 32610 million yen
R & D Spending	:	1620 million yen Capital Investment: 1745 million yen
Sales Breakdown	:	

SIC#	Product	Sales
2821	Polyesters	54 %
2843	Emulsifiers	30 %
2821	Phenolic resins	12 %
2851	Water-proof paints	4 %

Main Products	:	Plastics Materials and Resins: Polyesters, Phenolic resins; Surface Active Agents: Emulsifiers; Paints and Allied Products: Water-soluble paints
Banks	:	Kyowa Saitama Bank; Fuji Bank; Yasuda Trust and Banking; Long Term Credit Bank of Japan; Norinchukin Bank

Showa Shell Sekiyu K.K.

Japanese Name : ***Showa Sheru Sekiyu K.K.***

TSE Number : 5002

Description : Leading oil refiner and distributor affiliated with Royal Dutch Shell. The company was formed in 1985 through the merger of Shell Sekiyu and Showa Oil.

Address : Kasumigaseki Bldg., 3-2-5, Kasumigaseki, Chiyoda-ku, Tokyo 100

Tel & Fax : Tel: 03-580-0123 Fax: 03-580-3811

Statistics : Capital : 13654 million yen
Employees : 2379

Officers : CEO : Mr. Takeshi Henmi/President

Sales : December, 1988 - 1231060 million yen
December, 1989 - 1366270 million yen
December, 1990 - 1652000 million yen

Capital Investment : 41000 million yen

Sales Breakdown :

SIC#	Product
2911	Naptha, KerGas
2992	Lubricants

Main Products : Petroleum Refining: Naphtha, Fuel oil, Kerosene and gas oil; Lubricating oils and Greases

Shareholders : Shell Petroleum - 34.4% ; Anglo-Saxon Petroleum - 12.4% ; Dai-Ichi Kangyo Bank [8311] - 4.0% ; Sumitomo Trust and Banking [8403] - 3.6% ; Mexican Eagle Oil Co., Ltd. - 3.3%

Subsidiaries : Over 75 domestic subsidiaries

Overseas Subsidiaries : 8 subsidiaries

Sony Corp.

Japanese Name	:	***Soni K.K.***
TSE Number	:	6758
Description	:	Comprehensive electronics maker with a worldwide network of distribution outlets. The company's audio equipment still yields the largest percentage of sales but the company is moving into higher value-added lines of business including optical disks and materials for the communications industry. Sony made headlines with the multi-million dollar acquisition of CBS Records and is pursuing market share in the movie and home entertainment industries. The company has recently developed a final version for the MD (mini disc) music recording format, and has entered licensing negotiations with EMI and Warner Music.
Address	:	6-7-35, Kita-Shinagawa, Shinagawa-ku, Tokyo 141
Tel & Fax	:	Tel: 03-3448-2111 Fax: 03-3448-2183
Statistics	:	Capital : 296482 million yen Employees : 18130
Officers	:	CEO : Mr. Akio Morita/Chairman Contact : Mr. Norio Ohga/President
Sales	:	March, 1988 - 1029890 million yen March, 1989 - 1258300 million yen March, 1990 - 1536500 million yen March, 1991 - 1880580 million yen
R & D Spending	:	64793 million yen Capital Investment: 155900 million yen
Sales Breakdown	:	

SIC#	Product	Sales
3651	VCRs	38 %
3651	Televisions	25 %
3674	Magnetic disks	15 %

Main Products	:	Household Audio and Video Equipment: Video cassettes, Walkmans, Televisions, Amplifiers, Tuners, Radios, Recorders, Compact disc players; Electronic Data Storage Media: Magnetic Optical disks
Banks	:	Mitsui Trust and Banking; Bank of Tokyo; Fuji Bank; Mitsubishi Bank; Tokai Bank; Dai-Ichi Kangyo Bank; Sanwa Bank; Mitsui Trust and Banking; Toyo Trust and Banking; Industrial Bank of Japan
Factories	:	Shibaura Plant (Minato-ku, Tokyo); Haneda Plant (Ohta-ku, Tokyo); Atsugi Plant (Atsugi-shi)
Laboratories	:	Central Research Center (Yokohama-shi)

Sumitomo 3M Ltd.

Japanese Name	:	***Sumitomo Suriiemu K.K.***
TSE Number	:	Not listed
Description	:	Manufactures a range of tapes and materials for adhesive, recording and insulation purposes.
Address	:	2-33-1, Yamagawadai, Setagaya-ku, Tokyo 158
Tel & Fax	:	Tel: 03-3709-8111 Fax: 03-3709-8752
Statistics	:	Capital : 9465 million yen Employees : 2258
Officers	:	CEO : Mr. Asao Ito/President
Sales	:	October, 1988 - 110911 million yen October, 1989 - 120169 million yen
Sales Breakdown	:	SIC# Product 3842 Adhesive tapes 3695 Magnetic media 2842 Polishing materials 2241 Electric insulating materials
Main Products	:	Adhesive Products: Adhesive tapes, Adhesive films; Polishing Materials; Magnetic Media: Audio tapes, Floppy disks; Electric Insulation Materials
Banks	:	Sumitomo Bank; Sumitomo Trust and Banking; Dai-Ichi Kangyo Bank; Bank of Tokyo; Mitsubishi Bank; Yokohama Bank; Continental Bank; Citibank; Morgan Guarantee; Chase Manhattan Bank
Branches	:	Tokyo Branch (Minato-ku); Yokohama Branch (Yokohama-shi); Nagoya Branch (Nagoya-shi); Osaka Branch (Osaka-shi); Fukuoka Branch (Fukuoka-shi)
Factories	:	Sagamihara Factory (Sagami-shi); Fujikoyama Factory (Shunto-gun)
Shareholders	:	Minnesota Mining and Manufacturing Co.
Subsidiaries	:	Yamagata 3M; 3M Business Systems; Iwate 3M; 3M Yakuhin

Sumitomo Bakelite Co., Ltd.

Japanese Name	:	***Sumitomo Beekuraito K.K.***
TSE Number	:	4203
Description	:	Major producer of processed synthetic resin products, most notably IC laminates and plastic molding materials. A member of the Sumitomo group of companies, Sumitomo Bakelite also produces a variety of other plastic intermediate and finished products and has been emphasizing medical materials and instruments as a potential avenue for diversification. R&D continues to stress mainline electrical products, and the company is involved in technology exchanges with two American companies in the areas of silicon electronics materials and medical instruments.
Address	:	Hibiya Dai Bldg., 1-2-2, Uchisaiwai-cho, Chiyoda-ku, Tokyo 100
Tel & Fax	:	Tel: 03-3506-7245 Fax: 03-3506-7336
Statistics	:	Capital : 23600 million yen Employees : 2689
Officers	:	CEO : Mr. Masao Nomura/President
Sales	:	December, 1988 - 133172 million yen March, 1989 (3 months) - 33768 million yen March, 1990 - 137361 million yen March, 1991 - 152760 million yen
Capital Investment	:	9829 million yen

Sales Breakdown :

SIC#	Product	Sales
367	IC substrates and copper-clad laminates	32 %
2821	Synthetic resin molding materials	27 %
2821	Industrial resins	14 %
30	Molded plastic products	12 %
3081	Plastic films and sheets	14 %

Main Products	:	IC Substrates and Copper-Clad Laminates: Phenol laminates and substrates, Epoxy laminates and substrates, Melamine laminates; Synthetic Molding Materials: Phenol molding materials, Epoxy molding materials, Vinyl molding materials, Polyethylene molding materials, Polyester molding materials; Industrial Resins: Phenol resins, Urea resins, Melamine resins, Semiconductive resins, Formalin; Molded Plastic Products: Injection molded products, Compression molded products, PFC products, Polyethylene boards and other products, PVC containers, Engineering materials, Tennis court surfaces, Medical and scientific products; Plastic Films and Sheets: Vinyl sheets, Compound resin sheets, Polypropylene sheets, Styrol sheets; Other Products: Precision metal molds, Medical instruments, Wrapping and packaging equipment
Banks	:	Sumitomo Bank; Sumitomo Trust and Banking; Industrial Bank of Japan
Branches	:	Osaka Branch (Kita-ku, Osaka); Nagoya Branch (Nagoya-shi, Aichi-ken)
Factories	:	Amagasaki Plant (Amagasaki-shi, Hyogo-ken); Shizuoka Plant (Fujieda-shi, Shizuoka-ken); Utsunomiya Plant (Utsunomiya-shi, Tochigi-ken); Tsu Plant (Tsu-shi, Mie-ken)
Laboratories	:	Basic Research Center (Yokohama-shi, Kanagawa-ken); Utsunomiya Laboratory (Utsunomiya-shi, Tochigi-ken); Amagasaki Laboratory (Amagasaki-shi, Hyogo-ken); Shizuoka Laboratory (Fujieda-shi, Shizuoka-ken); Tsu Laboratory (Tsu-shi, Mie-ken)
R&D Areas	:	Electronics Materials: Conductive films, Magnetic sheilds and pastes, Polymid IC coatings; Medical Products: Precision Electrical Instruments:

Sumitomo Chemical Co., Ltd.

Japanese Name : ***Sumitomo Kagaku Kogyo K.K.***

TSE Number : 4005

Description : One of Japan's largest and most diversified chemical companies, serving markets in more than 110 countries. Sumitomo Chemical puts strong emphasis on R&D especially in value-added areas including fine chemicals, semiconductors, carbons, and engineering plastics. New business lines include photoresistors for ULSI circuits, high-purity aluminum compounds for optoelectronic products, carbon fibers for aerospace and leisure goods, and 'EKONOL', and 'ESPEX' super-engineering plastics for electrical components, bearings and piston rings. The company operates a major petrochemicals plant in Singapore.

Address : 4-5-33, Kitahama, Chuo-ku, Osaka 541

Tel & Fax : Tel: 06-220-3272 Fax: 06-220-3345

Statistics : Capital : 81459 million yen
Employees : 6773

Officers : CEO : Mr. Hideo Mori/President

Sales : December, 1988 - 556113 million yen
December, 1989 - 601653 million yen
December, 1990 - 716760 million yen

R & D Spending : 37100 million yen Capital Investment: 58800 million yen

Sales Breakdown :

SIC#	Product	Sales
2869	Ethylene, Toluene	30 %
2812	Caustic soda, hydrochloric acid	30 %
2821	Synthetic resins	15 %
2865	Pigments	5 %
3334	High-purity aluminium	5 %
2879	Pesticides, insecticides	5 %
3295	High-purity silicon	5 %
3624	Carbon fibers	5 %

Main Products : Industrial Organic Chemicals: Ethylene, Toluene, Xylene, Styrene monomer, Acrylic acid, Acetic acid, Acetone, Acid dyes, r-acid, Acrylic esters, Acrylonitrile, Alkybenzene, Phenol antioxidant, Sulfur antioxidant, Phosphorus antioxidant, Antistatic agent, Bactericides/preservatives for foods, Petroleum-based benzene, Brilliant carmine 6B, Ammonia cooling agent, Dimethoate, Epoxy resin curing agent, Ethylene-vinyl acetate copolymer, Emulsion ethylene-vinyl acetate copolymer, Powder ethylene-vinyl acetate copolymer, Halogen-based flame retardant, Non-inflammable polymer flame retardant, Flocculant, Formalin, Methacryl resin monomer, Organic pigments, Vulcanization accelerator, Organic antioxidant, Scorch retarder, Vulcanizing agent, Propylene, Rust inhibitor, Sea surface pollution prevention absorbent, Corrosion inhibitor, Toluene, Vinyl chloride monomer, Xylene, Salicyclic acid derivatives, Benzophenone-based UV absorbent, Benzotriazole-based UV absorbent, Beta-carotene; Alkalies and Chlorine, Caustic Soda, Hydrochloric acid; Plastics Materials and Resins: Synthetic resins; Cyclic Crudes and Intermediates: Pigments; Primary Aluminum: High-purity aluminum; Agricultural Chemicals: Pesticides, Insecticides; Minerals, Ground or Treated: High-purity silicon, Gallium; Carbon and Graphite Products: Carbon fibers and products

Banks : Sumitomo Bank; Sumitomo Trust and Banking; Norinchukin Bank; Industrial Bank of Japan; Long Term Credit Bank of Japan

Factories : Ehime Works (Niihama-shi, Ehime-ken); Chiba Works (Ichihara-shi, Chiba-ken); Osaka Works (Osaka-shi); Oita Works (Oita-shi); Misawa Works (Misawa-shi, Aomori-ken)

Sumitomo Corp.

Japanese Name	:	***Sumitomo Shoji K.K.***
TSE Number	:	8053
Description	:	Major general trading firm largely dependent on steel and chemical industries. The company experienced a drop in sales in 1990 due to the effects of unsecured investment in Iraq. Profits are expected to bounce back with increased sales of machinery and data communications equipment.
Address	:	1-2-2, Hitotsubashi, Chiyoda-ku, Tokyo 100
Tel & Fax	:	Tel: 03-3217-5000 Fax: 03-3636-7220
Statistics	:	Capital : 166991 million yen Employees : 8630
Officers	:	CEO : Mr. Tomiichi Akiyama/President
Sales	:	March, 1989 - 14570700 million yen March, 1990 - 21403600 million yen March, 1991 - 19212600 million yen
Sales Breakdown	:	

SIC#	Product	Sales
5051	Metals	35 %
5082	Machinery	30 %
1629	Construction	10 %
1541	Industrial plant	5 %
3731	Shipbuilding and aerospace	5 %
6531	Real estate	5 %
5149	Foodstuffs	5 %
506	Electrical goods	3 %
513	Textiles	2 %

Main Products	:	Metals: Iron, Steel, Non-ferrous metals; Petroleum and Petrochemicals: Carbon, Fuels; Machinery and Electric Systems; Shipbuilding: Ship repair and services; Industrial Machinery: Industrial plants; General Construction: Warehouses; Real Estate Services: Real estate development; Construction and Transportation equipment: Motor vehicles, Aerospace products, Shipbuilding; Foodstuffs; Chemical products: Fertilizers, Chemicals, Plastics; Textiles: Apparel and fabrics; Communications: Cable and satellite TV
Banks	:	Sumitomo Bank; Bank of Tokyo; Sumitomo Trust and Banking; Industrial Bank of Japan; Tokai Bank
Branches	:	46 domestic offices; 88 overseas offices
Subsidiaries	:	Summit, Inc.; Otto-Sumisho, Inc.; SC Hightech Center Corp.; Naf Naf Co., Ltd.; 92 other domestic subsidiaries
Overseas Subsidiaries	:	SC Steel Investment, Inc. (U.S.A.); Summit Agriculture Co., Ltd. (U.S.A.)

Sumitomo Electric Industries, Ltd.

Japanese Name : ***Sumitomo Denki Kogyo K.K.***

TSE Number : 5802

Description : Japan's leading manufacturer of electric wires and cables. A key member of the Sumitomo group of companies, Sumitomo Electric can boast a number of technological achievements in areas such as new materials, superconductivity and optoelectronic equipment. Current R&D continues to stress these areas, and the company has earmarked approximately five billion yen for the creation of a synchrotron orbital radiation research center to be completed sometime after 1992.

Address : 4-5-33, Kitahama, Chuo-ku, Osaka 541

Tel & Fax : Tel: 03-3423-5711 Fax: 03-3423-5093

Statistics : Capital : 65400 million yen
Employees : 14348

Officers : CEO : Dr. Noritaka Kurauchi/President

Sales : March, 1988 - 550115 million yen
March, 1989 - 605005 million yen
March, 1990 - 700375 million yen
March, 1991 - 781775 million yen

R & D Spending : 25000 million yen Capital Investment: 38000 million yen

Sales Breakdown :

SIC#	Product	Sales
3357	Electronic wires and cables	53 %
36	Electronics and communications systems	12 %
3399	Powder alloys	10 %
3357	Special wires	6 %
3714	Automobile brake components	6 %

Main Products : Electronic Wires and Related Materials: Electric cables, Fiber-optic cables, Ceramic wires, Various conductive materials; Electronics and Communications Systems: Public transportation controlling systems, Medical information systems, Band scan OCR systems, Optical LAN systems, Other data link systems, Automobile electronic systems and components; Powder Alloys; Special Wires: PC ferrous wires, Steel cord, Piano wire, Construction wires and cables, Automobile wires and cables; Automobile Brake Components: Disk brakes and pads, Anti-lock brake systems

Banks : Sumitomo Bank; Sumitomo Trust and Banking; Industrial Bank of Japan; Mitsubishi Bank; Mitsubishi Trust and Banking

Branches : Tokyo Branch (Minato-ku, Tokyo); Chubu Branch (Nagoya-shi, Aichi-ken); Kyushu Branch (Fukuoka-ken)

Factories : Osaka Plant (Konohana-ku, Osaka); Itami Plant (Itami-shi, Hyogo-ken); Nagoya Plant (Nagoya-shi, Aichi-ken); Yokohama Plant (Yokohama-shi, Kanagawa-ken); Kanto Plant (Kanuma-shi, Tochigi-ken); Sennan Plant (Sennan-gun, Osaka)

Sumitomo Heavy Industries, Ltd.

Japanese Name	:	***Sumitomo Jukikai Kogyo K.K.***
TSE Number	:	6302
Description	:	Major shipbuilder belonging to Sumitomo group. Heavy industries other than shipbuilding have accounted for an ever increasing share of the companies activities, so much so that shipbuilding is no longer a main line.
Address	:	2-2-1, Otemachi, Chiyoda-ku, Tokyo 100
Tel & Fax	:	Tel: 03-3245-4321 Fax: 03-3245-4337
Statistics	:	Capital : 30872 million yen Employees : 6263
Officers	:	CEO : Mr. Masataka Kubo/President
Sales	:	March, 1988 - 200026 million yen March, 1989 - 258214 million yen March, 1990 - 267678 million yen

Sales Breakdown :

SIC#	Product	Sales
3569	Centrifuges	30 %
3443	Process vessels	30 %
3559	Desalination equipment	19 %
3531	Crushing machines	10 %
3731	Shipbuilding	10 %

Main Products	:	General Industrial Machinery: Centrifuges, Heat-exchangers; Fabricated Plate Work; Process Vessels, Kilns; Construction Machinery: Crushers; Shipbuilding and Repair: Ship and frames; Special Industry Machinery: Desalination equipment; Atomic Energy Equipment
Banks	:	Sumitomo Bank; Sumitomo Trust and Banking; Industrial Bank of Japan; Yokohama Bank; Bank of Tokyo
Branches	:	Osaka Branch (Chuo-ku); Overseas Branches: London;
Laboratories	:	Hiratsuka Laboratory (Hiratsuka-shi, Kanagawa-ken)

Sumitomo Light Metal Industries, Ltd.

Japanese Name	:	***Sumitomo Keikinzoku Kogyo K.K.***
TSE Number	:	5738
Description	:	Manufacturer of wrought aluminum and copper products.
Address	:	5-11-3, Shibashi, Minato-ku, Tokyo 105
Tel & Fax	:	Tel: 03-3436-9700 Fax: 03-3434-6464
Statistics	:	Capital : 13418 million yen Employees : 3350
Officers	:	CEO : Mr. Katsumi Uchida/President Contact : Mr. Shigeki Tsubokura/General Manager, Overseas
Sales	:	March, 1988 - 173222 million yen March, 1989 - 199510 million yen March, 1990 - 209197 million yen March, 1991 - 229637 million yen
Sales Breakdown	:	SIC# Product 3354 Wrought aluminum products 3351 Wrought copper products
Main Products	:	Wrought Metal Products: Wrought aluminum products, Wrought copper products
Banks	:	Sumitomo Bank; Sumitomo Trust and Banking; Export-Import Bank of Japan; Industrial Bank of Japan; Long Term Credit Bank of Japan; Nippon Credit Bank
Branches	:	Osaka Branch (Osaka-shi); Nagoya Branch (Nagoya-shi); Hokkaido Branch (Sapporo-shi); Tohoku Branch (Sendai-shi); Shizuoka Branch (Shizuoka-shi); Hokuriku Branch (Takaoka-shi); Chugoku Branch (Hiroshima-shi); Kyushu Branch (Fukuoka-shi)
Factories	:	Nagoya Factory (Nagoya-shi); Chiba Factory (Kashiwa-shi)
Laboratories	:	Technical Research Laboratory (Nagoya-shi)
Shareholders	:	Sumitomo Metal [5401] - 17.3% ; Sumitomo Trust and Banking [8403] - 6.4% ; Sumitomo Bank [8318] - 4.9% ; Sumitomo Corp. [8053] - 4.2% ; Sumitomo Mutual Life - 4.2% ; Nippon Life - 1.8%

Sumitomo Metal Industries, Ltd.

Japanese Name :	***Sumitomo Kinzoku Kogyo K.K.***
TSE Number :	5405
Description :	Sumitomo Metal is one of Japan's top four integrated steel makers. Currently the company is diversifying into titanium, electronic materials and carbon fiber fields.
Address :	4-5-33, Kitahama, Chuo-ku, Osaka 541
Tel & Fax :	Tel: 06-220-5091 Fax: 06-220-5082
Statistics :	Capital : 202685 million yen Employees : 20573
Officers :	CEO : Mr. Yasuo Shingu/President
Sales :	March, 1988 - 909271 million yen March, 1989 - 1054580 million yen March, 1990 - 1119610 million yen March, 1991 - 1156850 million yen
R & D Spending :	37700 million yen Capital Investment: 111200 million yen

Sales Breakdown :

SIC#	Product	Sales
3312	Sheets	41 %
3317	Pipes, tubes	25 %
3324	Rolling stock	15 %
3324	Castings	15 %
3356	Titanium	4 %

Main Products :	Blast Furnaces and Mills: Steel sheets and plates; Steel Pipes and Tubes; Steel Investment Foundries: Rolling-stock parts, Steel casting and forging; Nonferrous Rolling and Drawing: Titanium products
Banks :	Sumitomo Bank; Sumitomo Trust and Banking; Industrial Bank of Japan; Long Term Credit Bank of Japan; Norinchukin Bank
Branches :	19 domestic branches
Factories :	Osaka Plant; Hyogo Plant; Wakayama Plant; Fukuoka Plant; Ibaraki Plant
Laboratories :	Future Technical Research Laboratory (Kashima-gun, Ibaraki-ken)
Subsidiaries :	15 consolidated subsidiaries; Over 100 unconsolidated subsidiaries
Overseas Subsidiaries :	10 consolidated subsidiaries in the U.S.A.; 15 unconsolidated subsidiaries worldwide

Sumitomo Metal Mining Co., Ltd.

Japanese Name	:	***Sumitomo Kinzoku Kozan K.K.***
TSE Number	:	5713
Description	:	Sumitomo Metal Mining is a top producer of gold and nickel, third in copper. The company is diversifying into electronic materials, construction materials, housing and chemicals.
Address	:	5-11-3, Shinbashi, Minato-ku, Tokyo 105
Tel & Fax	:	Tel: 03-3436-7701 Fax: 03-3436-7734
Statistics	:	Capital : 65467 million yen Employees : 3852
Officers	:	CEO : Mr. Akihiko Shinozaki/President
Sales	:	March, 1988 - 429239 million yen March, 1989 - 437972 million yen March, 1990 - 536642 million yen March, 1991 - 505609 million yen
R & D Spending	:	7704 million yen Capital Investment: 26103 million yen

Sales Breakdown :

SIC#	Product	Sales
3339	Gold	54 %
3331	Copper	15 %
3339	Nickel	10 %
3339	Zinc	6 %
3499	Electronic materials	6 %

Main Products	:	Primary Non-ferrous Metals: Gold, Nickel, Zinc; Primary Copper; Materials: Electronic materials, Antimony-based flame retardant
Banks	:	Sumitomo Bank; Sumitomo Trust and Banking; Industrial Bank of Japan; Long Term Credit Bank of Japan; Japan Development Bank
Branches	:	Osaka Branch (Chuo-ku); Sapporo Branch; Nagoya Branch; Fukuoka Branch; Hiroshima Branch; Overseas Branches: Sydney; London
Factories	:	Ehime Plant; Hyogo Plant; Hokkaido Plant; Electronics Plant (Tokyo); Hishikari Mine; Yokohama Plant (Kanagawa-ken); Tochigi Plant (Tochigi-ken); Mie Plant (Mie-ken)
Laboratories	:	Central Research Laboratory (Chiba); Electronics Materials Laboratory (Tokyo)
Overseas Subsidiaries	:	Sumitomo Metal Mining Arizona Inc. (Arizona, U.S.A.); Sumitomo Metal Mining Canada Ltd. (Canada); Sumitomo Metal Mining Singapore Pte. Ltd. (Singapore); M-SMM Electronics Sdn. Bhd. (Malaysia); Sumitomo Metal Mining U.S.A. Inc. (U.S.A.)

Sumitomo Optical Glass Inc.

TSE Number	:	Not listed
Address	:	3-15-10, Uchikanda, Chiyoda-ku, Tokyo 101

Sun Arrow Chemical Co, Ltd.

Japanese Name	:	***San Aroo Kagaku K.K.***
TSE Number	:	Not listed
Description	:	Manufacturer of vinyl chloride products.
Address	:	1-2, Harumi-cho, Tokuyama-shi, Yamaguchi-ken 745
Tel & Fax	:	Tel: 0834-21-8010 Fax: 0834-21-7510
Statistics	:	Capital : 1000 million yen Employees : 176
Officers	:	CEO : Mr. Yoshio Ohashi/President Contact : Mr. Yoshio Ohashi/President
Sales Breakdown	:	SIC# Product 2821 Plastics
Main Products	:	Plastics Products: Vinyl chloride polymer and monomer
Banks	:	Sanwa Bank; Japan Development Bank; Industrial Bank of Japan; Hiroshima Bank; Yamaguchi Bank
Branches	:	Tokyo Head Office (Minato-ku); Osaka Branch (Osaka-shi)
Factories	:	Main Factory (Tokuyama-shi)
Shareholders	:	Tokuyama Soda Co., Ltd.

Suzuki Metal Industry Co., Ltd.

Japanese Name	:	***Suzuki Kinzoku Kogyo K.K.***
TSE Number	:	5657
Description	:	Suzuki Metal Industry is a large-scale wire manufacturer, affiliated with Nippon Steel, marketing to the construction, electronics, and automotive industries. Established in 1938 to produce piano wire, the company has diversified into many wire-related fields and is now expanding abroad with a joint venture in Thailand, and is advancing into the production of materials for the leisure goods and office equipment industries.
Address	:	Dai-ichi Tekko Bldg. 5F, 1-8-2, Marunouchi, Chiyoda-ku, Tokyo 100
Tel & Fax	:	Tel: 03-3214-1681 Fax: 03-3214-7256
Statistics	:	Capital : 1200 million yen Employees : 710
Officers	:	CEO : Mr. Akira Akagawa/President Contact : Mr. Akira Muramoto/Export Manager
Sales	:	March, 1988 - 25410 million yen March, 1989 - 27501 million yen March, 1990 - 29223 million yen March, 1991 - 30500 million yen

Sales Breakdown :

SIC#	Product	Sales
3429	Piano wire	25 %
3315	Oil tempered wire	24 %
3315	PC wire and PC strands	17 %
3315	Hard drawn steel wire	16 %
3315	Stainless steel wire	15 %
3357	Titanium wire	2 %
3315	Fishing wire	1 %

Main Products	:	Steel Wire and Related Products: Steel wire, Stainless wire, Magnetic wires, Water-treated wires, Printer Dot wires; Cordage and Twine: Fishing lines; Piano Hardware: Piano wires
Banks	:	Industrial Bank of Japan; Mitsubishi Bank; Kyowa Saitama Bank
Branches	:	Nagoya Branch (Nagoya-shi); Osaka Branch (Osaka)
Factories	:	Narashino Plant (Narashino-shi, Chiba-ken)
Laboratories	:	Research and Development Center (Narashino-shi, Chiba-ken)
Subsidiaries	:	Nishikawa Seisen Co., Ltd.; Nitsutou Kinzoku Co., Ltd.

Tabuchi Electric Co., Ltd.

Japanese Name : ***Tabuchi Denki Kogyo***

TSE Number : 6624

Description : Manufacturer of transformers and power supplies for electronics and other industrial products. The company is venturing abroad with plants in the U.K., Thailand and Mexico, and is expanding into new markets including power sources for cordless phones and flat power sources.

Address : 1-12-22, Mitejima, Nishi-Yodogawaku, Osaka 555

Tel & Fax : Tel: 06-475-5111 Fax: 06-472-2487

Statistics : Capital : 2059 million yen
Employees : 387

Officers : CEO : Mr. Teruhisa Tabuchi/President
Contact : Mr. Teruhisa Tabuchi/President

Sales : March, 1989 - 28400 million yen
March, 1990 - 26600 million yen
March, 1991 - 29387 million yen

Sales Breakdown :

SIC#	Product	Sales
3669	Power sources for microwaves	30 %
3612	High-voltage transformers	30 %
3679	Power supplies	30 %

Main Products : Communications Equipment: Power sources for microwaves, Cordless phones; Transformers: High-voltage transformers for electronics; Electronic Components: Power supplies for industrial use

Banks : Fuji Bank; Dai-Ichi Kangyo Bank; Daiwa Bank; Sanwa Bank; Sumitomo Trust and Banking

Taiko Refractories Co., Ltd.

Japanese Name	:	***Taiko Rozai Co., Ltd.***
TSE Number	:	Not listed
Description	:	Manufacturer of industrial furnaces, ceramics and special cements.
Address	:	1-1, Mikiyama Shinmachi, Tobata-ku, Kitakyushu-shi, Fukuoka 804
Tel	:	093-871-1631
Statistics	:	Capital : 25 million yen Employees : 260
Sales	:	March, 1990 - 6709 million yen March, 1991 - 6142 million yen
Sales Breakdown	:	SIC# Product 3567 Industrial furnaces 3299 Fine ceramics 324 Fire resistant cement
Main Products	:	Industrial Furnaces; Ceramic Products: Fine ceramics; Cement Products: Fire resistant cement
Banks	:	Bank of Fukuoka; Yamaguchi Mutual Trust; Fukuoka Mutual Trust; Kyushu Mutual Trust
Factories	:	Fukuoka Plant (Fukuoka-ken); Oita Plant (Oita-ken); Kimitsu Plant (Chiba-ken)

Taisei Corp.

Japanese Name : ***Taisei Kensetsu K.K.***

TSE Number : 1801

Description : Taisei is the largest integrated construction firm in Japan and is rated fifth worldwide in terms of construction order bookings. The company is achieving high growth with the Japanese construction boom and is developing civil engineering projects boosted by technology in large-caliber shields, which was used in the construction of the Teito Rapid Transit Subway and in dam construction (Mikuni River). The company has expanded abroad and is facing unpaid debts related with Iraq projects but is increasing investment in Asia including a new plant in Indonesia. The company is advancing in the areas of factory automation and chemical research, and is a strong competitor in all areas of general construction including intelligent structure design and engineering.

Address : 1-25-1, Nishi-Shinjuku, Shinjuku-ku, Tokyo 163

Tel & Fax : Tel: 03-3348-1111 Fax: 03-3345-0481~2

Statistics : Capital : 94111 million yen
Employees : 12383

Officers : CEO : Mr. Yasuo Satomi/President
Contact : Mr. Hiroyoshi Miyazaki/Manager of Sales Promotion

Sales : March, 1988 - 1033600 million yen
March, 1989 - 1273300 million yen
March, 1990 - 1406000 million yen
March, 1991 - 1548900 million yen

Sales Breakdown :

SIC#	Product	Sales
1542	Building construction	69 %
162	Civil engineering	23 %
655	Real estate development	8 %

Main Products : Non-residential Construction: Commercial buildings and structures; Industrial Buildings: Warehouses, Plants, Sports domes; Civil Engineering Projects: Bridge, Tunnel and elevated buildings; Architectural Services: Architectural design and engineering; Subdividers and Developers: Real estate development

Banks : Fuji Bank; Yasuda Trust and Banking; Daiwa Bank; Industrial Bank of Japan; Mitsubishi Trust and Banking; Norinchukin Bank; Nippon Credit Bank

Laboratories : Technology Research Center (Yokohama-shi); Biotechnology Research Center (Narashino-shi)

Subsidiaries : Yuraku Real Estate Co., Ltd. (Chuo-ku); Taisei Road Construction Co., Ltd. (Chuo-ku); Taisei Prefab Construction Co., Ltd. (Shinagawa-ku)

Overseas Subsidiaries : Taisei America Corp. (New York, U.S.A.); Taisei Europe Ltd. (London, U.K.); Taisei (Malaysia) Sdn. Bhd. (Malaysia); Taisei Holland B.V. (Amsterdam, Netherlands); Taisei Hawaii Corp. (Hawaii, U.S.A.); Taisei Properties (Hong Kong) Ltd. (Hong Kong)

Taiyo Kagaku Co., Ltd.

Japanese Name	:	***Taiyo Kagaku K.K.***
TSE Number	:	2902
Description	:	Manufacturer of raw materials for foods, drugs and cosmetics. Currently manufactures emulsifiers, stabilizers, processed-egg products, extracts from natural substances. Began developing a technique in 1989 for isolating antibodies from eggs. New products include coffee concentrates and nutritional drinks, and the company has also diversified into hotel management.
Address	:	9-5, Akahorishin-machi, Yokkaichi-shi, Mie-ken 510
Tel & Fax	:	Tel: 0593-52-2555 Fax: 0593-52-9312
Statistics	:	Capital : 7049 million yen Employees : 546
Officers	:	CEO : Mr. Nagataka Yamazaki/President Contact : Mr. Nagataka Yamazaki/President
Sales	:	March, 1988 - 27092 million yen March, 1989 - 30332 million yen March, 1990 - 32155 million yen March, 1991 - 34476 million yen
R & D Spending	:	901 million yen Capital Investment: 1174 million yen

Sales Breakdown :

SIC#	Product	Sales
0259	Egg products	30 %
2099	Food additives	25 %
2836	Biological extracts (Lethicin, etc.)	10 %
2836	Dietary fiber	5 %
7011	Hotel management	5 %

Main Products	:	Raw Materials For: Food products, Pharmaceuticals, Cosmetics; Specialty Chemicals: Acetyl monoglyceride; Food-Related Products: Emulsifiers, Stabilizers, Processed egg products, Animal and plant extracts, Fruit sauces, Lecithin extracted from soya beans, Dietary fiber; Hotel Management
Banks	:	Dai-Ichi Kangyo Bank; Tokai Bank; Hyakugo Bank; Bank of Mie; Dai-san Bank
Branches	:	Tokyo Office (Tokyo); Osaka Office (Osaka); Sapporo Office (Sapporo); Nagoya Office (Nagoya); Fukuoka Office (Fukuoka)
Factories	:	Shiohama Factory (Yokkaichi-shi, Mie-ken); Yokkaichi Factory (Mie-ken); Nanbu Factory; Akahori Factory (Mie-ken); Suei Factory (Yokkaichi-shi, Mie-ken)
Laboratories	:	Central Research Laboratory (Yokkaichi-shi)

Takeda Chemical Industries, Ltd.

Japanese Name : ***Takeda Yakuhin Kogyo K.K.***

TSE Number	:	4502
Description	:	Top manufacturer of pharmaceuticals noted for its 'Alinamin' vitamin tablet.
Address	:	2-3-6, Dosho-machi, Chuo-ku, Osaka 541
Tel & Fax	:	Tel: 06-204-2111 Fax: 06-204-2880
Statistics	:	Capital : 47625 million yen Employees : 10919
Officers	:	CEO : Mr. Yoshimasa Umemoto/President
Sales	:	March, 1988 - 539754 million yen March, 1989 - 585015 million yen March, 1990 - 588761 million yen March, 1991 - 550871 million yen
R & D Spending	:	49400 million yen Capital Investment: 49400 million yen

Sales Breakdown :

SIC#	Product	Sales
2834	Pharmaceutical products	64 %
28	Chemical products	15 %
20	Foodstuffs	8 %
287	Agricultural chemicals	7 %
283	Veterinarian drugs	2 %
0000	Others	3 %

Main Products	:	Pharmaceutical Products: Drugs for the nervous system, Circulatory and respiratory drugs, Preparation streptomycin, Vitamin drugs and antibiotics, Digestive drugs, Hormone preparations, Vitamin preparations, Metabolic drugs, Biological drugs; Chemicals: Synthetic rubber latex, Polyureatane resins, Polyester resins, Activated carbon, Adermine hydrochloride, Amino acid, Phosphorous antioxidant, Dimethoate, Lactic acid, Monosodium L-glutamate, Acephate, Citric acid, Nicotine sulphate, Surface coating agent, Aliphatic ester plasticizers, Anhydrous Maleic acid; Food Products: Sodium glutamate, Vitamin C, Bactericides/preservatives; Agrichemicals; Veterinarian Drugs
Banks	:	Sumitomo Bank; Fuji Bank; Mitsubishi Bank; Sumitomo Trust and Banking
Factories	:	Osaka Factory (Osaka); Takasago Factory (Takasago-shi, Hyogo-ken); Hikari Factory (Hikari-shi, Yamaguchi-ken); Shimizu Factory (Shimizu-shi, Shizuoka-ken); Kashima Factory (Kashima-shi, Saga-ken); Shonan Factory (Kanagawa-ken)
Laboratories	:	Central Research Laboratory; Tsukuba Research Lab; Agricultural Science Research Laboratory (Kyoto); Bio Engineering Research Laboratory (Osaka-shi); Production Technology Laboratory;Applied Technology Laboratory;

Takiron Co., Ltd.

Japanese Name	:	***Takiron K.K.***
TSE Number	:	4215
Description	:	Manufacturer of plastic materials for use in construction, industry, agriculture and civil engineering. An affiliate of C. Itoh & Co., Takiron specializes in the production of plastic plates and related structural materials and is a market leader in PVC plates. R&D continues to stress plastic structural products and, acting through its London-based subsidiary, the company is involved in plastics technology tie-ups with British firms Wiggins Teape Co., Ltd. and I.C.I. Co., Ltd.. The company also provides engineering services related to sewer and drainage system construction.
Address	:	Osaka Kokusai Bldg. 15F, 2-3-13, Azuchi-machi, Chuo-ku, Osaka 541
Tel & Fax	:	Tel: 06-267-2800 Fax: 06-267-2655
Statistics	:	Capital : 6984 million yen Employees : 1361
Officers	:	CEO : Mr. Hitoshi Yamaguchi/President
Sales	:	March, 1988 - 47892 million yen March, 1989 - 52649 million yen March, 1990 - 54878 million yen March, 1991 - 59301 million yen
R & D Spending	:	897 million yen Capital Investment: 4761 million yen
Sales Breakdown	:	

SIC#	Product	Sales
308	Plastic building materials	48 %
308	Plastic industrial materials	29 %
308	Agricultural and engineering plastics	20 %
8711	Engineering services	1 %

Main Products	:	Plastic Products: PVC sheets, Glass net reinforced plastic plates, Synthetic floor tiles, Other misc. plastic building materials, PVC plates, Polycarbonate plates, ABS plates Acrylic plates, Channel plates, Conductive plates, PVC pipes and related products , Polyethylene drip pipes and drip nets for agricultural applications, LED display panels, Electricity conductive adhesives and related materials, Misc. other plastic agricultural and engineering materials; Engineering Services: Sewer and drainage systems
Branches	:	Tokyo Branch (Chuo-ku, Tokyo); Osaka Branch (Osaka-shi); Sendai Branch (Sendai-shi, Miyagi-ken); Nagoya Branch (Nagoya-shi, Aichi-ken); Fukuoka Branch (Fukuoka-shi, Fukuoka-ken)
Factories	:	Aboshi Factory (Ibo-gun, Hyogo-ken); Yasutomi Factory (Shisou-gun, Hyogo-ken); Tokyo Factory (Niihari-gun, Ibaraki-ken); Yokaichi Factory (Yokaichi-shi, Shiga-ken)
R&D Areas	:	Synthetic Resins and Plastics: Acrylic Resins, Transparent plastic sheets and panel, Super engineering plastics
Overseas Subsidiaries	:	Takiron (U.K.) Ltd (London, U.K);

Takuma Co., Ltd.

Japanese Name	:	***Takuma Kogyo***
TSE Number	:	6013
Description	:	Takuma is a boiler maker which has expanded into environmental systems and industrial machinery using boiler technology. The company has experienced financial difficulties and was the subject of a takeover bid by Cosmopolitan.
Address	:	1-3-23, Dojimahama, Kita-ku, Osaka-shi 530
Tel & Fax	:	Tel: 06-346-5161 Fax: 06-347-9146
Statistics	:	Capital : 8663 million yen Employees : 958
Officers	:	CEO : Mr. Junkichi Fukuda/President Contact : Mr. Junkichi Fukuda/President
Sales	:	March, 1988 - 48418 million yen March, 1989 - 50688 million yen March, 1990 - 56595 million yen March, 1991 - 73100 million yen
R & D Spending	:	439 million yen Capital Investment: 840 million yen
Sales Breakdown	:	

SIC#	Product	Sales
4953	Industrial waste systems	40 %
3589	Water treatment systems	30 %
3822	Garbage incinerators	10 %
3083	Laminated boards and plates	10 %
3443	Boiler plant	10 %

Main Products	:	Refuse Systems and Environmental Controls: Industrial waste systems, Waste-treatment systems, Garbage incinerators; Laminated Plastics: Laminated boards and plates; Fabricated Plate Work: Boiler plant supplies
Banks	:	Dai-Ichi Kangyo Bank; Sumitomo Bank; Kyowa Saitama Bank; Mitsui Taiyo Kobe Bank; Takugin Bank; Sanwa Bank
Branches	:	Nagoya Branch (Nagoya-shi); Fukuoka Branch (Fukuoka-shi); Sapporo Branch (Sapporo-shi); Sendai Branch (Sendai-shi); Hiroshima Sales Branch (Hiroshima-shi); Yokohama Sales Branch (Edogawa-ku)
Factories	:	Kyoto Plant (Minami-ku, Kyoto); Hyogo Plant (Takasaga-shi, Hyogo-ken)
R&D Areas	:	Boiler Technology: Diversification of energy supplies for boilers; New Materials: Void-free prepreg impregnating equipment
Overseas Subsidiaries	:	Taiwan Takuma Co., Ltd. (Taipei, Taiwan); TAC Environmental Engineering Co., Ltd. (Los Angeles, U.S.A.)

Tamura Corp.

Japanese Name	:	***Tamura Seisakusho K.K.***
TSE Number	:	6768
Description	:	Tamura Corp. is a small-sized firm which focuses on the production of communications equipment and computer-related parts and materials.
Address	:	1-19-41, Higashi-Oizumi, Nerima-ku, Tokyo 178
Tel & Fax	:	Tel: 03-3978-2111 Fax: 03-3923-0230
Statistics	:	Capital : 6999 million yen Employees : 1071
Officers	:	CEO : Mr. Itsuya Tamura/President
Sales	:	March, 1988 - 38229 million yen March, 1989 - 44257 million yen March, 1990 - 49288 million yen March, 1991 - 49826 million yen
R & D Spending	:	1400 million yen Capital Investment: 1984 million yen
Sales Breakdown	:	SIC# Product 3661 Communications equipment 3679 Power supplies 3674 Electronic materials
Main Products	:	Telephone and Telegraph Apparatus: Communications equipment; Electronic Components: Power supplies; Semiconductors and Related Devices: Electronic materials
Banks	:	Sumitomo Bank; Saitama Bank; Fuji Bank; Industrial Bank of Japan, Sumitomo Trust and Banking
Branches	:	Osaka Branch (Suita-shi); Electronic Materials Office (Iruma-shi); Kanagawa Office (Yamato-shi); Suzuka Office (Suzuka-shi)
Factories	:	Machine Factory (Nerima-ku); Electronic Parts Factory (Sakado-shi)

Tanaka Kinzoku Kogyo

Japanese Name : ***Tanaka Kinzoku Kogyo K.K.***

TSE Number : Not listed

Description : Tanaka Kinzoku is a producer of precious metals for application in the electronics, industrial machines, computers, and dental products industries.

Address : 2-6-6, Nihonbashi-Kayabacho, Chuo-ku, Tokyo 103

Tel & Fax : Tel: 03-3668-0111 Fax: 03-3667-7498

Statistics : Capital : 25000 million yen
Employees : 1000

Officers : CEO : Mr. Tadahiko Fukami/President
Contact : Mr. Tadahiko Fukami/President

Sales : March, 1988 - 453100 million yen
March, 1989 - 371700 million yen
March, 1990 - 307500 million yen

Sales Breakdown :

SIC#	Product	Sales
3341	Ingots	30 %
3339	Precious metals	30 %
3577	Computer materials	10 %
3499	Nozzles	10 %
2819	Catalysts	10 %
3843	Dental use materials	10 %

Main Products : Non-ferrous Metals: Precious metals, Ingots; Computer Peripheral Equipment: Computer parts and materials; Fabricated Metal Products: Chemical fiber nozzles, glass fiber nozzles; Industrial Chemicals: Petrochemical manufacture process catalysts, Nitric acid catalysts, Deodorizing catalysts; Dental Equipment and Supplies: Materials for dental products

Banks : Fuji Bank; Yasuda Trust and Banking; Long Term Credit Bank of Japan; Tokai Bank

Factories : Hiratska Plant (Hiratska-shi); Ichikawa Plant (Ichikawa-shi); Hiratska Plant No. 2 (Hiratska-shi); Ise Plant (Ise-shi); Tsuruoka Plant (Tsuruoka-shi)

Shareholders : Tanaka Precious Metals Sales - 47% ; Tanaka Precious Metal Jewelry - 12%

Tateyama Aluminum Industry Co., Ltd.

Japanese Name :	***Tateyama Aluminum Kogyo K.K.***
TSE Number :	5910
Description :	Mid-sized manufacturer of aluminum fixtures and related products for domestic, retail and commercial use.
Address :	550, Hayakawa, Takaoka-shi, Toyama-ken 933
Tel & Fax :	Tel: 0766-20-3131 Fax: 0766-20-3713
Statistics :	Capital : 6180 million yen Employees : 3438
Officers :	CEO : Mr. Eiji Takehira/President
Sales :	March, 1988 - 89005 million yen March, 1989 - 105796 million yen March, 1990 - 115029 million yen March, 1991 - 129270 million yen
R & D Spending :	730 million yen Capital Investment: 10500 million yen

Sales Breakdown :

SIC#	Product	Sales
3442	Housing construction materials	44 %
3442	Commercial construction materials	40 %
3499	Aluminum products for retail use	13 %
3499	Misc. aluminum shapes	3 %
3499	Misc. aluminum retail products	1 %

Main Products :	Aluminum Housing Construction Materials and Fixtures: 'Super Sash', Combination sashes, High quality sashes, Heat conserving sashes, Combination bay window sashes, Shutter sashes, Bathroom doors, Sliding doors, Screen doors, House fittings, Terrace and balcony fittings, Gates, Fences, etc.; Aluminum Commercial Construction Materials and Fixtures: Custom made doors and door sahses, Partitions, Curtain walls, Aluminum moldings, Railings, Panels; Aluminum Products for Use in Retail Establishments: Showcases, Furniture, Display fixtures, Ad-frames, etc.; Aluminum Shapes: Pressed shapes, Alumite shapes; Misc. Aluminum Retail Products: Stepladders, Ladders, Garden tools and supplies, Dish racks
Banks :	Long Term Credit Bank of Japan; Norinchukin Bank; Hokuriku Bank; Sumitomo Trust and Banking; Nippon Credit Bank; Fukui Bank; Hokkaido Bank; Sumitomo Bank; Yasuda Trust and Banking; Fijiyama Dai-ichi Bank
Branches :	Tokyo Branch (Shinjuku-ku, Tokyo); Osaka Branch (Moriguro-shi, Osaka); 12 other branches throughout Japan
Factories :	Takaoka Plant (Takaoka-shi, Toyama-ken); Fukuoka Plant (Fukuoka-shi, Fukuoka-ken); West Fukuoka Plant (Fukuoka-ken); Oyabe Plant (Oyabe-shi, Toyama-ken); Ibaraki Plant (Ushiku-shi, Ibaraki-ken)
R&D Areas :	Production Technologies: 'Product Information System', CAD (computer aided design) system; Products: Aluminum interior fixtures, Retail display fixtures
Subsidiaries :	Tateyama Gyokin Kogyo K.K. (Takaoka-shi, Toyama-ken)

TDK Corp.

Japanese Name	:	***TDK K.K.***
TSE Number	:	6762
Description	:	Major manufacturer of magnetic recording devices and electronic components. TDK is the world's largest manufacturer of magnetic tapes and has expanded its business lines to include a variety of information storage mediums including optical disks. Electronic parts, most notably IC components, have been making increasing contributions to company profits.
Address	:	1-13-1, Nihombashi, Chuo-ku, Tokyo 103
Tel & Fax	:	Tel: 03-3278-5111 Fax: 03-3278-5330
Statistics	:	Capital : 17858 million yen Employees : 8102
Officers	:	CEO : Mr. Hiroshi Sato/President
Sales	:	March, 1989 (4 months) - 118026 million yen March, 1990 - 379286 million yen March, 1991 - 413974 million yen
R & D Spending	:	23200 million yen Capital Investment: 32800 million yen
Sales Breakdown	:	

SIC#	Product	Sales
3695	Magnetic recording tape	30 %
3264	Ferrite magnetic products	30 %
3679	Electronic parts	20 %
3572	Floppy disks and diskettes	
3572	Optical disks	

Main Products	:	Magnetic Tapes; Ferrite Magnetic products: Ferrite cores; Electronic Parts: IC components, Thin magnetic heads, OA equipment components, Broadcasting satellite components; Computer storage devices: Floppy disks, Diskettes, Optical disks
Banks	:	Saitama Bank; Sumitomo Bank
Branches	:	Osaka Branch (Osaka-shi)
Laboratories	:	Technical Center (Ichikawa-shi)
Shareholders	:	Mitsubishi Bank [8315] - 8.0%; Sumitomo Trust and Banking [8403] - 7.0%; Matsushita Electric Industrial [6752] - 4.9%; Mitsui Trust and Banking [8401] - 4.2%; Yasuda Trust and Banking [8404] - 4.1%

Techmate

TSE Number	:	Not listed
Description	:	Manufacturer of fiber reinforced plastic. Techmate is in the process of completing a new plant in Hakodate (Hokkaido) which will boost the company's productive capacity significantly.
Address	:	
Tel	:	06-683-6163
Sales Breakdown	:	SIC# Product 3089 Fiber reinforced plastics
Main Products	:	Fiber Reinforced Plastics: CFRP (carbon fiber reinforced plastic)

Teijin Ltd.

Japanese Name :	***Teijin K.K.***
TSE Number :	3401
Description :	A leading polyester producer in Japan. The company is currently emphasizing the development of aramid fibers, new pharmaceuticals, and other new materials.
Address :	1-6-7, Minamihon-machi, Chuo-ku, Osaka 541
Tel & Fax :	Tel: 06-266-2131 Fax: 06-266-2133
Statistics :	Capital : 70137 million yen Employees : 6511
Officers :	CEO : Mr. Hiroshi Itagaki/President
Sales :	March, 1988 - 309666 million yen March, 1989 - 312264 million yen March, 1990 - 305700 million yen March, 1991 - 325727 million yen
R & D Spending :	21100 million yen Capital Investment: 37700 million yen

Sales Breakdown :

SIC#	Product	Sales
2281	Polyester fibers, new materials	51 %
2819	Industrial inorganic chemicals	22 %
2836	Biotechnological products	12 %
2284	Nylon	9 %
2284	Other fibers	6 %

Main Products :	Fibers: Tetoron, Nylon, Aramid fibers, Teviron, Acetate; Functional Products: Films, Floppy disks, Engineering plastics; Medical Products: Pharmaceuticals; Specialty Chemicals: Halogen-based flame retardant, Organic phosphorus-based flame retardant, Non-inflammable polymer flame retardant, Sea surface pollution prevention absorbent
Banks :	Sanwa Bank; Fuji Bank; Daiwa Bank; Sumitomo Trust and Banking; Toyo Trust and Banking
Branches :	Tokyo Branch (Chiyoda-ku); Kanagawa Branch (Yokohama-shi); Osaka Branch (Osaka-shi) and 2 other domestic branches; 6 domestic sales offices; Europe Office (Dusseldorf); Athens Representative Office, Hong Kong Representative Office
Factories :	Iwakuni Plant (Iwakuni-shi); Mihara Plant (Mihara-shi); Matsuyama Plant (Matsuyama-shi) and 4 other plants
Laboratories :	Tokyo Research Center (Hino-shi); Yokohama Research Center (Yokohama-shi); Sagamihara Research Center (Sagamihara-shi); Osaka Research Center (Osaka-shi); Iwakuni Reseach Center (Iwakuni-shi)
Shareholders :	Nippon Mutual Life - 6.2% ; Sanwa Bank [8320] - 4.7% ; Toyo Trust and Banking [8407] - 4.0% ; Sumitomo Mutual Life - 3.5% ; Sumitomo Trust and Banking [8403] - 3.1% ; Fuji Bank [8317] - 2.3%
Subsidiaries :	Teijin Bio Laboratories, Inc. (Minato-ku); Teijin Agrochemicals Ltd. (Mihara-shi) and 45 other domestic subsidiaries

Toa Planning

Japanese Name	:	***Toa Puraningu K.K.***
TSE Number	:	Not listed
Description	:	Engaged in the development and manufacture of ceramic composite materials.
Address	:	NK Bldg., 2 Fl., 7-2-4, Shinjuku, Shinjuku-ku, Tokyo 160
Tel	:	03-3202-6311
Sales Breakdown	:	SIC# Product 3299 Ceramic composite materials
Main Products	:	Ceramic Composite Materials: 'Ceranode' ceramic polymer coating material

Toagosei Chemical Industry Co., Ltd.

Japanese Name	:	***Toa Gosei Kagaku Kogyo K.K.***
TSE Number	:	4045
Description	:	Toagosei Chemical is a medium-ranking chemicals manufacturer which is diversifying into fine chemicals, including high-purity chemicals for semiconductor manufacture, high-purity gases, functional adhesives, paints and bio-agrochemicals.
Address	:	1-14-1, Nishi-Shinbashi, Minato-ku, Tokyo 105
Tel & Fax	:	Tel: 03-3597-7221 Fax: 03-3597-7217
Statistics	:	Capital : 13861 million yen Employees : 1797
Officers	:	CEO : Mr. Toshiaki Kametani/President
Sales	:	December, 1988 - 105408 million yen December, 1989 - 105400 million yen December, 1990 - 116017 million yen
Sales Breakdown	:	

SIC#	Product	Sales
2821	Synthetic resins	60 %
2812	Soda	27 %
2891	Instant adhesives	10 %
2873	Fertilizers	3 %

Main Products	:	Plastics Materials: Synthetic resins; Alkalies and Chlorine: Sodium chloride products; Specialty Chemicals: Acrylic esters, Ammonia cooling agent, Carbon dioxide cooling agent, Flocculant, Sequestering agent, Vinyl chloride monomer; Adhesives and Sealants: Instant adhesive 'Alone Alpha'; Nitrogenous Fertilizers
Banks	:	Mitsui Taiyo Kobe; Tokai Bank; Industrial Bank of Japan; Mitsui Trust and Banking; Norinchukin Bank

Toho Kinzoku Co., Ltd.

Japanese Name : ***Toho Kinzoku Kogyo***

TSE Number : 5781

Description : Comprehensive manufacturer of tungsten and molybdenum parts for the electronics, automotive and civil engineering industries. The company is showing slack earnings from mounting competition and will increase production of tungsten products, including fishing lines.

Address : 2-6-17, Kitahama, Chuo-ku, Osaka 541

Tel & Fax : Tel: 06-202-3376 Fax: 06-202-1390

Statistics : Capital : 2413 million yen
Employees : 346

Officers : CEO : Mr. Yasuhiro Iwai/President
Contact : Mr. Yasuhiro Iwai/President

Sales : March, 1988 - 6134 million yen
March, 1989 - 6804 million yen
March, 1990 - 6804 million yen
March, 1991 - 7218 million yen

Capital Investment : 352 million yen

Sales Breakdown :

SIC#	Product	Sales
3313	Molybdenum	35 %
3313	Tungsten	27 %
3341	Super hard alloys	21 %
3714	Automotive joints	15 %
3313	Tungsten fishing nets	2 %

Main Products : Electrometallurgical Products: Molybdenum, Tungsten, Tungsten fishing nets; Secondary Non-ferrous Metals: Super-hard alloys; Motor Vehicle Parts: Automotive joints

Banks : Sanwa Bank; Mitsui Taiyo Kobe Bank; Toyo Trust and Banking; Norinchukin Bank; Industrial Bank of Japan; Iyo Bank

Shareholders : Taiyo Mining & Industrial - 24.3% ; Kyoei Fire & Marine Insurance - 8.5% ; Nissho Iwai [8063] - 5.5% ; Sanwa Bank [8320] - 2.4% ; Mitsui Taiyo Kobe Bank [8314] - 2.1% ; Nissho Iwai Real Estate - 2.1% ; Toyo Trust and Banking [8407] - 1.9%

Toho Rayon Co., Ltd.

Japanese Name : ***Toho Reeyon K.K.***

TSE Number : 3403

Description : Toho Rayon is one of the largest PAN carbon fiber producers in the world. The company has tie-ups with BASF and is expanding into bicycles and other types of consumer goods.

Address : Nishikawa Bldg., 3-3-9, Nihonbashi, Chuo-ku, Tokyo 103

Tel & Fax : Tel: 03-3278-7615 Fax: 03-3278-7734

Statistics : Capital : 8877 million yen
Employees : 1753

Officers : CEO : Mr. Keizo Kanee/President

Sales : March, 1988 - 71926 million yen
March, 1989 - 71024 million yen
March, 1990 - 74149 million yen
March, 1991 - 77893 million yen

R & D Spending : 850 million yen Capital Investment: 6624 million yen

Sales Breakdown :

SIC#	Product	Sales
2824	Beslon	34 %
2211	Cotton	27 %
2221	Rayon	15 %
0279	Raw silk	10 %
3624	Carbon and graphite	10 %

Main Products : Organic Fibers: Beslon; Broadwoven Fabric Mills: Cotton; Man-made Fabric: Rayon; Animal Specialties: Raw silk

Banks : Fuji Bank; Mitsubishi Bank; Dai-Ichi Kangyo Bank; Yasuda Trust and Banking

Branches : Osaka Branch (Osaka-shi); Nagoya Office (Naka-ku)

Factories : Tokushima Factory (Itano-gun); Ogaki Factory (Ogaki-shi); Ibigawa Factory (Anpachi-gun); Kawase Factory (Hikone-shi); Mishima Factory (Sunto-gun)

Toho Titanium Co., Ltd.

Japanese Name	:	***Toho Chitaniumu K.K.***
TSE Number	:	5727
Address	:	Shinagawa NSS Bldg., 2-13-31, Konan, Minato-ku, Tokyo 108
Tel & Fax	:	Tel: 03-3458-4411 Fax: 03-3458-4431
Statistics	:	Capital : 3461 million yen Employees : 703
Officers	:	CEO : Mr. Shunichi Yashima/President Contact : Mr. Koji Suzuki/Sales Manager
Sales	:	March, 1989 - 15074 million yen March, 1990 - 18500 million yen March, 1991 - 19622 million yen
R & D Spending	:	1119 million yen
Sales Breakdown	:	

SIC#	Product	Sales
3356	Metallic titanium	70 %
3339	Titanium trichloride	20 %
2816	Titanium pigments	10 %

Banks	:	Industrial Bank of Japan; Mitsui Taiyo Kobe Bank; Nippon Credit Bank; Yokohama Bank
Factories	:	Chigasaki Factory (Chigasaki-shi, Kanagawa-ken)
Laboratories	:	R&D Laboratory (Kanagawa-ken)
Shareholders	:	Nippon Mining [5712] - 40.5% ; Mitsui & Co. [8031] - 17.9% ; Nomura International S.J.F. - 2.6% ; S.G. Warburg & Co. - 2.1% ; Government of Brunei - 2.1% ; Chuo Trust and Banking [8408] - 1.1% ; Mitsui Taiyo Kobe Bank [8314] - 1.1%

Tohoku Electric Power Co., Inc.

Japanese Name :	***Tohoku Denryoku K.K.***
TSE Number :	9506
Description :	Monopoly power supplier to Tohoku region.
Address :	3-7-1, Ichiban-cho, Aoba-ku, Sendai-shi, Miyagi-ken 980
Tel & Fax :	Tel: 022-225-2111 Fax: 022-222-2881
Statistics :	Capital : 246812 million yen Employees : 13398
Officers :	CEO : Mr. Teruyuki Akema/President Contact : Mr. Isao Wakabayashi/Manager, Research & Policy
Sales :	March, 1988 - 1037930 million yen March, 1989 - 1033790 million yen March, 1990 - 1067930 million yen March, 1991 - 1083400 million yen
Capital Investment :	282741 million yen
Sales Breakdown :	SIC# Product 4911 Power supply
Main Products :	Power Supply: Commercial power, Industrial power, Residential power
Banks :	Japan Development Bank; Industrial Bank of Japan; Long Term Credit Bank of Japan
Branches :	Aomori Branch (Aomori-shi); Iwate Branch (Morioka-shi); Akita Branch (Akita-shi); Yamagata Branch (Yamagata-shi); Niigata Branch (Niigata-shi); Tokyo Branch (Chiyoda-ku); New York Office (New York)
Factories :	One nuclear power station; 12 thermal power stations; 211 hydroelectric power stations
Laboratories :	Electricity Technology Research & Development Center (Sendai-shi); Applied Technology Research & Development Center (Sendai-shi)
Shareholders :	Nippon Life Insurance - 4.5% ; Industrial Bank of Japan [8302] - 4.2% ; Asahi Life Insurance - 3.1% ; Long Term Credit Bank of Japan [8303] - 3.1% ; Chiyoda Life Insurance - 2.3% ; Sumitomo Life Insurance - 2.1%
Subsidiaries :	Kitanihon Electric Cable Co., Ltd. (Sendai-shi); Aizu Gaishi Co., Ltd. (Sendai-shi); Higashinihon Industrial Co., Ltd. (Onuma-gun, Fukushima-ken)

Tohto Kasei Co., Ltd.

Japanese Name	:	***Toto Kasei K.K.***
TSE Number	:	Not listed
Address	:	2-15-4, Funabori, Edogawa-ku, Tokyo 134
Tel & Fax	:	Tel: 03-3686-6611 Fax: 03-3686-6624
Statistics	:	Capital : 2006 million yen Employees : 95
Officers	:	CEO : Mr. Tobei Tezuka/President Contact : Mr. Tobei Tezuka/President
Sales	:	March, 1989 - 20150 million yen March, 1990 - 20473 million yen March, 1991 - 26584 million yen

Sales Breakdown :

SIC#	Product	Sales
2821	Epoxy resins	60 %
2821	Polyamide resins	

Banks	:	Long Term Credit Bank of Japan; Fuji Bank; Norinchukin Bank; Mitsui Taiyo Kobe Bank; Sumitomo Bank
Branches	:	Osaka Office (Osaka-shi)
Factories	:	Tokyo Factory (Edogawa-ku)
Laboratories	:	Central Laboratory (Edogawa-ku)
Subsidiaries	:	Tohto Engineering Co., Ltd. (Osaka-shi); Sento Kasei Co,. Ltd. (Chiba-ken); Daito Kasei Co., Ltd. (Osaka-shi)
Overseas Subsidiaries	:	Kuk Do Chemical Industries Co., Ltd. (Korea); Jung Do Chemicals Co., Ltd. (Korea); Thai Epoxy and Allied Products Co., Ltd. (Thailand)

Tokin Corp.

Japanese Name	:	***Tokin K.K.***
TSE Number	:	6759
Description	:	Manufacturer of electronic components and materials. An affiliate of NEC Corp., Tokin is noted for its ferrite products as well as its activity in the growing market for telephone cards, other memory encoded cards and related machinery. R&D fields include YAG laser technology and related opto-electronic materials and technologies.
Address	:	2-5-8, Kita Aoyama, Minato-ku, Tokyo 107
Tel & Fax	:	Tel: 03-3402-6166 Fax: 03-3497-9756
Statistics	:	Capital : 8994 million yen Employees : 2030
Officers	:	CEO : Mr. Yukio Sato/President Contact : Mr. Tomomichi Akita/General Manager, Overseas Sales
Sales	:	March, 1988 - 61252 million yen March, 1989 - 66682 million yen March, 1990 - 67851 million yen March, 1991 - 74713 million yen
R & D Spending	:	5065 million yen
Sales Breakdown	:	see table below

SIC#	Product	Sales
367	Electronic components	43 %
3699	Electronic materials	30 %
3699	Misc. electronic products	27 %

Main Products	:	Electronic Components and Devices: Electronic coils, Transformers, Heat sensors, Magnetic sensors, Current sensors, Magnetic heads, Wave absorbers, Optical devices; Electronic Materials: Permalloys, Shape memory alloys, Ferrite cores, Ferrite magnets, Rare earth oxides, Rare earth magnets, Peizoceramics; Misc. Electronics Products: Electric switches, Magnetic card registration and reading devices, Pre-paid cards (telephone etc.) and related devices, Floppy disks, Static induction transistors (SIT) and SIT applied devices
Banks	:	Sumitomo Bank; Seventy-Seven Bank; Joyo Bank; Sumitomo Trust and Banking; Industrial Bank of Japan; Shokusan Bank
Branches	:	Osaka Branch (Chuo-ku, Osaka); Nagoya Branch (Nagoya-shi, Aichi-ken); Tsukuba Office (Tsukuba-shi, Ibaraki-ken); Kawasaki Office (Kawasaki-shi, Kanagawa-ken)
Factories	:	Suwa Factory (Sendai-shi, Miyagi-ken); Nishiura Factory (Sendai-shi, Miyagi-ken); Shiroishi Factory (Shiroishi-shi, Miyagi-ken); Hiroshima Factory (Hiroshima-shi, Hiroshima-ken)
Laboratories	:	Sendai Research Laboratories (Sendai-shi, Miyagi-ken); Tsukuba Laboratories (Tsukuba-shi, Ibaraki-ken); Tokyo Research Laboratories (Kawasaki-shi, Kanagawa-ken)
R&D Areas	:	Large-size YAG Laser Crystals and Technologies; Optoelectronics; Advanced Materials; Floppy Disks and Related Information Storage Mediums
Subsidiaries	:	Tokin Shiroishi K.K. (Shiroishi-shi, Miyagi-ken); Hokko Denshi K.K. (Minami-Akita-gun, Akita-ken); Tohoku Kakko K.K. (Shinagawa-ku, Tokyo); Tokin Shoji K.K. (Minato-ku, Tokyo)
Overseas Subsidiaries	:	Taiwan Tokin Electronics Ltd. (Taiwan); Tokin Electronics Ltd. (Hong Kong); Tokin America Inc. (California, U.S.A.); Tokin Magnetics Co., Inc. (California, U.S.A.); Tokin Europe GmbH (Germany)

Tokiwa Electric Co., Ltd.

Japanese Name	:	***Tokiwa Denki K.K.***
TSE Number	:	Not listed
Description	:	Manufacturer of industrial robots and auxiliary systems. Tokiwa is also active in the area of new materials including a new magnesium silicate paper which the company developed recently.
Address	:	1-46, Himegaoka, Shimogiri, Kani-shi, Gifu-ken 509-02
Tel	:	0574-62-6221
Sales Breakdown	:	SIC# Product 3599 Industrial robots and related products 26 Magnesium silicate papers 2819 Magnesium silicate compounds
Main Products	:	Industrial Robots and Related Products: Robot auxiliary systems; Composite Material Products: Magnesium silicate non-combustable paper

Tokuyama Soda Co., Ltd.

Japanese Name	:	***Tokuyáma Soda K.K.***
TSE Number	:	4043
Description	:	Manufacturer of a broad range of chemical products. Stepping up R&D into fine chemical products such as polysilicon resin lenses, dental materials, new ceramics and electronic materials.
Address	:	1-4-5, Nishi-shinbashi, Minato-ku, Tokyo 105
Tel & Fax	:	Tel: 03-3597-5049 Fax: 03-3597-5146
Statistics	:	Capital : 19274 million yen Employees : 2855
Officers	:	CEO : Mr. Kaoru Tsuji/President
Sales	:	March, 1988 - 134403 million yen March, 1989 - 144111 million yen March, 1990 - 153106 million yen March, 1991 - 169738 million yen
R & D Spending	:	10000 million yen Capital Investment: 16100 million yen

Sales Breakdown :

SIC#	Product	Sales
3241	Cement and related products	36 %
28	Chemicals	34 %
2821	Synthetic resins	16 %
0000	Misc. products	14 %

Main Products	:	Cement and Related Products: Portland cement, Quick drying cement, Blast furnace cement, Raw concrete, etc.; Industrial Chemicals: Soda ash, Caustic soda, Calcium chloride, Silicic acid soda, White carbon, Methylene chloride, Propylene chloride, Vinyl chloride monomer, Methylene chloride cooling agent, Epychlorohydrin; Synthetic Resins: Polypropylene, Polypropylene films, Polysilicon resins; Misc. Products: Polycrystaline silicon, Dental materials, Pharmaceutical intermediates, Agricultural chemical intermediates, Plastic monomers for use in lenses and related products, Ion exchange resins
Banks	:	Sanwa Bank; Dai-Ichi Kangyo Bank; Yamaguchi Bank; Hiroshima Bank; Sumitomo Trust and Banking; Toyo Trust and Banking; Mitsui Trust and Banking
Factories	:	Tokuyama Plant (Tokuyama-shi, Yamaguchi-ken); Nanyo Plant (Shinnanyo-shi, Yamaguchi-ken); East Plant (Tokuyama-shi, Yamaguchi-ken)
Laboratories	:	Tsukuba Research Center (Tsukuba-shi, Tsukuba-ken); Fujisawa Research Center (Fujisawa-shi, Kanagawa-ken)
R&D Areas	:	Electronics; Medical; Fine Chemicals: Plastic monomer lens material, Photochromic plastic lens materials, Photochromic materials, Agricultural chemical intermediates
Subsidiaries	:	Sun Arrow Chemical Co., Ltd. (Tokuyama-shi, Yamaguchi-ken); Tokuyama Raw Concrete Co., Ltd. (Minato-ku, Osaka); Seibu Tokuyama Raw Concrete Co., Ltd. (Tokuyama-shi, Yamaguchi-ken)
Overseas Subsidiaries	:	General Ceramics Inc. (New Jersey, U.S.A.); Figaro USA Inc. (Illinois, U.S.A.); Tokuyama America Inc.(California, U.S.A.); Tokuyama Europe GmbH. (Germany); Pornpat Chemicals Co., Ltd. (Bangkok, Thailand)

Tokyo Bi-Tech Laboratories

TSE Number	:	Not listed
Description	:	Manufacturer of metal alloy dental materials and other alloy products.
Address	:	
Tel & Fax	:	Tel: 03-3423-1219 Fax: 03-423-1948
Sales Breakdown	:	SIC# Product 3843 Metal alloy dental materials
Main Products	:	Dental Materials: Titanium alloy prosthetic dental root

Tokyo Denpa Co., Ltd.

Japanese Name	:	***Tokyo Denpa K.K.***
TSE Number	:	Not listed
Description	:	Manufacturer of industrial and scientific instruments specializing in artificial quartz crystal production technologies and applications.
Address	:	5-6-11, Chuo, Ota-ku, Tokyo 143
Tel	:	03-3751-6171
Statistics	:	Capital : 200 million yen Employees : 110
Sales	:	March, 1990 - 7720 million yen March, 1991 - 10500 million yen
Sales Breakdown	:	SIC# Product 3299 Artificial quartz crystals 3829 Measuring and controlling devices 3559 Quartz production technologies 3663 Communications equipment
Main Products	:	Measuring and Controlling Devices: Quartz oscillators, Surface acoustic wave bandpass filters, Thermal sensors; Quartz Production Technologies; Communications Equipment
Banks	:	Tokai Bank; Hokkaido Takugin Bank; Chuo Trust and Banking; Long Term Credit Bank of Japan
Factories	:	Tokyo Plant (Tokyo); Gunma Plant (Gunma-ken); Yamanashi Plant (Yamanashi-ken); Hokkaido Plant (Hokkaido); Overseas Production Facilities (Korea)
Shareholders	:	Myojo Denki; Tokai Bank [8321]; Tokyo Denpa Kiki ; Long Term Credit Bank of Japan [8303]; Chuo Trust and Banking; Chiyoda Life

Tokyo Electric Power Co., Inc.

Japanese Name	:	***Tokyo Denryoko***
TSE Number	:	9501
Description	:	Tokyo Electric is the main producer of electric power in Japan and is ranked first in the world. The company is striving to develop new power sources including nuclear reactors and has recently gained government approval to construct two new plants.
Address	:	1-1-3, Uchi-Saiwaicho, Chiyoda-ku, Tokyo 100
Tel & Fax	:	Tel: 03-3501-8111 Fax: 03-3592-1795
Statistics	:	Capital : 670410 million yen Employees : 39640
Officers	:	CEO : Mr. Sho Nasu/President
Sales	:	March, 1988 - 3940300 million yen March, 1989 - 3938500 million yen March, 1990 - 4086400 million yen March, 1991 - 4383690 million yen
Capital Investment	:	1357500 million yen

Sales Breakdown :

SIC#	Product	Sales
4911	Electrical power generation	60 %
4922	Gas transmission to residences	30 %
3443	Nuclear plants	10 %

Main Products	:	Electric Services: Electric power generation and transmission; Fabricated Plate Work: Nuclear power plants
Banks	:	Mitsui Taiyo Kobe Bank; Industrial Bank of Japan; Long Term Credit Bank of Japan; Dai-Ichi Kangyo Bank; Mitsubishi Bank; Fuji Bank

Tokyo Gas Co., Ltd.

Japanese Name	:	***Tokyo Gasu K.K.***
TSE Number	:	9531
Description	:	Tokyo Gas is Japan's largest city gas enterprise, supplying city gas to the Kanto area. Currently conducting R&D in new energy sources, communications and biotechnology.
Address	:	1-5-20, Kaigan, Minato-ku, Tokyo 105
Tel & Fax	:	Tel: 03-3433-2111 Fax: 03-3437-9130
Statistics	:	Capital : 141812 million yen Employees : 12519
Officers	:	CEO : Mr. Kunio Anzai/President
Sales	:	March, 1988 - 677621 million yen March, 1989 - 694268 million yen March, 1990 - 707168 million yen March, 1991 - 730580 million yen
R & D Spending	:	16100 million yen Capital Investment: 127611 million yen
Sales Breakdown	:	

SIC#	Product	Sales
4925	Gas	73 %
363	Gas appliances	14 %
871	Contract work	9 %

Main Products	:	Gas Production: Gas; Gas Appliances: Heaters, Air-conditioners, Central heating systems, Water heaters, Stoves, etc.; Other Products: Power transmission equipment, Coke and other by-products, Contract sevices
Banks	:	Dai-Ichi Kangyo Bank; Mitsubishi Bank; Mitsui Taiyo Kobe Bank; Long Term Credit Bank of Japan; Fuji Bank
Branches	:	6 Tokyo municipal branches; Yokohama Branch (Yokohama-shi); Kawasaki Branch (Kawasaki-shi); Fujisawa Branch (Fujisawa-shi, Kanagawa-ken); Saitama Branch (Saitama-ken); Chiba Branch (Chiba-ken)
Factories	:	Tsurumi Plant (Yokohama-shi); Toyosu Plant (Tokyo); Negishi Plant (Yokohama-shi); Kamagaya Plant (Chiba-ken)
Laboratories	:	Frontier Technology Laboratory (Yokohama-shi); Research Center (Tokyo)
R&D Areas	:	Power Generation and Transmission; New Energy Sources: Information Technologies: Biotechnology:
Overseas Subsidiaries	:	Tokyo Gas Engineering Singapore Ltd. (Singapore)

Tokyo Ohka Kogyo Co., Ltd.

Japanese Name	:	***Tokyo Ohka Kogyo K.K.***
TSE Number	:	4186
Description	:	Manufactures a range of fine chemicals and chemicals.
Address	:	Kosugi Bldg., 1-403, Kosugi-cho, Nakahara-ku, Kawasaki-shi, Kanagawa-ken 211
Tel & Fax	:	Tel: 044-722-7181 Fax: 044-733-2771
Statistics	:	Capital : 14640 million yen Employees : 1147
Officers	:	CEO : Mr. Hisashi Nakane/President
Sales	:	March, 1988 - 12422 million yen March, 1989 - 43914 million yen March, 1990 - 50497 million yen March, 1991 - 55225 million yen
R & D Spending	:	2342 million yen Capital Investment: 5949 million yen
Sales Breakdown	:	SIC# Product 289 Chemical products for electronics industry 386 Photosensitive materials 27 Printing
Main Products	:	Fine Chemical Products: Acetone, Chemical products for the electronics industry; Printing and Publishing: Printing materials; Photosensitive Materials: Photoresists
Banks	:	Mitsubishi Bank; Yokohama Bank; Mitsubishi Trust and Banking
Branches	:	Sagamihara Office (Koza-gun)
Factories	:	Kawasaki Factory (Kawasaki-shi); Aso Factory (Aso-gun); Utsunomiya Factory (Utsunomiya-shi); Gotenba Factory (Gotenba-shi); Yamanashi Factory (Minamikoma-gun); Sutama Factory (Kitakoma-gun); Kumagaya Factory (Kumagaya-shi)

Tokyo Photon Ltd.

TSE Number	:	Not listed
Description	:	Manufacturer of industrial and scientific instruments specializing in artificial quartz crystal production technologies and applications.
Address	:	Alusu Otowa II, #316, 1-22-18, Otowa, Bunkyo-ku, Tokyo 112
Tel	:	03-3941-6009

Tokyo Seimitsu Co., Ltd.

Japanese Name :	***Tokyo Seimitsu K.K.***
TSE Number :	7729
Description :	Manufacturer of measuring devices and semiconductor production machinery. Though measuring devices and related forms of industrial instrumentation still comprise the largest sales category, semiconductor production equipment has captured an increasing percentage of gross sales in recent years and it is in this area that the company is concentrating most of its R&D efforts.
Address :	9-7-1, Shimo-Renjaku, Mitaka-shi, Tokyo 181
Tel & Fax :	Tel: 0422-48-1011 Fax: 0422-45-3636
Statistics :	Capital : 1637 million yen Employees : 714
Officers :	CEO : Mr. Shozaburo Karube/President Contact : Mr. Shozaburo Karube/President
Sales :	March, 1988 - 13753 million yen March, 1989 - 19489 million yen March, 1990 - 21534 million yen March, 1991 - 23446 million yen
R & D Spending :	1105 million yen

Sales Breakdown :

SIC#	Product	Sales
382	Measuring and controlling devices	40 %
3559	Semiconductor production equipment	37 %
382	Automatic measuring instruments	19 %

Main Products :	Measuring and Controlling Devices: Cylindrical measuring devices, Electronic micrometers, Triple-axis coordinate measuring devices, Surface analysis devices; Semiconductor Production Equipment: Wafer probing devices, Wafer dicing machines, Wafer slicing machines, Micro forming machines; Automatic Measuring and Controlling Devices: Machine control guages, Automatic selection devices, Automatic assembling devices, Misc. automatic controlling devices
Banks :	Fuji Bank; Sanwa Bank; Toyo Trust and Banking; Mitsui Taiyo Kobe Bank; Kyowa Saitama Bank
Factories :	Hachioji Plant (Hachioji-shi, Tokyo); Tsuchiura Plant (Tsuchiura-shi, Ibaraki-ken)
Laboratories :	Laboratory (Hachioji-shi, Tokyo)
R&D Areas :	Industrial Machinery: Semiconductor production technology (hardware & software); Industrial Robots; Measuring and Controlling Instruments
Shareholders :	Masaaki Hayayama - 6.1% ; Makoto Takagi - 5.8% ; Nippon Seiko K.K. [5729] - 5.4% ; Fuji Bank [8317] - 5.0% ; Toyo Trust and Banking [8407] - 4.3% ; Mitsui & Co., Ltd. [8031] - 3.7%
Subsidiaries :	Tosei Engineering Corp. (Chofu-shi, Tokyo); Daiichi Seiki Co., Ltd. (Hino-shi, Tokyo); Tosei Systems Co., Ltd. (Hachioji-shi, Tokyo); Japan Physitec Co., Ltd. (Tokyo)
Overseas Subsidiaries :	Tokyo Seimitsu Europe GmbH (Germany); Tokyo Seimitsu America, Inc. (Michigan, U.S.A.)

Tokyo Tungsten Co., Ltd.

Japanese Name : ***Tokyo Tungsten K.K.***

TSE Number	:	5728
Description	:	Tokyo Tungsten is a full-fledged smelter of tungsten and molybdenum. Products are mainly for the electrical and office equipment industries. With sales sluggish, the company is attempting diversification into tungsten powder for semiconductors and is seeking collaboration with Sumitomo Electric.
Address	:	2-6-1, Kaji-cho, Chiyoda-ku, Tokyo 101
Tel & Fax	:	Tel: 03-3258-5316 Fax: 03-3256-6453
Statistics	:	Capital : 2476 million yen Employees : 567
Officers	:	CEO : Mr. Akira Arimoto/President Contact : Mr. Junji Matsubara/Overseas Section
Sales	:	March, 1988 - 10196 million yen March, 1989 - 11900 million yen March, 1990 - 12100 million yen March, 1991 - 13000 million yen

Sales Breakdown :

SIC#	Product	Sales
3313	Molybdenum and tungsten	44 %
3339	Smelted products	30 %
3341	Super-hard alloys	13 %

Main Products	:	Electrometallurgical Products: Molybdenum and tungsten products; Non-ferrous Metals: Smelted products; Secondary Non-ferrous Metals: Super-hard alloys
Banks	:	Sumitomo Trust and Banking; Sumitomo Bank; Industrial Bank of Japan; Long Term Credit Bank of Japan; Hokuriku Bank
Branches	:	Osaka Branch (Osaka); Fukuoka Office (Fukuoka-shi, Fukuoka-ken); Nagoya Office (Nagoya-shi, Aichi-ken)
Factories	:	Tokyo Factory (Tokyo); Toyama Factory (Toyama-ken)
Shareholders	:	Sumitomo Electric Industries [5802] - 50% ; Hiroshi Fukushima - 4.4% ; Sumitomo Trust and Banking [8403] - 3.7% ; Kanto Denzai - 2.7% ; Sumitomo Life Insurance - 1.7% ; Midori Sangyo - 1.1% ; Employees' Stockholding - 0.7%
Subsidiaries	:	Advanced Material Corp. (Sakata-shi, Yamagata-ken)

Tonen Chemical Corp.

Japanese Name : ***Tonen Kagaku K.K.***

TSE Number	:	Not listed
Description	:	Manufacturer of synthetic resins and related products.
Address	:	Togeki Bldg., 4-1-1, Tsukiji, Chuo-ku, Tokyo 104
Tel & Fax	:	Tel: 03-3542-7361 Fax: 03-5565-1402
Statistics	:	Capital : 4500 million yen Employees : 1540
Officers	:	CEO : Mr. Shigeru Matsumura/President
Sales	:	December, 1988 - 166321 million yen December, 1989 - 188480 million yen
Sales Breakdown	:	SIC# Product 2821 Synthetic resins
Main Products	:	Petrochemicals: Ethylene, Propylene, Butadiene, Petroleum-based benzene, iso-butylene, Toluene
Banks	:	Fuji Bank; Industrial Bank of Japan; Yasuda Trust and Baking Company; Mitsubishi Bank; Dai-Ichi Kangyo Bank; Tokai Bank; Kyowa Bank; Yokohama Bank; Suruga Bank
Branches	:	Osaka Branch (Osaka); Nagoya Branch (Nagoya-shi, Aichi-ken); Overseas Offices: Houston Office (Texas, U.S.A.)
Factories	:	Kawasaki Factory (Kawasaki-shi, Kanagawa-ku); Chidori Factory (Kawasaki-shi, Kanagawa-ku); Wakayama Factory (Arita-shi, Wakayama-ken)
Laboratories	:	General Laboratory (Iruma-gun, Saitama-ken); Technical Center (Kawasaki-shi, Kanagawa-ku)
Shareholders	:	Tonen Corp. [5005]

Toppan Printing Co., Ltd.

Japanese Name : ***Toppan Insatsu K.K.***

TSE Number : 7911

Description : The second largest printing company in Japan. The company is aggressively diversifying into electronic parts, color filters for liquid crystals, high-precision shadow masks and investing capital into technology for high definition television and IC cards. Efforts to become an integrated firm include linkages abroad with data transceiving services in the U.K. and a production plant in the U.S.. Toppan Moore, a domestic subsidiary, is also registering healthy sales.

Address : 1, Kanda-Izumicho, Chiyoda-ku, Tokyo 101

Tel & Fax : Tel: 03-3835-5111 Fax: 03-3837-7675

Statistics : Capital : 90054 million yen
Employees : 12378

Officers : CEO : Mr. Hiromichi Fujita/President

Sales : March, 1989 - 734000 million yen
March, 1990 - 781600 million yen
March, 1991 - 846200 million yen

Sales Breakdown :

SIC#	Product	Sales
2731	General publishing	50 %
2721	Periodical publishing	20 %
2621	Wrapping papers	10 %
2759	Securities and special papers	10 %
3545	Precision tools	10 %

Main Products : Publishing: Books, Periodicals, General publishing, Wrapping papers; Machine Tool Accessories: Precision tools, IC leadframes

Banks : Long Term Credit Bank of Japan; Mitsui Taiyo Kobe Bank; Mitsubishi Bank; Dai-Ichi Kangyo Bank; Sumitomo Bank

Branches : Precision Electronics Sales Branch (Kawagoe); Kyushu Branch (Fukuoka-shi); Sengoku Sales Branch (Hiroshima-shi); Shikoku Sales Branch (Takamatsu-shi); Hokkaido Sales Branch (Sapporo-shi)

Factories : Kannabe Plant (Hiroshima); Nagoya Plant (Nagoya-shi)

Subsidiaries : Toppan Moore K.K.

Toray Industries, Inc.

Japanese Name	:	***Toray K.K.***
TSE Number	:	3402
Description	:	Toray is a top manufacturer of synthetic fibers, ranking first in artificial leather and polyester films. The company is rapidly diversifying into reverse osmosis membranes, engineering plastics, and pharmaceuticals. New products include a Polyacrylonitrile (PAN)-based carbon fiber used in the medical and aerospace industries and a series of medical products, including CA72-4 Cancer Diagnostic Kit, 'Plasmax' — a plasma separator, Natural Beta-Interferon — a drug treatment against tumors and viral diseases and 'Filtryzer' — a PMMA fiber membrane artificial kidney. The company recently developed an organic optical material which generates green laser light.
Address	:	2-2-1, Nihonbashi-Muromachi, Chuo-ku, Tokyo 103
Tel & Fax	:	Tel: 03-3245-5111 Fax: 03-3245-5555
Statistics	:	Capital : 94734 million yen Employees : 10047
Officers	:	CEO : Mr. Katsunosuke Maeda/President
Sales	:	March, 1988 - 541511 million yen March, 1989 - 553069 million yen March, 1990 - 549140 million yen March, 1991 - 585434 million yen
R & D Spending	:	30100 million yen Capital Investment: 47700 million yen

Sales Breakdown :

SIC#	Product	Sales
2823	Synthetic fiber 'Tetron'	31 %
3081	Synthetic films	17 %
2823	Nylon	16 %
2821	Resins	11 %
0000	New businesses	11 %
28	Chemicals	7 %
2823	'Essene' and other synthetic resins	6 %
2823	'Toraylon'	3 %

Main Products	:	Fibers: Nylon, Polyester, Acrylic fibers, Non-woven fabrics, Artificial leathers, CAD/CAM systems for garment manufacture; Chemicals: Nylon, Polyamide resin, ABS resin, PBT resin, PPS resin, Polybutylene terephthalate, Polyester films 'Lumirror', Polypropylene films, Polyolefin foam, Synthetic fiber raw materials, Plastics raw materials, Essential amino acids and their derivatives, Pharmaceutical intermediates, Agricultural intermediates, Fiber anti-static agent, Marine pollution absorbtive agent, Textile finishing agents, Terephthalic acid, Xylene, Resins for paints, Industrial adhesives, Other specialty chemicals; Medical Products: Artificial kidney 'Filtryzer', Plasma separator 'Plasmax', HPV DNA recombination kit, Natural interferon-beta 'Feron', Monoclonal antibody cancer diagnostic kit 'CA72-4', Catheters, Other pharmaceuticals; Materials: Electronic and imaging materials, Electron beam-resistant material 'EBR-9', Optical fibers, Building materials, Optical products, Fine ceramics, (polyacrylonitrile) PAN - based carbon fiber, Functional membranes; Other Equipment: Environmental control equipment, Liquid crystal displays, Lens cleaners, Computer peripheral equipment
Banks	:	Mitsui Taiyo Kobe Bank; Dai-Ichi Kangyo Bank; Long Term Credit Bank of Japan; Mitsui Trust and Banking
Branches	:	Osaka Head Office (Osaka-shi); Nagoya Branch (Nagoya-shi); Hokuriku Branch (Fukui-shi); Kyushu Branch (Fukuoka-shi); Tohoku Branch (Sendai-shi); Seoul Office; Beijing Office; European Headquarters (London); U.K. Branch (Nottingham)

Toray Research Center, Inc.

Japanese Name	:	***Tore Risachi Senta K.K.***
TSE Number	:	Not listed
Description	:	Toray Research Center has established its reputation in chemical analysis and R&D. The company has recently entered the bio field with such services as DNA base sequence analysis.
Address	:	3-1-8, Nihonbashi-Muromachi, Chuo-ku, Tokyo 103
Tel & Fax	:	Tel: 03-3245-5665 Fax: 03-3245-5804
Statistics	:	Capital : 250 million yen Employees : 310
Officers	:	CEO : Mr. Fumio Nishikawa/President Contact : Mr. Fumio Nishikawa/President
Sales	:	March, 1988 - 3624 million yen March, 1989 - 4264 million yen March, 1990 - 4431 million yen March, 1991 - 4864 million yen
Sales Breakdown	:	SIC# Product 8731 Chemical analysis 3823 Computerized analysis systems
Main Products	:	Analytical Services: Chemical analysis, DNA base sequencing analysis; Computer Systems: Computerized analysis systems
Banks	:	Mitsui Taiyo Kobe Bank; Dai-Ichi Kangyo Bank; The Industrial Bank of Japan
Branches	:	Tokyo Sales Office (Chuo-ku); Kansai Sales Office (Otsu-shi); System Sales Office (Urayasu-shi, Chiba-ken)
Laboratories	:	Shiga Research Division (Otsu-shi, Shiga-ken); Kamakura Research Division (Kamakura-shi, Kanagawa-ken); Nagoya Research Division (Nagoya-shi, Aichi-ken)
Shareholders	:	Toray Industries Inc. - 100%

Toshiba Ceramics Co., Ltd.

Japanese Name	:	***Toshiba Seramikkusu K.K.***
TSE Number	:	5213
Address	:	Shinjuku Nomura Bldg., 1-26-2, Nishi-Shinjuku, Shinjuku-ku, Tokyo 163
Tel & Fax	:	Tel: 03-3348-7411 Fax: 03-3343-8627
Statistics	:	Capital : 17851 million yen Employees : 2022
Officers	:	CEO : Mr. Kiyohiko Kasuya/President
Sales	:	March, 1988 - 48579 million yen March, 1989 - 53356 million yen March, 1990 - 56116 million yen March, 1991 - 59309 million yen
R & D Spending	:	3207 million yen

Sales Breakdown :

SIC#	Product	Sales
3674	Silicon wafers	31 %
3295	Quartz glass	27 %
3255	Refractories	20 %
2819	Carbon and silicon carbide	11 %

Banks	:	Mitsui Taiyo Kobe Bank; Mitsui Trust and Banking; Tokai Bank; Kyowa-Saitama Bank
Branches	:	Osaka Branch (Osaka); Nagoya Branch (Nagoya-shi, Aichi-ken)
Factories	:	Oguni Manufacturing Facility (Nishiokitama-gun); Hadano Plant (Hadano-shi); Kariya Plant (Kariya-shi); Togane Plant (Togane-shi)
Laboratories	:	Research and Development Center (Hadano-shi)
R&D Areas	:	Photomasks for liquid crystal display units; SiC coated carbon fiber-reinforced carbon; Refractories for the refining of steel; Super-high temperature electric furnaces; Silicon wafer evaluation technology
Subsidiaries	:	Toshiba Monoflax; Kawatana Toshiba Ceramics; Tokuyama Ceramics; Tokai Ceramics; Tocera Engineering; Kyushu Ceramics; Niigata Toshiba Ceramics
Overseas Subsidiaries	:	Toshiba Ceramics America; Toshiba Ceramics Europe; Quartz International

Toshiba Chemical Corp.

Japanese Name	:	***Toshiba Kagaku K.K.***
TSE Number	:	4232
Description	:	Manufactures a range of plastic sheeting and chemical products, and related production machinery.
Address	:	3-3-9, Shimbashi, Minato-ku, Tokyo 105
Tel & Fax	:	Tel: 03-3502-3212 Fax: 03-3501-9876
Statistics	:	Capital : 10172 million yen Employees : 1033
Officers	:	CEO : Mr. Hideo Osawa/President
Sales	:	March, 1988 - 39188 million yen March, 1989 - 40538 million yen March, 1990 - 45257 million yen March, 1991 - 47638 million yen
R & D Spending	:	1100 million yen Capital Investment: 6000 million yen
Sales Breakdown	:	SIC# Product 3083 Laminated sheets 3089 Resin products 3644 Insulating materials 3544 Molds 354 Machinery
Main Products	:	Laminated Sheets: Sheeting for use in electronic manufacturing, Industrial use laminated sheets, Consumer electronics-use laminated sheets; Non-current Carrying Materials: Insulating materials; Plastics Products: Synthetic resin products; Molding Equipment: Metal molds; Metalworking Machinery
Banks	:	Mitsui Taiyo Kobe Bank; Tokai Bank; Kyowa Bank
Branches	:	Chubu Branch (Nagoya-shi); Kansai Branch (Osaka-shi)
Factories	:	Kawaguchi Factory (Kawaguchi-shi); Chidorimachi Factory (Kawasaki-shi)
Laboratories	:	Technical Development Center (Kawasaki-shi)
Shareholders	:	Toshiba Corp. [6502]-
Overseas Subsidiaries	:	Toshiba Chemical America, Inc. (California, U.S.A.)

Toshiba Corp.

Japanese Name : ***K.K. Toshiba***

TSE Number : 6502

Description : Toshiba is a major world electronics producer ranking second in Japan behind Hitachi. Through extensive R&D the company is advancing into high value-added information-related products. Toshiba is expanding into the communications field with overseas tieups with major foreign firms including Siemens, Olivetti, AT&T and recently, IBM. The company will be a major competitor in the development and marketing of HDTV and broadcast satellites.

Address : Toshiba Bldg., 1-1-1, Shibaura, Minato-ku, Tokyo 105-01

Tel & Fax : Tel: 03-3457-4511 Fax: 03-3456-1631~3

Statistics : Capital : 271834 million yen
Employees : 162000

Officers : CEO : Mr. Joichi Aoi/President

Sales : March, 1988 - 2682780 million yen
March, 1989 - 2921500 million yen
March, 1990 - 3060900 million yen
March, 1991 - 3227700 million yen

R & D Spending : 265300 million yen Capital Investment: 249800 million yen

Sales Breakdown :

SIC#	Product	Sales
357	Information/communication devices	57 %
3629	Heavy electrical apparatus	22 %
36	Consumer electronics	21 %

Main Products : Information Processing Equipment and Systems: Computers Workstations, Optical document filing systems, Word processors, Printers, Hard disk drives, Optical character readers, Plain-paper copiers, IC cards, CD-ROM drives; Telecommunications Equipment: Digital private branch exchanges, Key telephones, Packet switching systems, Virtual circuit switches, LAN systems, Master control panels, Radio and TV transmitters, Color cameras, Open array antennas, Radar systems, Airport systems, Facsimilie machines; Control Systems: Distribution and banking systems, Securities trading systems, Government systems, Industrial systems, Control and instrumentation systems; Medical Systems: Computed tomography scanners, Ultrasound and X-ray diagnostic equipment, Magnetic resonance imaging systems, Nuclear medicine equipment; Automated Equipment: Automatic fare collection equipment, Letter and package sorting systems, Banking machines, POS systems; Semiconductors: Advanced memories, Microprocessors, Gate arrays, Standard cells, Logic ICs, Bipolar ICs, Power devices, Diodes, LEDs, Transistors, Laser diodes, Charge-coupled devices; Electron Tubes: Color picture tubes, Display tubes, Magnetron tubes; Other Electronic Components: Printed circuit boards, Surface acoustic wave devices, Sensors, Liquid crystal displays; Power Plant Systems: Nuclear equipment, Thermal equipment, Geothermal equipment, Hydroelectric equipment, Computer controlled transmission systems, Fuel cells; Industrial Electrical Apparatus: Industrial control equipment, Indication/synchronous/DC motors, Process control instruments, Power transformers, Voltage regulators, Frequency converters; Transportation Equipment: Mass transportation control systems, Electric locomotives, Electrical equipment for railway coaches, Railway substation equipment, Elevator equipment, Escalator equipment; Machinery: Machine tools, Precision machinery, Semiconductor manufacturing equipment, Plastic-fabricating

		machines, Die-casting equipment; Video Products: Video cassette recorders and cameras, Camcorders, Video disc players, Color televisions, Digital still cameras, High-definition TV systems; Audio Products: Stereo systems, Compact disc players, Radio cassette recorders, Headphone type compact stereo cassette players; Household Products: Room air-conditioners, Heaters, Electric fans, Washing machines, Clothes dryers, Vacuum cleaners, Refrigerators, Microwave ovens, Food processors, Juicers, Blenders, Toasters, Rice cookers, Meat grinders; Lighting Equipment: Incandescent and fluorescent bulbs, Lamps, Lamp fixtures; Materials: Steel tubing, Alloys, Quartz glass, Ceramics, Insulating materials, Resins, Magnetic and Non-magnetic materials, Silicone, Tungsten and other molybdenum products; Other Products: Optical equipment, Lightwave telemeters, Clinical thermometers
Banks	:	Mitsui Taiyo Kobe Bank; Tokai Bank; Long Term Credit Bank of Japan; Industrial Bank of Japan; Mitsui Trust and Banking
Factories	:	Ome Factory (Ome-shi); Fuchu Factory (Fuchu-shi); Yanagimachi Factory (Kwasaki-shi); Hino Factory (Hino-shi); Nasu Factory (Otawara-shi, Tochigi-ken); Fukaya Factory (Fukaya-shi) and 20 other domestic factories
Laboratories	:	Research and Development Centre (Kawasaki-shi); Systems Engineering Center (Kawasaki-shi); Environmental Engineering Laboratory (Kawasaki-shi) and 12 other domestic laboratories; 1 overseas laboratory

Toshiba Tungaloy Co., Ltd.

Japanese Name :	***Toshiba Tangaroi K.K.***
TSE Number :	6139
Address :	1-7, Tsukagoshi, Saiwai-ku, Kawasaki-shi, Kanagawa-ken 210
Tel & Fax :	Tel: 044-548-8704 Fax: 044-548-8795
Statistics :	Capital : 10183 million yen Employees : 1753
Officers :	CEO : Mr. Kenichiro Ando/President Contact : Mr. Kozo Ohara/Manager, Overseas Dept.
Sales :	March, 1988 - 30641 million yen March, 1989 - 36743 million yen March, 1990 - 40527 million yen March, 1991 - 44753 million yen
R & D Spending :	1746 million yen Capital Investment: 7305 million yen

Sales Breakdown :

SIC#	Product	Sales
3313	Super-hard metal alloys	42 %
3541	Cutting tools	31 %
3541	Wear-resistant tools	11 %
3291	Friction materials and parts	6 %
3531	Construction tools	3 %

Banks :	Mitsui Taiyo Kobe Bank; Kyowa Bank; Yokohama Bank
Branches :	Tokyo Branch (Chiyoda-ku); Osaka Branch (Osaka-shi); Nagoya Branch (Nagoya-shi); Fukuoka Branch (Fukuoka-shi)
Shareholders :	Toshiba [6502] - 40.15% ; Mitsubishi Trust and Banking [8402] - 3.95% ; Sumitomo Trust and Banking [8403] - 3.28% ; Mitsui Trust and Banking [8402] - 2.44% ; Chuo Trust and Banking [8407] 2.14% ; Toyo Trust and Banking [8407] - 2.04%
Overseas Subsidiaries :	Tungaloy Singapore (Pte) Ltd. (Singapore); Toshiba Tungaloy (Europe) GmbH

Tosoh Corp.

Japanese Name	:	***Tosoh K.K.***
TSE Number	:	4042
Description	:	Major vertically-integrated chemical manufacturer and front runner in domestic production of chlorine and caustic soda. Also produces cement, which it sells through Onoda Cement, and polyolefins, which it sells through the Union Polymer Co.
Address	:	1-7-7, Akasaka, Minato-ku, Tokyo 107
Tel & Fax	:	Tel: 03-3585-6707 Fax: 03-3588-8129
Statistics	:	Capital : 40591 million yen Employees : 5069
Officers	:	CEO : Mr. Toshiaki Yamaguchi/President Contact : Mr. Masahiro Shimizu/International Business Division
Sales	:	March, 1988 - 210254 million yen March, 1989 - 226697 million yen March, 1990 - 237570 million yen March, 1991 - 325903 million yen
R & D Spending	:	16900 million yen Capital Investment: 21111 million yen

Sales Breakdown :

SIC#	Product	Sales
2821	Olefins	19 %
2821	Polyolefins	15 %
281	PVC and rubber	13 %
2819	Inorganic chemicals	13 %
28	Fine chemicals	12 %
3241	Cement	8 %
2869	Organic chemicals	6 %
33	Electrolytic metals	5 %
33	Materials	4 %
382	Analytical equipment	3 %

Main Products	:	Petrochemicals Division: Olefins, Polyolefins, Petroleum-based benzene; Basic Chemicals Division: Chlorine, Caustic soda, Other inorganic chemicals, Organic chemicals, Chemical fertilizers, Polyvinyl chloride, Rubber, Cement; Specialty Chemicals Division: Epoxy resin curing agent, Ethylene-vinyl acetate copolymer, Halogen-based flame retardant, Styrene monomer, Toluene, Vinyl chloride monomer, Xylene, Electrolytes, Inorganic fine chemicals, Organic fine chemicals, Analytical equipment, Functional materials, Electronic materials
Banks	:	Industrial Bank of Japan; Norinchukin Bank; Tokai Bank; Sumitomo Trust and Banking; Fuji Bank; Yamaguchi Bank; Chuo Trust and Banking
Branches	:	Osaka Offices (Kita-ku and Taisho-ku, Osaka); Nagoya Office (Nagoya-shi, Aichi-ken); Fukuoka Office (Fukuoka-shi, Fukuoka-ken); Sendai Office (Sendai-shi, Miyagi-ken); Tsukuba Office (Tsukuba-shi, Ibaraki-ken)
Laboratories	:	Tokyo Research Center (Ayase-shi, Tokyo); Chemistry Laboratory (Shinnanyo-shi, Yamaguchi-ken); High Polymer Laboratory (Yokkaichi-shi, Mie-ken); Petrochemical Laboratory (Yokkaichi-shi, Mie-ken); Tsukuba Laboratory (Tsukuba-shi, Ibaraki-ken)

Toto Ltd.

Japanese Name	:	***Toto Kiki K.K.***
TSE Number	:	5332
Address	:	2-1-1, Nakashima, Kokurakita-ku, Kitakyushu-shi, Fukuoka-ken 802
Tel & Fax	:	Tel: 093-951-2707 Fax: 093-922-6789
Statistics	:	Capital : 35264 million yen Employees : 10457
Officers	:	CEO : Mr. Yoshine Koga/President Contact : Mr. Masaaki Sakamoto/Manager, International Sales
Sales	:	November, 1988 - 293099 million yen March, 1989 (4 months) - 116243 million yen March, 1990 - 345725 million yen March, 1991 - 394640 million yen
R & D Spending	:	11267 million yen Capital Investment: 27055 million yen
Sales Breakdown	:	

SIC#	Product	Sales
3261	Sanitaryware and metal fittings	46 %
3088	Plastic bathroom and kitchen fixtures	28 %
3088	Toilet seats	22 %
326	Bathroom tiles and other accessories	4 %

Banks	:	Sumitomo Bank; Mitsubishi Bank; Dai-Ichi Kangyo Bank; Fuji Bank; Daiwa Bank; Tokai Bank; Sawa Bank; Mitsui Taiyo Kobe Bank
Laboratories	:	Laboratory (Chigasaki-shi)
R&D Areas	:	Production Technologies: FA technologies, CIM (computer integrated manufacturing) technologies, CAD design systems; Consumer Products: Bathroom fixtures, Tiles, etc.; Ceramics: New ceramics
Overseas Subsidiaries	:	Toto Industries (Atlanta), Inc. (Georgia, U.S.A.); Toto Kiki U.S.A., Inc. (California, U.S.A.); Toto Gerate GmbH. (Germany); Toto Kiki (H.K.) Ltd. (Hong Kong)

Toyo Engineering Corp.

Japanese Name :	***Toyo Enjiniaringu Kooporeishon***
TSE Number :	6330
Description :	One of Japan's top three plant engineering firms. The company exports its products to over 40 countries, and is known for its advanced urea technology.
Address :	Kasumigaseki Bldg., 3-2-5, Kasumigaseki, Chiyoda-ku, Tokyo 100
Tel & Fax :	Tel: 03-3592-7411 Fax: 03-3593-0749
Statistics :	Capital : 12219 million yen Employees : 1487
Officers :	CEO : Mr. Uzuhiko Uwatoko/President
Sales :	March, 1989 - 117371 million yen March, 1990 - 105905 million yen March, 1991 - 142428 million yen

Sales Breakdown :

SIC#	Product	Sales
162	Petrochemical plant	73 %
162	Factory automation	11 %
162	Petroleum refining plant	6 %
162	Chemical fertilizer plant	4 %
162	Nuclear power plant	2 %

Main Products :	Plant Engineering: Petrochemical plant, Oil and gas refining plant, Factory automation, Fertilizer plant, Nuclear power plant
Banks :	Mitsui Taiyo Kobe Bank
Branches :	Peking Branch (China); Berlin Branch (Germany); Bangkok Branch (Thailand); Jakarta Branch (Indonesia); Moscow Branch (U.S.S.R)
Laboratories :	Engineering Center (Narashino-shi); Research Center (Mobara-shi)
Shareholders :	Mitsui Toatsu Chemicals [4001] - 22.8% ; Mitsui & Co. [8031] - 17%
Overseas Subsidiaries :	Subsidiaries in New York; London; Bangkok; Bombay; Luxembourg; The Bahamas; Kuala Lumpur; Seoul; Hong Kong

Toyo Tanso Co., Ltd.

TSE Number	:	Not listed
Description	:	Engaged in the development and manufacture of graphite materials and non-ferrous alloys.
Address	:	10-1, Nihonbashi, Chuo-ku, Tokyo 103-91
Tel	:	03-3814-7561
Sales Breakdown	:	SIC# Product 3299 Graphite composites and other materials 3339 Non-ferrous alloys
Main Products	:	Graphite Materials and Related Products: Pure (99.9999%) graphite, Graphite sheets, Graphite material for use in fission reactors, Heat resistant carbon materials; Non-ferrous Alloys: Copper alloys

Toyo Tire & Rubber Co., Ltd.

Japanese Name	:	***Toyo Gomu Kogyo K.K.***
TSE Number	:	5105
Description	:	A major manufacturer of tires for automobiles, trucks and buses. Also involved in the production of industrial chemicals.
Address	:	1-17-8, Edobori, Nishi-ku, Osaka 550
Tel & Fax	:	Tel: 06-441-8801 Fax: 06-445-2225
Statistics	:	Capital : 22275 million yen Employees : 4000
Officers	:	CEO : Mr. Kenichi Katori/President Contact : Mr. Yuji Hosoda/General Manager
Sales	:	March, 1988 - 180832 million yen March, 1989 - 194384 million yen March, 1990 - 206673 million yen March, 1991 - 219688 million yen
R & D Spending	:	8300 million yen
Sales Breakdown	:	SIC# Product 3011 Tires 3069 Fabricated rubber products 28 Industrial chemicals
Main Products	:	Automotive Tires: Tires for automobiles, Tubes for automobiles; Fabricated Rubber Products: Industrial rubber products
Banks	:	Sanwa Bank; Long Term Credit Bank of Japan; Toyo Trust and Banking; Daiwa Bank; Mitsui Taiyo Kobe Bank
Branches	:	Sapporo Office (Sapporo-shi); Sendai Office (Sendai-shi); Tokyo Office (Shibuya-ku); Nagoya Office (Nagoya-shi); Osaka Office (Osaka-shi); Hiroshima Office (Hiroshima-shi); Fukuoka Office (Fukuoka-shi)
Factories	:	Itami Plant (Itami-shi); Sendai Plant (Iwanuma-shi); Hyogo Plant (Kako-gun); Fukushima Plant (Fukushima-shi)
Laboratories	:	Merchandise Development Center (Ibaraki-shi)
Shareholders	:	Toyobo Co. [5105] - 6.7% ; Toyota Motor Corp. [7203] - 4.7% ; Sanwa Bank [8320] - 4.3% ; Nippon Life - 3.9% ; Long Term Credit Bank of Japan [8303] - 3.6% ; Japan Securities Clearing - 3.3% ; Nissan Motor Co., Ltd. [7201] - 2.8%
Subsidiaries	:	Fukushima Rubber Co., Ltd.; Ryoto Tire Co., Ltd.; Toyo Soflan Co., Ltd.; Chubu Soflan Co., Ltd.; Saitama Seibu Toyo Rubber Co., Ltd.; T.G.K. Co., Ltd.
Overseas Subsidiaries	:	Toyo Tire (U.S.A) Corp. (U.S.A.); Toyo Reifen GmbH (Germany); Toyo Tyre Australia Ltd.; Toyo Tire Canada Inc. (Canada)

Toyobo Co., Ltd.

Japanese Name	:	***Toyo Bosei K.K.***
TSE Number	:	3101
Description	:	Major comprehensive textile manufacturer producing a broad range of synthetic and natural textile products. Toyobo Co. has a long history as an innovator in its field and has expanded its business operations to encompass a number of product lines outside of textiles. A leading supplier of plastic films for packaging and industrial use, the company has also established itself in the field of engineering plastics, carbon fibers, reverse-osmosis membranes and medical materials. The company has targeted pharmaceuticals as the next area for expansion.
Address	:	2-2-8, Dojimahama, Kita-ku, Osaka 530
Tel & Fax	:	Tel: 06-348-3111 Fax: 06-348-3192
Statistics	:	Capital : 43341 million yen Employees : 8360
Officers	:	CEO : Mr. Saburo Takizawa/President
Sales	:	March, 1988 - 298378 million yen March, 1989 - 315594 million yen March, 1990 - 328751 million yen March, 1991 - 338515 million yen
R & D Spending	:	9700 million yen Capital Investment: 18600 million yen

Sales Breakdown :

SIC#	Product	Sales
22	Synthetic fibers and textiles	47 %
2211	Cotton fibers and textiles	15 %
3081	Films	13 %
22	Textile secondary products	10 %
28	Functional and biomedical products	9 %
22	Wool	4 %

Main Products	:	Natural Fibers and Textile Goods: Cotton and cotton blends, Wool and woolen blends, Sportswear and other apparel; Synthetic Fibers and Textile Goods: Polyester staple fiber, Polyester yarns and fabrics, Nylon filament yarns and fabrics, Acrylic staple fiber, Acrylic yarns and fabrics, Protein-grafted fibers, Protein-grafted filament yarns and fabrics, Polynosic staple fiber, Polynosic yarns and fabrics; Films: Polypropylene film, PET films, Nylon film; Functional and Biomedical Products: Thermoplastic copolyester resin, Glass-fiber-reinforced nylon resin, Glass-fiber-reinforced PET resin, Thermosetting polyester resin, Polyester elastomer, Photosensitive nylon water-developable printing plates, ACF (activated carbon fiber) felts and papers, ACF solvent recovery equipment, ACF pollution control equipment, ACF electret filters, Hollow fiber reverse-osmosis modules, Holllow fiber artificial kidney units, Flexible copper-clad laminate and UV-curable inks and pastes, Enzymes, growth factors, Restriction enzymes, Diagnostic kits, Enzyme immunoassay kits, Ventricular assist system, Autologous transfusion system, Alkali protease
Banks	:	Dai-Ichi Kangyo Bank; Mitsubishi Bank; Sumitomo Bank; Long Term Credit Bank of Japan; Industrial Bank of Japan; Fuji Bank; Sanwa Bank
Branches	:	Tokyo Branch (Chuo-ku, Tokyo); Nagoya Branch (Nagoya-shi, Aichi-ken); Overseas Offices: Boston Office (U.S.A.); San Francisco Office (U.S.A.); Paris Office (France); Hong Kong Office (Hong Kong)
Laboratories	:	Research Institute (Otsu-shi, Shiga-ken)

Toyoda Automatic Loom Works

Japanese Name	:	***Toyoda Jido Shokki Seisakusho***
TSE Number	:	6201
Description	:	Manufacturer of textiles and industrial machinery, automobiles and automobile parts. The founding member of the Toyota group of companies, Toyoda Automatic Loom now earns close to half of its income from producing parts and assembling automobiles for its affiliate, Toyota Motor Corp.
Address	:	2-1, Toyoda-cho, Kariya-shi, Aichi 448
Tel & Fax	:	Tel: 0566-22-2511 Fax: 0566-23-3255
Statistics	:	Capital : 30895 million yen Employees : 7866
Officers	:	CEO : Mr. Yoshitoshi Toyoda/President
Sales	:	March, 1988 - 367828 million yen March, 1989 - 437443 million yen March, 1990 - 491514 million yen March, 1991 - 583686 million yen
R & D Spending	:	16756 million yen

Sales Breakdown :

SIC#	Product	Sales
3711	Automobiles and automobile parts	47 %
3537	Industrial vehicles	23 %
3563	Air compressors	20 %
3552	Textile machinery	10 %

Main Products	:	Automobiles and Automobile Parts: Car assembly, Car engines, Air compressors for use in car air conditioners, Misc. car parts; Industrial Vehicles: Fork lifts, Shovel loaders, Tractors; Textile Machinery: Spinning machines, Ring and open ended precision spinning machines, Air jet looms
Banks	:	Tokai Bank; Mitsui Taiyo Kobe Bank; Sumitomo Bank; Daiwa Bank; Fuji Bank; Bank of Tokyo; Kyowa Saitama Bank; Sanwa Bank
Branches	:	Tokyo Branch (Chiyoda-ku, Tokyo); Osaka Branch (Osaka)
Factories	:	Kariya Plant (Kariya-shi, Aichi-ken); Nagakusa Plant (Obu-shi, Aichi-ken); Kyowa Plant (Obu-shi, Aichi-ken); Takahama Plant (Takahama-shi, Aichi-ken); Hekinan Plant (Hekinan-shi, Aichi-ken)
R&D Areas	:	Information Processing Technology; Textile Machinery
Overseas Subsidiaries	:	Toyoda Industrial Equipment Manufacturing Inc. (Indiana, U.S.A.); Michigan Automotive Compressor Inc. (Michigan, U.S.A.); Toyoda Textile Machinery Inc. (North Carolina, U.S.A.)

Toyota Central Research and Development Laboratories Inc.

TSE Number	:	Not listed
Description	:	Engaged in R&D activities for parent company Toyota Motor Corp.
Address	:	41-1, Nagakute-cho, Oaza, Yokomichi, Aichi-gun, Aichi-ken 480-11
Tel & Fax	:	Tel: 0561-62-6111 Fax: 0561-62-6245
Sales Breakdown	:	SIC# Product 8731 Research and development
Main Products	:	Research and Development Activities: Bio-engineering, New materials, Other (tantalum pentoxide films etc.)
R&D Areas	:	Automotive Engineering and Materials
Shareholders	:	Toyota Motor Corp. [7203]

Toyota Motor Corp.

Japanese Name :	***Toyota Jidosha K.K.***
TSE Number :	7203
Description :	Major integrated car manufacturer with a worldwide distribution and production network. Ranked number one in Japan, with a 40% market share, and second in the world in terms of units manufactured. Toyota has production facilities in Canada, U.S., Mexico and Europe, and initiated a joint venture with General Motors in the U.S. and with Volkswagen in Germany. While car sales in the U.S. have been levelling off, sales of automotive parts have increased. The company is aggressively moving into the European market to uplift sales that have slowed in the U.S. and Japan, and is continually upgrading production processes including the well-known, Just-in-Time, production system.
Address :	1, Toyota-cho, Toyota City, Aichi-ken 471
Tel & Fax :	Tel: 0565-28-2121 Fax: 0565-23-5721
Statistics :	Capital : 254400 million yen Employees : 69071
Officers :	CEO : Mr.Shoichiro Toyoda/President
Sales :	June, 1989 - 7190600 million yen June, 1990 - 7999000 million yen June, 1991 - 8600000 million yen

Sales Breakdown :

SIC#	Product	Sales
3711	Cars	50 %
3714	Vehicle parts and supplies	33 %
3713	Trucks	10 %
3444	Construction materials	5 %
6411	Car insurance	5 %

Main Products :	Motor Vehicles: Passenger cars, Sports cars, Trucks, Jeeps, Motorcycles, Vehicle parts and supplies; Sheet Metalwork: Construction materials; Insurance Agents: Car insurance
Banks :	Mitsui Taiyo Kobe Bank; Tokai Bank; Sanwa Bank; Long Term Credit Bank of Japan
Branches :	Osaka Sales Branch (Osaka-shi); Yokosuka Branch (Kanagawa-ken); Yokohama Sales Branch (Yokohama-shi)
Factories :	Motomachi Plant (Toyota-shi); Takaoka Plant (Toyota-shi); Tohara Plant (Toyobashi-shi)

Tsuda Kenkyusho

TSE Number	:	Not listed
Description	:	Independent research laboratory engaged in the development of non-ferrous alloys and related metals materials.
Address	:	
Tel	:	0265-52-5994
Sales Breakdown	:	SIC# Product 8731 Research and development 3339 Non-ferrous alloys and related materials
Main Products	:	Nonferrous Alloys: Molybdenum plating for aluminum

Tsutsunaka Plastic Industry Co., Ltd.

Japanese Name : ***Tsutsunaka Purasuchikku Kogyo K.K.***

TSE Number : 4225

Description : Engaged in the manufacture of plastic boards for lighting equipment and signboards. Production of resin boards and water treatment systems is being intensified. The company is a member of the Sumitomo group and has a 30% share of the domestic PVC board market.

Address : Nihon Itagarasu Bldg., 3-5-11, Dosho-machi, Chuo-ku, Osaka 541

Tel & Fax : Tel: 06-229-5000 Fax: 06-229-5091

Statistics : Capital : 9635 million yen
Employees : 786

Officers : CEO : Mr. Takuji Akagaki/President
Contact : Mr. Tetsuo Inaba/Manager, Overseas Divison

Sales : March, 1989 - 34384 million yen
March, 1990 - 37378 million yen
March, 1991 - 41920 million yen\

Sales Breakdown :

SIC#	Product	Sales
2821	Plastic boards	35 %
2821	Styrene boards	25 %
2821	PVC resin boards	20 %
2899	Water treatment products	15 %
2911	High-density polyethylene	5 %

Main Products : Plastic Boarding: Hard PVC board, Styrene board, Polymer alloy, Polycarbonate board; Environmental Hygiene Products: Water treatment equipment, Water treatment chemicals; Resin Products: Waterproofing compounds for sheeting, High density polyethylene

Banks : Sumitomo Bank; Sumitomo Trust and Banking; Long Term Credit Bank of Japan; Joyo Bank; Ashikaga Bank; Bank of Kyoto; Hokkaido Takushoku Bank

Branches : Tokyo Branch (Bunkyo-ku); Nagoya Branch (Nagoya-shi); Fukuoka Branch (Fukuoka-shi)

Factories : Osaka Factory (Kashiwara-shi); Kanto Factory (Kanuma-shi); Nara Plant (Nara-shi)

Laboratories : Osaka Laboratory (Kashiwara-shi)

Shareholders : Sumitomo Bakelite [4203] - 26.0%; Nippon Sheet Glass [5202] - 14.2%; Sumitomo Trust [8403] - 5.2%; Sumitomo Bank [8318] - 3.3%; Long Term Credit Bank of Japan [8303] - 2.4%

Subsidiaries : Tsutsunaka Shiito Bosui K.K.; Kyodo Purasuchikku K.K.; Chubu Jushi Hanbai K.K.; K.K. Kyodo

Ube Industries, Ltd.

Japanese Name	:	***Ube Kosan K.K.***
TSE Number	:	4208
Description	:	All-round maker of a wide range of chemical, raw material and related products. The company has diversified into pharmaceuticals and is a leading manufacturer of nylon materials.
Address	:	Ark Mori Bldg., 1-12-32, Akasaka, Minato-ku, Tokyo 107
Tel & Fax	:	Tel: 03-3505-9271 Fax: 03-3505-9277
Statistics	:	Capital : 41805 million yen Employees : 7191
Officers	:	CEO : Mr. Motoo Nakahigashi/President Contact : Mr. Tetsuya Hironaka/General Manager, International
Sales	:	March, 1988 - 364989 million yen March, 1989 - 402887 million yen March, 1990 - 426190 million yen March, 1991 - 467848 million yen
R & D Spending	:	16000 million yen Capital Investment: 29500 million yen

Sales Breakdown :

SIC#	Product	Sales
3241	Cement, concrete and related products	36 %
355	Plant and machinery	19 %
291	Petrochemicals	17 %
282	Functional resins	12 %
28	Industrial reagents	5 %
287	Fertilizers	3 %
28	Fine chemicals	3 %
2999	Coke	3 %

Main Products	:	Petrochemical Products: Polyethylene, Polypropylene, Polybutadine; Chemicals: Phenol, Caprolactam, Ammonia, Ammonia cooling agent, Sulphuric acid, Nitric acid, Oxalic acid, Ammonium Nitrate; Building Materials and Products: Cement, Concrete, Prefabricated housing, Civil engineering materials, Housing materials, Landscaping materials, Bridge sections and other steel structures; Machinery and Plant Engineering: Plastics processing machinery, Metals processing machinery, Chemical plant, Other production equipment; Electronic Products: Electronic devices, Electronic parts, Electronic materials; Synthetic Resin products: Nylon resins, Polyamide resins, Other functional resins; Fine Chemicals: Catechol, Hydroquinone, Pharmaceuticals, Agrichemicals, Food additives, Catalysts, High purity industrial gases; Advanced Materials: Fine ceramics; Chemical Fertilizers: Carbamide, Ammonium Sulphate, Compound fertilizers; Metal Products: Steel, Iron, Ferrosilicon, Magnesium; Coal Products: Coal, Cokes, Talc; Real Estate Services
Banks	:	Sanwa Bank; Industrial Bank of Japan; Norinchukin Bank; Yamaguchi Bank
Laboratories	:	Ube Research Laboratory (Ube-shi, Yamaguchi-ken); Cement & Cokes Research Laboratory (Ichihara-shi, Chiba-ken); Hirakata Research Laboratory (Hirakata-shi, Osaka); Ube Plastics & Processing Machines Research Laboratory (Ube-shi)

Uchihashi Estec Co., Ltd.

Japanese Name	:	***Uchihashi Esutekku K.K.***
TSE Number	:	Not listed
Description	:	Engaged in the manufacture and sale of thermal cut-offs, thermal cut-off/resistor assemblies and solder.
Address	:	2-9-14, Kita-Imazu, Tsurumi-ku, Osaka 538
Tel & Fax	:	Tel: 06-962-6661 Fax: 06-962-6669
Statistics	:	Capital : 92 million yen Employees : 250
Officers	:	CEO : Mr. Hisao Uchihashi/President Contact : Mr. Toshihiko Nakano/Manager, International Trade
Sales Breakdown	:	SIC# Product 33 Metals and metal products 367 Electronic components
Main Products	:	Components: Thermal cut-offs, Thermal cut-off/resistor assemblies; Metal Products: Solder
Banks	:	Sanwa Bank
Branches	:	Tokyo Domestic Business Office (Tokyo); Nagoya Domestic Business Office (Aichi-ken)
Factories	:	Asahi Factory; Nanjo Factory

Ultex Corp., Inc.

TSE Number	:	Not listed
Description	:	Engaged in the development and manufacture of automated welding equipment and related products.
Address	:	Kyobashi Takano Bldg., 7 Fl., 1-6-7, Kyobashi, Chuo-ku, Tokyo 104
Tel	:	03-3535-3573
Sales Breakdown	:	SIC# Product 3548 Welding equipment
Main Products	:	Welding Equipment: CAW (computer assisted welders) for metals and ceramics

Unitika Ltd.

Japanese Name :	***Unichika K.K.***
TSE Number :	3103
Description :	A major textile producer. Currently expanding into bioproducts and plastics, and developing several enzymes for clinical diagnosis and artificial skin using chitin. Strengthening environmental equipment such as water/waste water treatment and garbage incineration equipment.
Address :	4-1-3, Kyutaro-machi, Chuo-ku, Osaka 541
Tel & Fax :	Tel: 06-281-5695 Fax: 06-281-5697
Statistics :	Capital : 23798 million yen Employees : 5596
Officers :	CEO : Mr. Keita Taguchi/President Contact : Mr. Hiroshi Ashida/General Manager, International
Sales :	March, 1988 - 248183 million yen March, 1989 - 249841 million yen March, 1990 - 264470 million yen March, 1991 - 294129 million yen
R & D Spending :	6300 million yen Capital Investment: 19500 million yen

Sales Breakdown :

SIC#	Product	Sales
2299	Man-made textiles	41 %
2241	Natural fibers, textiles	25 %
2821	Plastics products	14 %
8711	Plant engineering	10 %
2836	Biotechnological products	2 %

Main Products :	Fiber Products: Nylon fiber, Polyester fiber, Cotton fiber; Construction: Housing; Biotechnology Products: Enzymes, Artificial skin; Plastics Products: Spunbonded materials; Plant Engineering: Water treatment equipment, Municipal refuse incinerator
Banks :	Sanwa Bank
Branches :	Tokyo Head Office (Tokyo); Nagoya Office (Aichi-ken); Dusseldorf Office (Germany)
Laboratories :	Research and Development Center (Uji-shi)
R&D Areas :	Fiber Products: PET filament, Thermal fibers for marine sportswear, Nylon 46 fiber and related plant

Veetech Japan Co., Ltd.

Japanese Name	:	***Nippon Bitekku K.K.***
TSE Number	:	Not listed
Description	:	Vacuum products manufacturer with mainlines in ultra-high vacuum systems, vacuum deposition systems, sputtering systems and surface analysis equipment.
Address	:	1316-3, Katsukura, Katsuta-shi, Ibaraki 312
Tel & Fax	:	Tel: 0292-72-8161 Fax: 0292-72-9814
Statistics	:	Capital : 20 million yen Employees : 25
Officers	:	CEO : Mr. Tateo Ken Saito/President Contact : Mr. Tateo Ken Saito/President
Sales	:	March, 1988 - 650 million yen March, 1989 - 870 million yen March, 1990 - 910 million yen March, 1991 - 1000 million yen

Sales Breakdown :

SIC#	Product	Sales
3821	UHV systems	40 %
3559	Vacuum deposition system	40 %
3829	Surface analysis equipment	20 %

Main Products	:	Vacuum Equipment: Ultra-high vacuum systems, Vacuum evaporator, Sputtering deposition system; Analytical Equipment: Reflected high energy electron diffraction system
Banks	:	Joyo Bank; Kanto Bank; Ashikaga Bank
Branches	:	Katsuta Branch (Katsuta-shi, Ibaraki-ken); Mito Branch (Mito-shi, Ibaraki-ken)
Factories	:	Main Factory (Katsuta-shi, Ibaraki-ken)
Laboratories	:	R&D Department (Katsuta-shi, Ibaraki-ken)
Shareholders	:	Mr. T. Saito - 43% ; Mr. E. Saito - 7.5% ; Mr. T. Komori - 8.5% ; B-Dec Ltd. - 5.0%
Subsidiaries	:	B-Dec Co., Ltd. (Tsukuba-shi, Ibaraki-ken); TNK International Co., Ltd., (Tokyo)

Victor Co. of Japan, Ltd.

Japanese Name : ***Nihon Bikutoru K.K.***

TSE Number : 6792

Description : Leading manufacturer of video cassette recorders and car audio equipment.

Address : 4-8-14, Nihonbashi-Honcho, Chuo-ku, Tokyo 103

Tel & Fax : Tel: 03-3246-9861 Fax: 03-3246-0780

Statistics : Capital : 34106 million yen
Employees : 13367

Officers : CEO : Mr. Takuro Bojo/President
Contact : Mr. Yutaka Uchiyama/Director, International Marketing

Sales : September, 1988 (6 months) - 308922 million yen
March, 1989 (6 months) - 312900 million yen
March, 1990 - 618500 million yen
March, 1991 - 638700 million yen

Sales Breakdown :

SIC#	Product	Sales
3651	Video	47 %
3651	Audio	19 %
3651	Television	16 %
357	Information related equipment and others	12 %
3652	Entertainment	6 %

Main Products : Video: VHS video cassette recorders, Camcorders, Video Cameras, Video disc players, Blank video tapes; Audio: Stereo components, CD players, DAT decks, Radio cassette recorders, Car audio equipment, Blank audio tapes; TV: Color TVs, Display monitors; Information Related Equipment and Others: Personal computers, Computer peripherals, Institutional and educational equipment, Cordless answering telephones, Dictation machines, Acoustic equipment, Furniture

Banks : Industrial Bank of Japan; Sumitomo Bank; Mitsui Taiyo Kobe Bank; Kyowa Saitama Bank; Fuji Bank; Bank of Yokohama

Wako Engineering Co., Ltd.

Japanese Name	:	***Wako Enjiniiringu K.K.***
TSE Number	:	Not listed
Description	:	Engineering firm engaged in the development and implementation of new industrial technologies.
Address	:	4-8-30, Kudan-Minami, Chiyoda-ku, Tokyo 102
Tel	:	03-3239-6158
Sales Breakdown	:	SIC# Product 8711 Plant engineering 3599 Industrial machinery and systems
Main Products	:	Plant Engineering; Industrial Machinery and Systems

Yokohama Rubber Co., Ltd.

Japanese Name	:	***Yokohama Gomu K.K.***
TSE Number	:	5101
Address	:	5-36-11, Shinbashi, Minato-ku, Tokyo 105
Tel & Fax	:	Tel: 03-3432-7111 Fax: 03-3432-5616
Statistics	:	Capital : 17967 million yen Employees : 6308
Officers	:	CEO : Mr. Kazuo Motoyama/President
Sales	:	December, 1988 - 254364 million yen December, 1989 - 264672 million yen December, 1990 - 285967 million yen
R & D Spending	:	153 million yen Capital Investment: 29800 million yen
Sales Breakdown	:	

SIC#	Product	Sales
5014	Vehicle tires and tubes	71 %
3011	Belt conveyers and auto hoses	22 %
3728	Aircraft fuel tanks and hoses	8 %

Banks	:	Dai-Ichi Kangyo Bank; Industrial Bank of Japan; Mitsubishi Trust Bank; Sumitomo Trust; Yokohama Bank
Factories	:	Hiratsuka Plant (Kanagawa-ken); Mie Plant (Mie-ken); Mishima Plant (Shizuoka-ken); Shinshiro Plant (Aichi-ken); Ageo Plant (Saitama-ken); Ibaraki Plant (Ibaraki-shi); Onomichi Plant (Hiroshima-ken)

PRODUCT INDEX

02 Agricultural Production — Livestock and Animal Specialities

Taiyo Kagaku Co., Ltd	Egg products
Toho Rayon Co., Ltd.	Raw silk

10 Metal Mining

Dowa Mining Co., Ltd.	Gold and Silver Zinc
Nittetsu Mining Co., Ltd.	Metallic minerals

12 Coal Mining

Mitsui Mining Co., Ltd.	Coal

14 Mining & Quarrying of Non-metallic Minerals, except Fuels

Nittetsu Mining Co., Ltd.	Mineral mining - limestone, etc.
Onoda Cement Co., Ltd.	Limestone & other

15 Building Construction & Operative Builders — General Contractors

JGC Corp.	Chemical plants
Kajima Corp.	Domes, commercial construction Warehouses and cleanrooms Dormitories, hotel construction
Obayashi Corp.	Building construction
Shimizu Corp.	Housing construction
Sumitomo Corp.	Industrial plant
Taisei Corp.	Building construction

16 Heavy Construction other than Building Construction — Contractors

Kajima Corp.	Civil engineering
Kawasaki Heavy Industries, Ltd.	Plant and steel structure engineering
Mitsubishi Heavy Industries, Ltd.	Power plants
Nippon Steel Corp.	Heavy construction
Shimizu Corp.	Road construction
Sumitomo Corp.	Construction
Taisei Corp.	Civil Engineering
Toyo Engineering Corp	Petrochemical plant Petroleum refining plant Factory automation Chemical fertilizer plant Nuclear power plant

20 Food & Kindred Products

Ajinomoto Co., Inc.	Seasonings Processed foods
Asahi Chemical Industry Co., Ltd.	Frozen yogurt
Honen Corporation	Corn starch Vegetable oils
Kanebo, Ltd.	Foods
Kao Corp.	Cooking oil
Kyowa Hakko Kogyo Co., Ltd.	Alcoholic spirits
Nihon Nohyaku Co., Ltd.	Chemical feedstuffs
Nippon Oil & Fats Co., Ltd.	Food products
Nitta Gelatin Inc.	Food additives
Otsuka Chemical Co., Ltd.	Foods and beverages
Taiyo Kagaku Co., Ltd.	Food additives
Takeda Chemical Industries, Ltd.	Foodstuffs

22 Textile Mill Products

Daiwabo Co., Ltd.	Textiles
Kanebo, Ltd.	Natural fibres Synthetic fibres

Komatsu Seiren Co., Ltd.	Synthetic fabrics processing Synthetic leather
Kurabo Industries Ltd.	Wollen fabrics Yarn, cotton garments
Mitsubishi Rayon Co., Ltd.	Synthetic fibres
Sakai Textile Manufacturing	Textile dyeing Apparel
Sumitomo 3M Ltd	Electric insulating materials
Teijin Ltd.	Other fibres Nylon Polyester fibres, new materials
Toho Rayon Co., Ltd	Cotton Rayon
Toyobo Co., Ltd.	Textile secondary products Cotton fibres & textiles Synthetic fibres & textiles Wool
Unitika Ltd.	Natural fibres, textiles Man-made textiles

23 Apparel & other finished products made from fabric & similar materials

Kanebo, Ltd.	Clothing
Nishikawa Rubber Co., Ltd.	Auto trimming

24 Lumber and Wood Products

Asahi Chemical Industry Co., Ltd.	Prefabricated buildings
Nitta Venee	Compressed laminated panels Plywood Misc. wooden building materials

25 Furniture and Fixtures

Daiwabo Co., Ltd.	Home furnishings
Otsuka Chemical Co., Ltd.	Furniture

26 Paper and Allied Products

Honshu Paper Co., Ltd.	Processed paper products Paper & paperboard products
Kao Corp.	Disposable diapers
Tokiwa Electric Co., Ltd.	Magnesium silicate papers
Toppan Printing Co., Ltd.	Wrapping Papers

27 Printing, Publishing & Allied Industries

Tokyo Ohka Kogyo Co., Ltd.	Printing
Toppan Printing Co., Ltd.	General Publishing Periodical Publishing Securities and special papers

28 Chemicals and Allied Products

Ajinomoto Co., Inc.	Specialty chemicals Pharmaceuticals
Asahi Chemical Industry Co., Ltd.	Acrylic rubber Acids Bacterins Rayon fibres Organic chemicals Polyester and vinyl fibres
Asahi Glass Co., Ltd.	Chemicals
Asahi Organic Chemicals Industry Co., Ltd.	Plastic piping materials Industrial-use resins
Baxter Ltd.	Blood component products

Chugai Pharmaceutical Co., Ltd.	Clinical reagents & equipment
	Chemotherapy drugs
	Nervous system agents
	Hormones
	Antibiotics
	Digestive agents
	Respiratory agents
	Metabolic agents
	Insecticides
Chugoku Kogyo Co., Ltd.	Inorganic chemicals
Chukoh Chemical Industries, Ltd.	Synthetic resin sealing materials
	Engineering plastics
	Synthetic resins & resin products
Cosmo Oil Co., Ltd., Group	Organic chemicals
Dai Nihon Glass Industries Ltd.	Specialized resin products
Dai-Ichi Kogyo Seiyaku Co., Ltd.	Resin additives
	Surface active agents
	Water soluble polymers
	Urethane chemicals
Daicel Chemical Industries, Ltd.	Organic chemicals
	Functionals
	Cellulosics
	Plastics and films
Daikin Industries, Ltd.	Fluorochemicals
Dainichiseika Color & Chemical Mfg. Co., Ltd.	Chemical products
	Coloring agents
Dainippon Ink & Chemicals, Inc.	Specialty plastics & compounds
	Printing supplies & equipment
	Polymers & related Products
Denki Kagaku Kogyo Co., Ltd.	Organic Chemicals
	Inorganic chemicals
Dow Corning Toray Silicone Co., Ltd.	Silicones
Du Pont Japan Ltd.	Acrylic fibres
	Diagnostic kits
	Herbicides
Furukawa Co., Ltd.	Chemicals
Hayashi Chemical Industries Co., Ltd.	Synthetic dyes
	Dyestuffs
	Soil Ameliorants & fertilizers
Hitachi Chemical Co., Ltd.	Inorganic materials
	Housing, construction material
	Pharmaceuticals
	Organic materials and resins
Hoechst Gosei Co., Ltd.	Specialty chemicals
	Synthetic resins
Hoechst Mitsubishi Kasei Co., Ltd.	Natural dye products
Honen Corporation	Specialty chemicals
Honshu Paper Co., Ltd.	Chemical products
Idemitsu Kosan Co., Ltd.	Monoclonal antibodies
	Agricultural chemicals
	Fine chemicals
Idemitsu Petrochemical Co., Ltd.	Biological products
	Synthetic resins & plastics
Japan Metals & Chemicals Co., Ltd.	Chemicals and related products
Japan Organo Co., Ltd.	Resin, carbon & chemicals
Japan REC Co., Ltd.	Synthetic resin composite materials
Japan Synthetic Rubber Co., Ltd.	ABS, AS AES plastics
	SBR, BR, IR, NBR, PDM
Kanebo, Ltd.	Cosmetics
	Pharmaceuticals
Kanebo-NSC Ltd.	Adhesives
Kaneka Corporation	Synthetic fibres
	PVCs & soda
	Pharmaceuticals & foodstuffs
	Synthetic resins products
	Special synthetic resins

Kankyo Sozai Kenkyusho K.K.	Organic solvents Pollution control materials
Kansai Paint Co., Ltd.	Oil paints Synthetic resin-based paints Lacquers Thinners
Kao Corp.	Fatty acids Shampoos, bath agents Polyurethanes Sulfonated oils Detergents Plasticizers
Kokusai Chemical K.K.	Synthetic resins
Kuraray Co., Ltd.	Medical and pharmaceuticals Fine chemicals Various synthetic fibres
Kureha Chemical Industry Co., Ltd.	Specialty chemicals Synthetic resins Inorganic chemicals
Kyowa Hakko Kogyo Co., Ltd.	Inorganic chemicals Pharmaceuticals
Matsumoto Yushi-Seiyaku Co., Ltd.	Fire retardants Metal cleaning agents Synthetic resin treatments Surfactants Synthetic materials
Mitsu Kagaku	Chemical products Silicone dissolving agent
Mitsubishi Gas Chemical Co., Inc.	Methanol Engineering plastics Urea Ammonia
Mitsubishi Kasei Corp.	Carbon and inorganic chemicals
Mitsubishi Pencil Co., Ltd.	Cosmetics
Mitsubishi Petrochemical Co., Ltd.	General chemicals Specialty chemicals Fine chemicals Synthetics
Mitsubishi Rayon Co., Ltd.	Resin products
Mitsui Petrochemical Industries, Ltd.	Synthetic resins Chemicals & raw materials Specialty chemicals
Mitsui Toatsu Chemicals, Inc.	Fine chemicals Industrial chemicals Synthetic resins & resin processing
Mochida Pharmaceutical Co., Ltd.	Pharmaceuticals
NGK Insulators, Ltd.	Ceramic products
Nihon Nohyaku Co., Ltd.	Disinfectants Herbicides Disinfectant pesticides Pesticides
Nippon Carbon Co., Ltd.	Carbon
Nippon GE Plastics Co., Ltd.	Synthetic resins
Nippon Mining Co., Ltd.	Pharmaceuticals
Nippon Nyukazai Co., Ltd.	Organic chemicals
Nippon Oil & Fats Co., Ltd.	Coatings Welding materials Oleo and specialty chemicals Chemicals and explosives
Nippon Paint Co., Ltd.	Surface treatment agents Paints and thinners
Nippon Petrochemicals Co., Ltd.	Specialty chemicals & materials Chemicals Plastics
Nippon Pillar Packing Co., Ltd.	'Pillaflon' fluorine resins
Nippon Sanso K.K.	Oxygen, Nitrogen, Argon

Nippon Shokubai Co., Ltd.	Catalysts
	Polymers and resins
	Fine and specialty chemicals
	Chemical products
Nippon Steel Chemical Co., Ltd.	Toluene
	Pigments
	Synthetic resins
	Argon
Nippon Steel Corp.	Aluminum and copper compounds
Nippon Tungsten Co., Ltd.	Tungsten carbide
Nippon Zeon Co., Ltd.	Synthetic rubber
	Synthetic resins
Nissin Chemical Industry Co., Ltd.	Copolymer resins
	Acrylic rubber
Nitivy Co., Ltd.	Ion exchange fiber
	Water soluble eliminable fiber "Solvron"
	Continuous alumina fiber
Nitta Gelatin Inc.	Adhesives
	Collagen
	Gelatine
Nittetsu Mining Co., Ltd.	Chemical products
Okamoto Industries, Inc.	Vinyl & vinyl products
Osaka Gas Co., Ltd.	Industrial gasses
Osaka Sanso Kogyo, Ltd.	Oxygen, Nitrogen, Argon
Otsuka Chemical Co., Ltd.	Agricultural chemicals
	Chemical products
Polyplastics Co., Ltd.	Synthetic resins
Sakai Chemical Industry Co., Ltd. (Osaka)	Zinc products
	Barium products
	Pharmaceuticals
	Stabilizers
	Catalysts
Sanyo Chemical Industries, Ltd.	High polymers
	Polyurethane products
	Surfactants
	Plastic specialty products
Sekaicho Rubber Co., Ltd.	Sealants
Sekisui Chemical Co., Ltd.	Adhesive tapes
	Plasticizers
	Moulding compounds
Sekisui Plastics Co., Ltd.	Synthetic resins
Shikoku Chemicals Corp.	Fine chemicals
	Inorganic chemical products
	Organic chemical products
Shin-Etsu Chemical Co., Ltd.	Organic chemicals
	Chemical fertilizers
	Inorganic chemicals
	Synthetic resins
Shin-Etsu Handotai Co., Ltd.	Silicon for semiconductors
Shin-Etsu Polymer Co., Ltd.	Construction & engineering materials
Showa Denko K.K.	Chemicals & carbons
	Olefins and organic chemicals
Showa Highpolymer Co., Ltd.	Polyesters
	Emulsifiers
	Water-proof paints
	Phenolic resins
Sumitomo 3M Limited	Polishing materials
Sumitomo Bakelite Co., Ltd.	Synthetic resin molding materials
	Industrial resins
Sumitomo Chemical Co., Ltd.	Caustic soda, hydrochloric acid
	Synthetic resins
	Ethylene, Toulene
	Pigments
	Pesticides, insecticides
Sun Arrow Chemical Co, Ltd.	Plastics

Taiyo Kagaku Co., Ltd.	Biological extracts (Lethicin, etc.) Dietary fiber
Takeda Chemical Industries, Ltd.	Agricultural Chemicals Pharmaceutical products Chemical products Veterinarian drugs
Tanaka Kinzoku Kogyo	Catalysts
Teijin Ltd.	Biotechnological products Industrial inorganic chemicals
Toagosei Chemical Industry Co., Ltd.	Fertilizers Instant adhesives Soda Synthetic resins
Toho Rayon Co., Ltd.	Beslon
Toho Titanium Company, Limited	Titanium pigments
Tohto Kasei Co., Ltd.	Polyamide resins Epoxy resins
Tokiwa Electric Co., Ltd.	Magnesium silicate compounds
Tokuyama Soda Co., Ltd.	Synthetic resins Chemicals
Tokyo Ohka Kogyo Co., Ltd.	Chemical products for electronic ind
Tonen Chemical Corp.	Synthetic resins
Toray Industries, Inc.	Nylon Synthetic fiber 'Tetron' 'Toraylon' 'Essene' and other synthetic resins Chemicals Resins
Toshiba Ceramics Co., Ltd.	Carbon and silicon carbide
Tosoh Corp.	PVC and rubber Inorganic chemicals Fine chemicals Olefins Polyolefins Organic chemicals
Toyo Tire & Rubber Co., Ltd.	Industrial chemicals
Toyobo Co., Ltd.	Functional & biomedical products
Tsutsunaka Plastic Industry Co., Ltd.	Styrene boards Plastic boards PVC resin boards Water treatment products
Ube Industries, Ltd.	Industrial reagents Functional resins Fertilizers Fine chemicals
Unitika Ltd.	Plastics products Biotechnological products

29 Petroleum Refining & Related Industries

Furukawa Co., Ltd.	Fuel and energy
Idemitsu Kosan Co., Ltd.	Petroleum products
JGC Corp.	Petrochemicals
Koa Oil Co., Ltd.	Fuel oil Kerosene Gasoline
Mitsubishi Kasei Corp.	Petrochemical products
Mitsubishi Materials Corp.	Petroleum & coal products
Mitsubishi Mining & Cement Co., Ltd.	Petrochemicals
Mitsui Mining Co., Ltd.	LPG & petroleum Coke
Nippon Mining Co., Ltd.	Petroleum products
Nippon Oil & Fats Co., Ltd.	Petrochemicals
Nippon Oil Co., Ltd.	Fuel oil Gasoline Other petroleum based products
Nippon Petroleum Refining Co., Ltd.	Petrochemical products

Nippon Shokubai Co., Ltd.	Petrochemicals
Nippon Steel Chemical Co., Ltd.	Coke
Showa Shell Sekiyu K.K.	Naptha, KerGas
	Lubricants
Tsutsunaka Plastic Industry Co., Ltd.	High-density polyethylene
Ube Industries, Ltd.	Coke
	Petrochemicals

30 Rubber and Miscellaneous Plastic Products

Asahi Organic Chemicals Industry Co., Ltd.	Moulded materials
C.I. Kasei Company, Ltd.	PVC sheeting
	PVC film
	PVC piping
Chukoh Chemical Industries, Ltd.	Fluororesin coated glass fiber fabric
Dai Nihon Glass Industries Ltd.	Fiber reinforced plastic (FRP) products
Idemitsu Petrochemical Co., Ltd.	Functional films
JSP Corporation	Polystyrene foam boards
	Polystyrene foam sheets
	Polyolefin foams
Kansai Paint Co., Ltd.	Marking film, others
Kureha Chemical Industry Co., Ltd.	Plastic products
Nippon GE Plastics Co., Ltd.	Misc. plastic products
Nippon Pillar Packing Co., Ltd.	Mechanical seals
	Vertical gaskets
	Grand packing
Nippon Polyester Co., Ltd.	FRP molded products
Nissin Chemical Industry Co., Ltd.	Prophylactics
	Insulation materials
Nitta Venee	Synthetic ivory
NOK Corporation	Oil seals
	Industrial rubber products
Okamoto Industries, Inc.	Condoms & sundry goods
	Footwear
	Automobile tires
	Building & industrial material
Sekaicho Rubber Co., Ltd.	Rubber footwear
Sekisui Chemical Co., Ltd.	Film
Sekisui Plastics Co., Ltd.	Building materials
	Plastic materials
	Foam sheets
Shin-Etsu Polymer Co., Ltd.	Plastic films & sheets
	Plastic industrial materials
Sumitomo Bakelite Co., Ltd.	Moulded plastic products
	Plastic films & sheets
Takiron Co., Ltd.	Plastic industrial materials
	Plastic building materials
	Agricultural and engineering plastics
Takuma Co., Ltd.	Laminated boards and plates
Techmate	Fibre reinforced plastics
Toray Industries, Inc.	Synthetic films
Toshiba Chemical Corporation	Resin products
	Laminated sheets
Toto Ltd.	Plastic bathroom & kitchen fixtures
	Toilet seats
Toyo Tire & Rubber Co., Ltd.	Fabricated rubber products
	Tires
Toyobo Co., Ltd.	Films
Yokohama Rubber Co., Ltd.	Belt conveyers and auto hoses

32 Stone, Clay, Glass and Concrete Products

Asahi Glass Co., Ltd.	Glass products
	Building materials
	Ceramics and refractories
Denki Kagaku Kogyo Co., Ltd.	Cement
Fuji Electrochemical Co., Ltd.	Ferrites, ceramics
Hayashi Chemical Industries Co., Ltd.	New ceramics

Hoya Corp.	Crystal and glassware Molded porcelain parts for electronics
Kanebo, Ltd.	New materials
Kurosaki Refractories Co., Ltd.	Fire bricks and refractories
Kyocera Corp.	Fine Ceramics Bioceramics
Matsumoto Yushi-Seiyaku Co., Ltd.	Synthetic diamonds
Mitsubishi Materials Corp.	Nonmetal & metal advanced materials Cement Construction materials
Mitsubishi Mining & Cement Co., Ltd.	Cement
Mitsui Mining Co., Ltd.	Cement & construction material
NGK Insulators, Ltd.	Electric power products
NGK Spark Plug Co., Ltd.	Ceramic electronics & other products
Nihon Cement Co., Ltd.	Cement and related products
Nikkiso Co., Ltd.	New materials
Nippon Electric Glass Co., Ltd.	Glass fibres Glass tubes & bulbs
Nippon Sheet Glass Co., Ltd.	Fine glass Sheet glass Glass building materials
Nippon Steel Chemical Co., Ltd.	Cement
Nippon Steel Corp.	Ferrite
Nippon Tungsten Co., Ltd.	Ceramics
Noritake Co., Ltd.	Industrial materials Tableware Ceramic products
Onoda Cement Co., Ltd.	Cement
Shibata Scientific Technology Ltd.	Industrial glass equipment
Shikoku Chemicals Corp.	Building materials
Showa Denko K.K.	Ceramics & other materials
Sumitomo Chemical Co., Ltd.	High-purity silicon
Taiko Refractories Co., Ltd.	Fine ceramics Fire resistant cement
TDK Corporation	Ferrite magnetic products
Toa Planning	Ceramic composite materials
Tokuyama Soda Co., Ltd.	Cement & related products
Tokyo Denpa Co., Ltd.	Artificial quartz crystals
Toshiba Ceramics Co., Ltd.	Refractories Quartz glass
Toshiba Tungaloy Co., Ltd.	Friction materials and parts
Tosoh Corp.	Cement
Toto Ltd.	Sanitaryware & metal fittings Bathroom tiles & other accessories
Toyo Tanso Co., Ltd.	Graphite composites & other materials
Ube Industries, Ltd.	Cement, concrete & related products

33 Primary Metal Industries

Alithium Ltd.	Aluminum-lithium alloys
Araya Industrial Co., Ltd.	Steel pipe
Chugoku Kogyo Co., Ltd.	Scrups & recoveries
Daido Steel Co., Ltd.	Titanium alloys Special and cast steel
Dowa Mining Co., Ltd.	Rolled copper
Furukawa Co., Ltd.	Metals
Furukawa Electric Co., Ltd. ; The	Rolled copper products Nonferrous products
Hitachi Cable, Ltd.	optical fibres and wires
Hitachi Metals Co., Ltd.	Steel pipe & fittings Specialty Steels
Ishikawajima Precision Casting	Precision castings
Iwate Iron Co., Ltd.	Iron & steel products Composite metal molding materials
Japan Metals & Chemicals Co., Ltd.	Ferroalloys Electrometallurgical materialsElectronic

Kobe Steel, Ltd.	Iron and steel
	Non-ferrous metals
	Welding rods
KSL Alcoa Aluminum	Aluminum can stock
Mitsubishi Materials Corp.	Copper
	Other nonferrous metals
	Gold & silver
Nakadai Kinzoku Co., Ltd.	Metal products
NGK Insulators, Ltd.	Metals
Nippon Light Metal Company, Ltd.	Aluminum sheets
	Aluminum ingots and alloys
	Aluminum products
	Aluminum products
Nippon Metal Industry Co., Ltd.	Primary steel products
Nippon Mining Co., Ltd.	Zinc
	Copper
Nippon Seisen Co., Ltd.	Metal fibres and filter elements
	Nonferrous wires
	Stainless steel wires
Nippon Stainless Steel Co., Ltd.	Steel castings
	Titanium products and others
	Stainless steel
Nippon Steel Corp.	Titanium products
	Steel products
Nippon Tungsten Co., Ltd.	Tungsten, Electrical contacts
Nippon Yakin Kogyo Co., Ltd.	Stainless steel sheets
Nisshin Steel Co., Ltd.	Stainless and surface-treated
	High frequency pipes
	Hot & cold-rolled steel prod
Nittetsu Micrometal Corp.	Micro wire products
NKK Corporation	Steel bars
	Steel tubing
	Steel plating
Omi Goukin Co., Ltd.	Aluminum alloys & related products
Osaka Titanium Co., Ltd.	Titanium
Sakai Chemical Industry Co., Ltd. (Osaka)	Titanium dioxide
Shin-Etsu Chemical Co., Ltd.	Alloys
Shinko Wire Co., Ltd.	Steel cables
	Steel wires
	Stainless steel wire products
Showa Electric Wire & Cable Co., Ltd.	Covered wires and windings
	Optical and glass fibres
	Aluminum wires
	Bare wires
Sumitomo Chemical Co., Ltd.	High-purity aluminium
Sumitomo Electric Industries, Ltd.	Special wires
	Powder alloys
	Electronic wires and cables
Sumitomo Light Metal Industries, Ltd.	Wrought aluminium products
	Wrought copper products
Sumitomo Metal Industries, Ltd.	Titanium
	Castings
	Sheets
	Pipes, tubes
	Rolling stock
Sumitomo Metal Mining Co., Ltd.	Zinc
	Copper
	Nickel
	Gold
Suzuki Metal Industry Co., Ltd.	Hard drawn steel wire
	Fishing wires
	Stainless steel wire
	Oil tempered wire
	Titanium wire
	PC wire and PC strands

Tanaka Kinzoku Kogyo	Ingots
	Precious metals
Toho Kinzoku Co., Ltd.	Tungsten fishing nets
	Molybdenum
	Tungsten
	Super hard alloys
Toho Titanium Company, Limited	Metallic titanium
	Titanium trichloride
Tokyo Tungsten Co., Ltd.	Super hard alloys
	Molybdenum and Tungsten
	Smelted products
Toshiba Tungaloy Co., Ltd.	Super-hard metal alloys
Tosoh Corp.	Electrolytic metals
	Materials
Toyo Tanso Co., Ltd.	Nonferrous alloys
Tsuda Kenkyusho	Nonferrous alloys & related materials
Uchihashi Estec Co., Ltd.	Metals & metal products

34 Fabricated Metal Products, except Machinery & Transportation Equipment

Aisin Seiki Co., Ltd.	Automobile parts
Daido Steel Sheet Corp.	Galvanized steel sheets
Hitachi Metals Co., Ltd.	Magnetic & electronic materials
	Machine & building frames
Japan Steel Works, Ltd.	Pressure vessels and reactors
	Generator components & engines
Mitsubishi Materials Corp.	Sintered alloy products
	Aluminum Cans
	Special alloys & industrial machinery
Nachi-Fujikoshi Corp.	Misc. steel products
	Tools
Nippon Metal Industry Co., Ltd.	Secondary steel products
Nippon Yakin Kogyo Co., Ltd.	Stainless steel fabrications
Nishikawa Rubber Co., Ltd.	Auto moldings
Sekisui Chemical Co., Ltd.	Building materials
	Housing
	Pipe fittings
Sumitomo Heavy Industries, Ltd.	Process Vessels
Sumitomo Metal Mining Co., Ltd.	Electronics materials
Suzuki Metal Industry Co., Ltd.	Piano wire
Takuma Co., Ltd.	Boiler Plant
Tanaka Kinzoku Kogyo	Nozzles
Tateyama Aluminium Industry Co., Ltd.	Comercial construction materials
	Housing construction materials
	Misc. aluminum retail products
	Misc. aluminum shapes
	Aluminum products for retail use
Tokyo Electric Power Co., Inc.	Nuclear Plants
Toyota Motor Corp.	Construction materials

35 Industrial & Commercial Machinery & Computer Equipment

Anelva Corp.	Other equipment and components
	Semiconductor manufacturing equipment
Canon Inc.	Computer peripherals
	Copiers
Daido Steel Co., Ltd.	Machinery
Daihen Corp.	Welding equipment & industrial robots
Daikin Industries, Ltd.	Air conditioning and refrigeration eqpt.
	Defense equipment
	Hydraulic and Lubrication Equipment
Denkoh Co., Ltd.	Infrared heating devices
	Semiconductor manufacturing equipment
	Belt kilns
Dowa Mining Co., Ltd.	Industrial machines
Du Pont Japan Ltd.	Electronic components
Fuji Electric Co., Ltd.	Heavy electrical equipment
Fujitsu Ltd.	Computer equipment

Furukawa Co., Ltd.	Machinery
Hitachi Cable, Ltd.	Semiconductor and rubber prod
	Electronics and leadframe prod
Hitachi Metals Co., Ltd.	Plant & equipment
Hitachi, Ltd.	Industrial machinery
	Industrial machinery
	Industrial machinery
	Office Automation equipment
Honda Engineering	Machine tools & dies
Honda Motor Co., Ltd.	Farm & misc. combustion powered products
Idemitsu Petrochemical Co., Ltd.	Optical disks
Ishikawajima-Harima Heavy Industries Co., Ltd	Industrial machinery and plant
Iwamoto Seisakusho Co., Ltd.	Precision machinery
Japan Aircraft Manufacturing Co., Ltd.	Industrial machinery & eqmnt.
Japan Organo Co., Ltd.	Customized Plant
	General plant
Japan Steel Works, Ltd.	Molding machines & compressors
	Circuit board machines
	Plastic processing machinery
JEOL Ltd.	Industrial equipment
JGC Corp.	Gas related equipment
Kawasaki Heavy Industries, Ltd.	Machinery
Komatsu Ltd.	Industrial machinery
	Contsruction machinery
Kurabo Industries Ltd.	Precision filtration equipment
Matsushita Industrial Equipment Co., Ltd.	Welding & cutting equipment
Matsushita Seiko Co., Ltd.	Industrial air-conditioning equipment
	Household air-conditioners
Minebea Co., Ltd.	Bearings
	Machinery parts
Mitsubishi Electric Corp.	Heavy Machinery
Mitsubishi Heavy Industries, Ltd.	Industrial mahinery
Mitsubishi Pencil Co., Ltd.	Office automation equipment
	Misc. office goods
Mitsui Mining Co., Ltd.	Industrial machinery
Nachi-Fujikoshi Corp.	Industrial equipment
	Machine tools
	Bearings
NEC Corporation	Computers & computer components
Nihon Cement Co., Ltd.	Industrial machinery
Nikkiso Co., Ltd.	Pumps & related equipment
Nippon Canon Inc.	Industrial machinery
Nippon Light Metal Company, Ltd.	Ice cream machines
Nippon Sanso K.K.	Gas-related equipment
Nippondenso Co., Ltd.	Automobile heaters & airconditioners
Nissan Motor Co., Ltd.	Industrial machinery and Marine
Nissei Electric Co., Ltd.	Metal die DNC system
Nissin Electric Co., Ltd.	Other electrical equipment
Nitta Gelatin Inc.	Pneumatic conveying system
Nittetsu Mining Co., Ltd.	Industrial machinery
NOK Corporation	Mechanical seals
NSK, Ltd.	Machinery parts & other
	Bearings
Oki Electric Industry Co., Ltd.	Information processing systems
Osaka Diamond Industrial	Drills, handtools
	Diamond-tipped tools
	Diamond wheels
Sanyo Electric Co., Ltd.	Information sysytems & electronics
Seiko Epson Corporation	Computers & printers
Shibata Scientific Technology Ltd.	Plastics production equipment
Shin-Etsu Polymer Co., Ltd.	Industrial machinery
Shizuoka NEC Ltd.	Electronics production machinery
	Information processing equipment
	Image processing equipment

Sumitomo Heavy Industries, Ltd.	Crushing machines Desalination equipment Centrifuges
Taiko Refractories Co., Ltd.	Industrial furnaces
Takuma Co., Ltd.	Water Treatment Systems
Tanaka Kinzoku Kogyo	Computer materials
TDK Corporation	Floppy disks & diskettes Optical disks
Tokiwa Electric Co., Ltd.	Industrial robots & related products
Tokyo Denpa Co., Ltd.	Quartz production technologies
Tokyo Seimitsu Co., Ltd.	Semiconductor production equipment
Toppan Printing Co., Ltd.	Precision tools
Toshiba Chemical Corporation	Moulds Machinery
Toshiba Corp.	Information/Communication devices
Toshiba Tungaloy Co., Ltd.	Wear-resistant tools Cutting tools Construction tools
Toyoda Automatic Loom Works	Industrial vehicles Textile machinery Air compressors
Ube Industries, Ltd.	Plant and machinery
Ultex Corp., Inc.	Welding equipment
Veetech Japan Co., Ltd.	Vacuum deposition system
Victor Co. of Japan, Ltd.	Information related equipment and others
Wako Engineering Co., Ltd.	Industrial machinery & systems

36 Electronic & Other Electrical Equipment & Components, except Computer Equipment

Advantest Corp.	Semiconductor testing devices
Asahi Glass Co., Ltd.	Electronic components and materials
Canon Inc.	Communications equipment
Chukoh Chemical Industries, Ltd.	Integrated Circuits & IC substrates
Daihen Corp.	Transformers & related products
Denkoh Co., Ltd.	Semicondutors and allied products
Dowa Mining Co., Ltd.	Electronic materials
Fuji Electric Co., Ltd.	Consumer electronics & other products Control/information systems & devices
Fuji Electrochemical Co., Ltd.	Dry cell batteries Magnetic heads and supplies
Fujitsu Ltd.	Electronic components Communications equipment
Furukawa Electric Co., Ltd. ; The	Wires and cables
Hitachi Cable, Ltd.	copper products and coils
Hitachi Chemical Co., Ltd.	Electronics materials
Hitachi Maxell, Ltd.	Electronic eqmnt. & batteries Magnetic recording tape Electronic appliances
Horiba Ltd.	Electronic equipment
Hosiden Corporation	Audio parts & components Electronic machine components Composite parts & other Electromagnetic parts
Hoya Corp.	Discs and tapes High-density IC's
Japan REC Co., Ltd.	Non-metal electrical materials
Kaneka Corporation	Electronics materials
Kao Corp.	Floppy discs
Kyocera Corp.	Electronic equipment Electronic parts Semiconductors and related products
Matsushita Electric Industrial Co., Ltd.	Audio equipment Batteries and kitchen products Home appliances Video equipment Communication and industrial equipment Electronic components

Tokin Corp.	Electronic components
	Electronics materials
	Misc. electronics products
Tokyo Denpa Co., Ltd.	Communications equipment
Tokyo Gas Co., Ltd.	Gas appliances
Toshiba Ceramics Co., Ltd.	Silicon wafers
Toshiba Chemical Corporation	Insulating materials
Toshiba Corp.	Heavy Electrical Apparatus
	Consumer electronics
Uchihashi Estec Co., Ltd.	Electronic components
Victor Co. of Japan, Ltd.	Audio
	Video
	Television
	Entertainment

37 Transportation Equipment

Araya Industrial Co., Ltd.	Bicycle wheel rims
	Bicycles
Daicel Chemical Industries, Ltd.	Propellants & propulsion systems
Hitachi Metals Co., Ltd.	Automobile parts
Honda Motor Co., Ltd.	Automobiles
	Parts and other
	Motorcycles
Ishikawajima-Harima Heavy Industries Co., Ltd	Shipbuilding and ocean structures
	Aerospace
Isuzu Motors Ltd.	Heavy trucks & buses
	Light trucks & buses
	Engines, parts & other
	Passenger cars & vans
Japan Aircraft Manufacturing Co., Ltd.	Motor boats
	Aircraft repairs
	Aircraft & parts
Kawasaki Heavy Industries, Ltd.	Consumer products and components
	Shipbuilding
	Aerospace
	Rolling stock
Kinugawa Rubber Industrial Co., Ltd.	Rubber interior parts & fittings
	Rubber vibration insulators
	Rubber weather strips
	Rubber hoses
	Rubber brake parts
Mazda Motor Corp.	Passenger automobiles
	Trucks
	Automotive parts
Minebea Co., Ltd.	Transportation equip. & parts
Mitsubishi Heavy Industries, Ltd.	Ship building
	Aircraft
Mitsubishi Motors Corp.	Trucks and Buses
	Passenger automobiles
	Minicars
	CKD kits for overseas assembly
Nippondenso Co., Ltd.	Automotive fuel management systems
	Automotive meters
	Automotive filters
	Automotive electronic equipment
	Radiators
Nissan Motor Co., Ltd.	Automobiles
	Aircraft & parts
	Automobile parts & accessories
Sumitomo Corp.	Shipbuilding and aerospace
Sumitomo Electric Industries, Ltd.	Automobile brake components
Sumitomo Heavy Industries, Ltd.	Shipbuilding
Toho Kinzoku Co., Ltd.	Automotive joints
Toyoda Automatic Loom Works	Automobiles & automobile parts

Toyota Motor Corp.	Trucks
	Vehicle Parts and Supplies
	Cars
Yokohama Rubber Co., Ltd.	Aircraft fuel tanks and hoses

38 Measuring, Analysing & Controlling Instruments; Photographic Medical & Optical Goods

Advantest Corp.	Electronic measuring devices
Asahi Chemical Industry Co., Ltd.	Artificial kidney
Baxter Ltd.	Surgical supplies
	Dialysis hardware
Canon Inc.	Optical products & others
	Cameras & video equipment
Du Pont Japan Ltd.	Mammography machines
Horiba Ltd.	Scientific analyzers
	Medical analyzers
	Automotive emissions analyzers
Hoya Corp.	Laser oscillators
	Eyecare products
J. Morita Corp.	Medical instruments
JEOL Ltd.	Analytical instruments
	Electromedical equipment
	Electron optical devices
Kyocera Corp.	Precision optical equipment
Mochida Pharmaceutical Co., Ltd.	Medical equipment
Nikkiso Co., Ltd.	Medical equipment & materials
	Dental equipment & materials
	Measuring devices
Nikkoshi Co., Ltd.	Analytical instruments
Nippon Mining Co., Ltd.	Optical goods
Nissin Electric Co., Ltd.	Control systems, charged ptcl beam eqpt.
Nitta Gelatin Inc.	Dental cement
Seiko Epson Corporation	Watches
Seiko Instruments Inc.	Processing equipment
	Alcohol sensors
	Timepieces
Shibata Scientific Technology Ltd.	Biological testing equipment
	Scientific instruments
	Lab glassware & glass apparatus
	Medical & sanitary equipment
	Pollution measuring equipment
Shimadzu Corp.	Precision medical apparatus
	Biotech equipment
	Aerospace-related equipment
	Industrial control equipment
	Measuring equipment
Sumitomo 3M Limited	Adhesive tapes
Takuma Co., Ltd.	Garbage incinerators
Tanaka Kinzoku Kogyo	Dental use materials
Tokyo Bi-Tech Laboratories	Metal alloy dental materials
Tokyo Denpa Co., Ltd.	Measuring & controlling devices
Tokyo Ohka Kogyo Co., Ltd.	Photosensitive materials
Tokyo Seimitsu Co., Ltd.	Automatic measuring instruments
	Measuring & controlling devices
Toray Research Center, Inc.	Computerized analysis systems
Tosoh Corp.	Analytical equipment
Veetech Japan Co., Ltd.	UHV systems
	Surface analysis equipment

39 Miscellaneous Manufacturing Industries

Mitsubishi Pencil Co., Ltd.	Pencils
	Marking Pens
	Ball-point pens

42 Motor Freight Transportation & Warehousing

Mitsui & Co., Ltd.	Warehousing

48 Communications

Nippon Telegraph and Telephone Corp.	Telegraph Digital data exchange Pagers Data communications facilities Telephone Leased circuits

49 Electric, Gas & Sanitary Services

Chubu Electric Power Co., Inc.	Electric power
Electric Power Development Co., Ltd.	Electricity supply
Japan Metals & Chemicals Co., Ltd.	Geothermal electricity generation
Kansai Electric Power Co., Inc.	Commercial & industrial use Residential use
Osaka Gas Co., Ltd.	Gas Gas supply work
Takuma Co., Ltd.	Industrial waste systems
Tohoku Electric Power Co., Inc.	Power supply
Tokyo Electric Power Co., Inc.	Electrical Power Generation Gas Transmission to residences
Tokyo Gas Co., Ltd.	Gas

50 Wholesale Trade — Durable Goods

C. Itoh & Co., Ltd.	General machinery and construction Metal compounds Timber products and commodities
Chugoku Kogyo Co., Ltd.	Nonferrous metals (wholesale)
Hitachi, Ltd.	Power generation plants
Marubeni Corp.	Metals Minerals Machinery Construction equipment
Mitsui & Co., Ltd.	Machinery and equipment Steel Nonferrous metals
Nissho Iwai Corp.	Machinery Metal products
Sumitomo Corp.	Metals Machinery Electrical goods
Yokohama Rubber Co., Ltd.	Vehicle tires and tubes

51 Wholesale Trade — Non-durable Goods

C. Itoh & Co., Ltd.	Chemicals and energy Textiles Foodstuffs and food products
Cosmo Oil Co., Ltd., Group	Lubricant oils Fuel oil Kerosene Gasoline
Daiya Reform Corp.	Concrete waterproofing agents
Marubeni Corp.	Fuels Textiles Chemicals
Mitsui & Co., Ltd.	Chemicals Foodstuffs Texitiles Petroleum
Nissho Iwai Corp.	General commodities Energy Foodstuffs and food products Textiles
Sumitomo Corp.	Textiles Foodstuffs

64 Insurance Agents, Brokers & Service
Toyota Motor Corp. . . . . .Car Insurance

65 Real Estate
Kajima Corp. . . . . .Real estate development
Nittetsu Mining Co., Ltd. . . . . .Real estate
Obayashi Corp. . . . . .Real estate
Sumitomo Corp. . . . . .Real estate
Taisei Corp. . . . . .Real estate development

70 Hotels, Rooming Houses, Camps & Other Lodging Places
Taiyo Kagaku Co., Ltd. . . . . .Hotel management

73 Business Services
Du Pont Japan Ltd. . . . . .Environmental-related software
Kanebo, Ltd. . . . . .Computer software
Nippon Steel Corp. . . . . .Software

87 Miscellaneous Services
Fujitsu Laboratories Ltd. . . . . .Personal systems R&D
Materials R&D
Telecommunications R&D
Space technology R&D
Information processing R&D
Electronic devices R&D
Electronic systems R&D
Kajima Corp. . . . . .Archictural design and engineering
Kurosaki Refractories Co., Ltd. . . . . .Industrial design
Kyushu Kogyo Gijitsu Shinkenjo . . . . .Industrial technology development
NGK Insulators, Ltd. . . . . .Engineering services
Nippon Steel Corp. . . . . .Engineering contracts
NKK Corporation . . . . .Engineering
NTT, Electron Application Institute .Electronics R&D
Obayashi Corp. . . . . .Civil engineering
Optoelectronics Technology Research Corp. . . . . .Optoelectonics R&D
Shinko Wire Co., Ltd. . . . . .Engineering services
Shizuoka NEC Ltd. . . . . .Communications equipment R&D
Showa Denko K.K. . . . . .Engineering services
Takiron Co., Ltd. . . . . .Engineering services
Tokyo Gas Co., Ltd. . . . . .Contract work
Toray Research Center, Inc. . . . . .Chemical analysis
Toyota Central Research and Development
Laboratories Inc. . . . . .Research & development
Tsuda Kenkyusho . . . . .Research & development
Unitika Ltd. . . . . .Plant engineering
Wako Engineering Co., Ltd. . . . . .Plant engineering

00 Miscellaneous
Asahi Glass Co., Ltd. . . . . .Others
Chugoku Kogyo Co., Ltd. . . . . .Electronics & misc. materials
Dai-Ichi Kogyo Seiyaku Co., Ltd. . . . . .Others
Dainichiseika Color & Chemical Mfg. Co., Ltd. . . . . .Electronics materials & others
Dainippon Ink & Chemicals, Inc. . . . . .Others
Daiwabo Co., Ltd. . . . . .Industrial Materials
Furukawa Co., Ltd. . . . . .Electronic materials
Japan Organo Co., Ltd. . . . . .Others
Mitsubishi Kasei Corp. . . . . .Functional products
Mitsui Toatsu Chemicals, Inc. . . . . .Functional products
Nihon Cement Co., Ltd. . . . . .Other
Nittetsu Mining Co., Ltd. . . . . .Fuels and other products
NKK Corporation . . . . .Ferroalloys, new materials, related prod
Takeda Chemical Industries, Ltd. . . . . .Others
Tokuyama Soda Co., Ltd. . . . . .Misc. products
Toray Industries, Inc. . . . . .New businesses